T0305843

Cambridge Studies in Biological and Evolutionary Anthropology 51

Monkeys of the Taï Forest

A great deal has been written about primates; however few volumes have focused on an entire community of sympatric monkeys at a single site. The primary purpose of this book is to provide a multi-thematic snapshot of the entire monkey community of the Taï forest (Ivory Coast), drawing upon diverse sets of data collected by researchers over the years. The major themes covered include the following: feeding ecology, social behavior, positional behavior and habitat use, predator-prey interactions, vocal communication, and conservation. In addition, portraits of all species are provided, summarizing the major behavioral characteristics of each, as little is known about these West African monkeys.

W. SCOTT MCGRAW is Associate Professor in the Department of Anthropology at The Ohio State University and Affiliated Research Scientist at Yerkes National Primate Research Center, Emory University in Atlanta, Georgia.

KLAUS ZUBERBÜHLER is a Lecturer in the School of Psychology at the University of St Andrews, Scotland.

RONALD NOË is a Professor at the University of Louis-Pasteur and the Department of Ecology, Physiology, and Ethology (IPHC-CNRS), Strasbourg, France.

Cambridge Studies in Biological and Evolutionary Anthropology 51

Series Editors

HUMAN ECOLOGY
C. G. Nicholas Mascie-Taylor, University of Cambridge
Michael A. Little, State University of New York, Binghamton
GENETICS
Kenneth M. Weiss, Pennsylvania State University
HUMAN EVOLUTION
Robert A. Foley, University of Cambridge
Nina G. Jablonski, California Academy of Science
PRIMATOLOGY
Karen B. Strier, University of Wisconsin, Madison

Monkeys of the Taï Forest

An African Primate Community

EDITED BY

W. S. McGRAW
Ohio State University

K. ZUBERBÜHLER
University of St. Andrews

and

R. NOË
Université Louis Pasteur

CAMBRIDGE
UNIVERSITY PRESS

Shaftesbury Road, Cambridge CB2 8EA, United Kingdom

One Liberty Plaza, 20th Floor, New York, NY 10006, USA

477 Williamstown Road, Port Melbourne, VIC 3207, Australia

314–321, 3rd Floor, Plot 3, Splendor Forum, Jasola District Centre, New Delhi – 110025, India

103 Penang Road, #05–06/07, Visioncrest Commercial, Singapore 238467

Cambridge University Press is part of Cambridge University Press & Assessment, a department of the University of Cambridge.

We share the University's mission to contribute to society through the pursuit of education, learning and research at the highest international levels of excellence.

www.cambridge.org
Information on this title: www.cambridge.org/9780521816335

© Cambridge University Press & Assessment 2007

First published 2007

A catalogue record for this publication is available from the British Library

Library of Congress Cataloging-in-Publication data
The Monkeys of the Taï Forest : an African primate community / edited
by W. S. McGraw, K. Zuberbühler, R. Noë .
    p.   cm. -- (Cambridge studies in biological and evolutionary
      anthropology ; 51)
Includes bibliographical references and index.
ISBN 978-0-521-81633-5 (hardback)
    1.   Cercopithecidae--Behavior--Côte d'Ivoire--Parc National
de Taï . 2.   Social behavior in animals--Côte d'Ivoire--Parc National
de Taï . I. McGraw, William Scott. II. Zuberbühler, K. (Klaus),
1964- III. Noë , Ronald, 1951- IV. Title. V. Series.
QL737.P93M654 2007
599.8´6096668--dc22

                               2007002289

ISBN   978-0-521-81633-5   Hardback

Contents

Part III   Habitat use

Part IV   Conservation

Color plate section appears between pages 178 and 179

Contributors

Cécile Benetton
le bois d'Amont, 39140 Arlay, France

Karin Bergmann
Taï Monkey Project, CSRS, Abidjan, Côte d'Ivoire and
Johann Wolfgang Goethe-Universität, Frankfurt am Main,
Theobald-Christ-Str. 21, 60316 Frankfurt am Main, Germany

Redouan Bshary
Université de Neuchâtel, Institut de Zoologie, Eco-Ethologie,
Rue Emile-Argand 11, Case postale 158, CH-2009 Neuchâtel, Switzerland

Paul Buzzard
Fauna Flora International, 95 Xinxiang Bei Sihuan Xi Lu # 25,
Haidian Qu, Beijing 100080, People's Republic of China

Christelle Deffernez
Taï Monkey Project, CSRS, Abidjan, Côte d'Ivoire and
Institut d'Ecologie, Université de Lausanne, Switzerland;
Route de Loëx, 47, CH-1213 Onex, Switzerland

Winnie Eckardt
Biological Science, University of Chester, Parkgate Road,
CH1 4BJ, UK

Tim Förderer
FILMplattform GmbH, Germany

Cécile Fruteau
Center for Economic Research, Universiteit Tilburg,
Nederland and Ethologie des Primates, CEPE (CNRS UPR 9010) and
Université Louis Pasteur, France

David Jenny
Zoologisches Institut, Universität Bern,
Switzerland and Centre Suisse de Recherches Scientifiques, Abidjan,
Côte d'Ivoire

Inza Koné
Laboratory of Zoology, University of Cocody, 22 BP 582 Abidjan 22, Côte
d'Ivoire and Centre Suisse de Recherches Scientifiques en Côte d'Ivoire,
Taï Monkey Project, 01 B.P. 1303 Abidjan 01 Côte d'Ivoire;
Correspondence to: C/O CSRS, 01 B.P. 1303 Abidjan 01, Côte d'Ivoire

Amanda H. Korstjens
Behavioural Biology Group, Utrecht University, The Netherlands;
Taï Monkey Project, CSRS, Abidjan, Côte d'Ivoire and Max Planck
Institut Seewiesen, Starnberg, Germany;
Correspondence to: Senior lecturer in Biological Anthropology,
School of Conservation Sciences, Bournemouth University,
Talbot Campus, Christchurch House (room C240), Poole BH12 5BB,
Dorset, UK

Melanie Krebs
Taï Monkey Project, CSRS, Abidjan, Côte d'Ivoire and
Umweltbildung Ostfriesland, Self-employed with a service for
environmental education and intercultural learning: environmental
education biologist; Hufschlag 2, 26524 Hagermarsch, Germany

W. Scott McGraw
Department of Anthropology, 124 W. 17th Ave., The Ohio State
University, Columbus, Ohio, 43210-1364

Estelle C. Nijssen
Behavioural Biology Group, Utrecht University, The Netherlands;
Taï Monkey Project, CSRS, Abidjan, Côte d'Ivoire; Trianonstraat 24,
6213 AC Maastricht, Netherlands

Boris A. M. van Oirschot
Taï Monkey Project, CSRS, Abidjan, Côte d'Ivoire and Nature Awareness
Counsellor, Boris' Wereldgift, Wandelmeent 56, 1218 CR Hilversum,
The Netherlands

Cornelia Paukert
Taï Monkey Project, CSRS, Abidjan, Côte d'Ivoire and Zoologisches
Institut, Universität Frankfurt, Germany; Würzburger Str. 27, 74078
Heilbronn, Germany

Friederike Range
Dep. für Verhaltens-, Neuro- und Kognitions-Biologie, Universität Wien
Althanstrasse 14, A-1091 Wien, Austria

Johannes Refisch
Centre Suisse de Recherches Scientifiques en Côte d'Ivoire, Taï Monkey
Project, 01 B.P. 1303 Abidjan 01 Côte d'Ivoire and International Gorilla
Conservation Project (AWF, FFI, WWF), P.O. Box 48177, 00100 Nairobi,
Kenya

Eva Ph. Schippers
Behavioural Biology Group, Utrecht University, The Netherlands and
Taï Monkey Project, CSRS, Abidjan, Côte d'Ivoire; AAP Sanctuary,
Kemphaanweg 16, Postbus 50313, 1305 AH Almere, The Netherlands

Susanne Shultz
Population Biology Research Group, Biosciences Building,
University of Liverpool, Liverpool L69 7ZB, England, UK

Yannick Storrer-Meystre
Department of Ethology, Zoology, University of Neuchâtel, Switzerland

Simon Thomsett
The Peregrine Fund, 5668 West Flying Hawk Lane, Boise, ID 83709, USA

Klaus Zuberbühler
School of Psychology, University of St Andrews, St Andrews, KY16 9JU,
Scotland, UK

Preface

The study of primate behavior and ecology has been an ongoing area of research for over 50 years, building on the pioneering work of such people as C. R. Carpenter, Washburn and DeVore, C. R. L. Hall, J. J. Petter, and the Altmanns. There are relatively few of the 300 plus species of living primates that have not been the object of at least a survey; many have been studied for a complete year; and a few taxa have been the subject of long-term efforts lasting decades (e.g. Strier *et al.* 2006). Primatology has grown to become an integral part of anthropology or zoology in most parts of the world and a discipline that is the focus of numerous national and international organizations, more than half a dozen specialist journals and numerous book series.

Although year-long studies of a single species have long been the standard research protocol in primatology, some of the greatest advances in our understanding of primate behavioral ecology have come from coordinated studies of numerous species at a single site. Because all of the species are living in the same habitat with identical climatic and phono-logical variations, they enable a clearer insight into species-specific differences and similarities in adaptive strategies. The comparative, synecological studies of the primate assemblages such as those conducted at Makokou in Gabon, Morondava in Madagascar, Kibale Forest in Uganda, Kuala Lompat in Malaysia, Raleighvallen-Voltsberg in Suriname, and Manu in Peru stand out as milestones in the history of primatology and have disproportionately advanced our understanding of the relationship between behavior and ecology in primate evolution. With this volume, the Taï Forest joins this pantheon and offers an in-depth, comparative view of the diurnal primate fauna of west-central Africa.

In addition, this volume clearly demonstrates the sophisticated and diverse nature of studies in primatology at the beginning of the twenty-first century. In addition to providing the critical baseline data on behavior and ecology of the monkey taxa at Taï, these papers use both observational and experimental methods to probe the nature of locomotion and posture, communication, and predator-prey interactions. Most significantly research on the Taï monkey is fully integrated with conservation work to make possible effort to ensure that this extraordinary assemblage of primates, and research to understand ever more about their behavior and ecology will continue for generations to come.

J. G. Fleagle
Stony Brook University
New York

1 The Monkeys of the Taï forest: an introduction

W. S. McGraw and K. Zuberbühler

Introduction

With several notable exceptions (e.g. Schaller 1963, Goodall 1965, 1968), early field primatology in Africa was practically equivalent to observing baboons on the savannah. Because of the prominence of open-country primates in models of human evolution as well as the difficulties of seeing and habituating cercopithecids in dense forest, many of the first studies of African primates focussed on terrestrial monkeys such as Olive baboons (Washburn & Devore 1961a, 1961b), Chacma baboons (Hall 1962), Hamadryas baboons (Kummer 1968), Yellow baboons (Altmann & Altmann 1970), Gelada baboons (Crook 1966, Crook & Aldrich-Blake 1968, Dunbar & Dunbar 1974), patas monkeys (Hall 1965) and vervet monkeys (Struhsaker 1967) (but see Haddow 1952, Rowell 1966, Aldrich-Blake 1968, 1970, Chalmers 1968a, 1968b, Gautier & Gautier-Hion 1969, Struhsaker 1969, Gartlan & Struhsaker 1972). Interest in arboreal primates eventually prompted more biologists to venture beneath the closed canopy and with Struhsaker's (1975) classic monograph on red colobus monkeys as a reference point, our knowledge of forest-dwelling African monkeys has grown significantly over the last 30 years. The result has been a burgeoning literature on African cercopithecoids including detailed treatments of guenons (e.g. Gautier-Hion *et al.* 1988, Glenn & Cords 2002), colobines (Davies & Oates 1994) and monkeys throughout the Congo Basin (Gautier-Hion *et al.* 1999). These and other contributions on both extant and extinct cercopithecoids (e.g. Whitehead & Jolly 2000) represent the state of the art in phylogeny reconstruction, functional morphology and behavioral biology and have provided significant insight into the habits of forest dwelling African monkeys that as recently as 35 years ago were largely unknown (Napier & Napier 1970).

Monkeys of the Taï Forest, ed. W. Scott McGraw, Klaus Zuberbühler and Ronald Noë. Published by Cambridge University Press. © Cambridge University Press 2007.

Much of what we know about African monkeys is based on work at several well-known sites including Kibale Forest, Uganda (e.g. Struhsaker 1978, Chapman & Chapman 1996), Tiwai Island, Sierra Leone (e.g. Oates & Whitesides 1990), Makokou Forest, Gabon (e.g. Gautier & Gautier-Hion 1969), Lope Reserve, Gabon (Tutin *et al.* 1997), and Kakamega Forest, Kenya (e.g. Cords 1984). The elegant research carried out at these localities has generated a wealth of long-term data for several monkey species, some of whom are among the best known – and thoroughly studied – of all primates (e.g. red colobus). At the same time, there have been few attempts to summarize the interactions between food, predators, habitat and social life for *all* members of any particular cercopithecid community living in sympatry. After reviewing research conducted by the ~40 students involved in our project since 1989, we felt we could describe some of the principle machinations within a single African monkey community in a manner similar to Terborgh's study of New World monkeys (Terborgh 1983). Ideally, the result would be a multi-disciplinary overview that could inform anthropological, psycho-logical (Gleitman 1999), philosophical (Allen & Bekoff 1997) and linguistic (Tallerman 2005) disciplines in ways a collection of papers scattered throughout specialty journals could not.

In this book, we report on a community of eight Old World monkeys living in the Taï forest of western Ivory Coast. We summarize results of approximately 15 years of research conducted by a large number of individuals, all of whom carried out fieldwork at Taï. From the start, it has been our intention to understand the behavior of these primates as determined by habitat characteristics, predators, food availability, other group members and neighbors. A volume summarizing the behavior of one Taï primate – the chimpanzee – already exists (Boesch & Boesch-Achermann 2000) and one aim of the present book is to complement information on the Taï ape with that on the lesser-known cercopithecids sharing the same forest. Eventually, we hope the third group of Taï primates – the nocturnal lorises and galagos – are similarly studied so that the entire Taï primate community can be examined collectively. Ultimately, data sets from additional sites can be used to compare communities so that the ecological, phylogenetic and historical factors responsible for the composition of faunas we observe – and con-serve – today are better understood (Fleagle *et al.* 1999). This chapter provides background information on the Taï forest, presents a brief history of the Taï Monkey Project, and introduces the eight monkey species with general remarks on their natural histories. We then discuss the content of subsequent chapters.

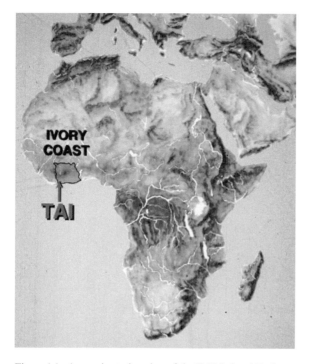

Figure 1.1. Approximate location of the Taï National Park.

Taï National Park

Taï National Park is the last substantial block of intact forest remaining in West Africa. The forest was once part of a large forest belt, the Upper Guinea Forest that covered a vast area from Ghana to Sierra Leone. Its decline in size has been dramatic, particularly in the twentieth century (Martin 1991). The official area today is 330,000 ha in addition to a 20,000 ha buffer zone, which is contiguous to the 73,000 ha "Réserve de Faune du N'Zo" to the north of the park. The park is located in the southwest corner of Ivory Coast near the Liberian border about 200 km south of Man and 100 km from the Gulf of Guinea coast in the districts of Guiglo and Sassandra (0°15′–6°07′N, 7°25′–7°54′W) (see Figure 1.1).

The park was declared a UNESCO World Heritage Site in 1982; detailed information about its features and history can be found on the UNESCO website (http://whc.unesco.org). Briefly, the Taï forest first obtained protection in 1927 when it was declared a "Forest and Wildlife Refuge." About half a century later, in 1972, it obtained National Park status. Five years later, 20,000 ha of buffer zone were added, and the park became

Figure 1.2. Under story of the Taï forest (Photo: Scott McGraw).

internationally recognized as a Biosphere Reserve under UNESCO's Man and the Biosphere Programme. In 1982 it became part of the UNESCO World Heritage List.

The park contains some 1,300 species of higher plants including 150 identified as endemic. Vegetation is predominantly dense evergreen ombrophilous forest of a Guinean type with 40–60 m high continuous canopy and large numbers of epiphytes and lianas (see Figure 1.2). The forest is recovering from commercial timber exploitation, which officially ceased in 1972. The park contains a fauna typical of West African forests. Some noteworthy non-primate mammals include giant pangolins (*Manis gigantean*), tree pangolins (*M. tricuspis*) and long-tailed pangolins (*M. tetradactyla*), golden cats (*Felis aurata*), leopards (*Panthera pardus*), elephants (*Loxodonta africana*), bushpigs (*Potamochoerus porcus*), giant forest hogs (*Hylochoerus meinertzhageni*), pygmy hippopotamus (*Choeropsis liberiensis*), water chevrotains (*Hyemoschus aquaticus*), bongos (*Tragelaphus euryceros*), buffalos (*Syncerus caffer*), and several species of forest duikers and rodents. Almost 1,000 species of vertebrates including over 230 bird species have been identified in the park. Altitudes range from 80 m to 396 m with Mount Niénokoué as the highest peak (see Figure 1.3). It comprises an ancient sloping granitic peneplain, broken by several inselbergs, which were formed by volcanic intrusions. The soils

Figure 1.3. Taï forest canopy viewed from top of Mount Nienokoue
(Photo: Scott McGraw).

are ferralitic of generally low fertility. There are two distinct climatic zones
with annual average rainfall of 1,700 mm in the north and 2,200 mm in
the south. The rains peak in June and September and there is a marked
dry season from December to February. Temperatures range from 24° C to
27° C and the relative humidity is constantly high at between 85 and
90 per cent.

The principal conservation problems facing the Taï National Park are
illegal poaching, logging, farming, and gold mining. There is increasing
degradation of and human encroachment into the forest, particularly
in the surrounding buffer zone that is generally not respected by
local farmers. Destabilization of the country following a failed 2002
military coup has led to an increase in poaching activity in and around
the park. The impact of these activities on the local fauna is likely to be
enormous.

Background of the Taï Monkey Project

Most early publications dealing with African monkeys were largely
taxonomic and contained little behavioral information, particularly on
the habits of West African cercopithecids (e.g. Pocock 1907, Elliot 1913,
Schwarz 1928, 1929, Rode 1937, Sanderson 1940, Dekeyeser 1955). Before
his tragic death at age 30, Angus Booth provided some of the earliest
observations on West African monkeys in a series of influential papers

(Booth 1954, 1955, 1956a, 1956b, 1957, 1958a, 1958b, 1960), but it was not until the 1970s that the behavior of the Taï Forest primates first came to light (e.g. Struhsaker & Hunkeler 1971). Intensive studies on the forest's chimpanzees began in 1976 (Boesch & Boesch-Achermann 2000) and work on monkeys began shortly thereafter (Galat 1978, Galat & Galat-Luong 1985). In addition to primates, there have been numerous studies on the forest's non-primate fauna including those on leopards (Hoppe-Dominik 1984, Jenny 1996), elephants (Alexandre 1978, Roth *et al.* 1984, Merz 1986, Roth & Hoppe-Dominik 1987), crocodiles (Waitkuwait 1981), pygmy hippos (Galat-Luong 1981), duikers (Newing 2001), and birds (Thiollay 1985, Balchin 1988, Gartshore 1989). These contributions have been vital in informing our research.

The Taï Monkey project was founded in 1989 when Ronald Noë and Bettie Sluijter, then at the University of Zurich, undertook a pilot study on red colobus monkeys. The eminent primatologist Hans Hummer (University of Zurich) had suggested to Noë and Sluijter that they investigate whether some of the peculiarities of red colobus monkeys, especially their large group size, male philopatry and tendency to form polyspecific associations, could be explained as adaptive responses to chimpanzee predation (see Figure 1.4). At the time, the Taï chimpanzees were already well-known monkey hunters and red colobus were their preferred prey (Boesch & Boesch 1989). Kummer envisioned a long-term cooperative endeavor in which one research group studied the predators while the other studied the prey.

A successful four-month pilot study led to additional funding and the project's first students, Klaus Zuberbühler and Kathy Holenweg, arrived in January 1991. Klaus and Kathy were responsible for habituating the first group of red colobus and Diana monkeys as well as establishing the primary study grid. They selected an area with a high density of monkeys near the field station of the "Institute d'Ecologie Tropicale" (IET) on the western border of the park. The IET research station is approximately 20 km from the nearest village and 25 km from the Cavally River that forms the border with Liberia. The grid established in 1991 has since been enlarged but still forms the core of the project's study site (see Figure 1.5).

The following three years witnessed significant expansion in research activity as a growing number of students travelled to Taï, primarily to investigate the anti-predation adaptations of monkeys. In January 1992, Redouan Bshary (Max Planck Institut für Verhaltensphysiologie) started a three-year study on the relationship between red colobus – Diana monkey associations and chimpanzee hunting behavior. In November 1992, Kauri Adachi (University of Kyoto) initiated a study of guenon socio-ecology

Figure 1.4. The Taï Monkey Project began as an attempt to determine whether peculiar features of red colobus monkeys – including large group sizes, male philopatry and frequent formation of polyspecific associations – were adaptations to predation by chimpanzees. Here, part of a red colobus monkey group rests and grooms during the late afternoon (Photo: Scott McGraw).

Figure 1.5. The Audrenisrou River near the research station of the Taï Monkey Project (Photo: Klaus Zuberbühler).

and five months later, Scott McGraw (SUNY Stony Brook) started work on comparative positional behavior and habitat use. As the number of students grew, so too did the number of field assistants. By the end of 1994, there were − on average − six students and six field assistants studying monkeys at any one time. The breadth of research has increased over the years, but the number of personnel in the forest has remained stable. Table 1.1 provides an overview of Masters and Ph.D. students including the general topics of study and date of thesis.

A community of West African monkeys

There are eight monkey species in the Taï forest. Seven occur throughout the park: the Diana monkey *Cercopithecus diana*, Campbell's monkey *Cercopithecus campbelli*, the lesser spot-nosed monkey *Cercopithecus petaurista*, the red colobus monkey *Procolobus badius*, the King (or Western black and white) colobus monkey *Colobus polykomos*, the olive colobus monkey *Procolobus verus* and the sooty mangabey *Cercocebus atys*. The eighth species, the putty-nosed monkey *Cercopithecus nictitans stampflii*, is found at significantly lower densities and mainly in northern portions of the forest. The low densities and patchy distribution of putty-nosed monkeys at Taï and elsewhere in West Africa can be explained by competitive exclusion from the Diana monkey *C. diana* (Oates 1988a, Eckardt & Zuberbühler 2004). Although we have studied *C. nictitans*, the species is not part of the monkey community near the IET research station; most of the project's research − and that comprising the majority of this book − is focused on the seven species found within our $2 \times 2 \, km^2$ study grid.

It is not clear how long each species has existed in the Taï region, nor is the exact order of their arrival known. Most authorities agree that Taï monkeys are early descendents from primates that migrated from central Africa and that they have been isolated in the Upper Guinea forest for a considerable period of time (Grubb 1978, 1982, Oates & Trocco 1983, Kingdon 1989, Disotell & Raaum 2002, Tosi *et al.* 2005). Some version of Holocene refuge theory is routinely used to explain the presence of multiple species at a single locality (Lonnberg 1929, Livingstone 1975, 1982, Kukla 1977, Grubb 1982, Hamilton 1988) and while the Taï forest is situated midway between the two proposed West African refugia − one in Sierra Leone/Liberia and the other in eastern Ivory Coast/western Ghana (Booth 1958a, 1958b, Hamilton 1988, Oates 1988b) − there is growing evidence that the guenons and mangabeys diverged as early as the late Miocene (see Figure 1.6). Such early divergence dates are problematic for arguments that rely on Pleistocene glaciers and concomitant forest oscillations to

Table 1.1. *Thesis research conducted in the Taï Monkey Project*

Topic	Reference
Anti-predator behavior	Mangabeys: Range (2004); Olive colobus: van der Hoeven (1996); Korstjens (2001); Red colobus: Zuberbühler (1993); Bshary (1995); Korstjens (2001); King colobus: Korstjens (2001); Diana monkeys: Zuberbühler (1993), Bshary (1995); Shultz (2003); Campbell's monkeys: Wolters (2001); all guenons: Hansen (1996)
Association behavior	Mangabeys: Bshary (1995); McGraw (1996); Olive colobus: Bergmann (1998); Korstjens (2001); King colobus: Bergmann (1998); Korstjens (2001); Red colobus: Blank (1997); Holenweg 1992; Bshary (1995); Höner (1993); Leumann (1994); Korstjens (2001); Diana monkeys; Bshary (1995); Wolters (2001); Holenweg (1992); Leumann (1994); Eckardt (2002); Höner (1993); Campbell's monkeys: Wolters (2001); Putty-nosed monkeys: Eckardt (2002); all guenons: Buzzard (2004); Adachi-Kanazawa (2004)
Feeding behavior	Mangabeys: Rutte (1998); Bergmüller (1998); Red colobus: Schabel (1993); Wachter (1993); King colobus: Nijssen (1999); all colobines: Schaaff (1995); Korstjens (2001); Diana monkeys: Schabel (1993), Wachter (1993); Eckardt (2002); putty-nosed monkeys Eckardt (2002); all guenons: Buzzard (2004)
Ranging behavior	Mangabeys: Janmaat (2006); Förderer (2001); Olive colobus: Schippers (1999); Korstjens (2001); Red colobus: Korstjens (2001); Höner (1993); King colobus: Paukert (2002); Bitty (2001); Korstjens (2001); Diana monkeys: Höner (1993); Eckardt (2002), Wolters (2001); Campbell's monkeys: Wolters (2001); putty-nosed monkeys: Eckardt (2002); all guenons: Buzzard (2004)
Vocal behavior	Mangabeys: Range (2004); Red colobus: Zuberbühler (1993); Bshary (1995); Olive colobus: Koffi (in prep); King colobus: Tranquilli (2003); Diana monkeys: Zuberbühler (1993, 1998); Bshary (1995); Uster (2000); Eckardt (2002); Campbell's monkeys: Wolters (2001); putty-nosed monkeys: Eckardt (2002)
Social behavior	Mangabeys: Benneton (2002); Range (1998, 2004), Meystre-Storrer (2002); Olive colobus: Deschner (1996); Schippers (1999); Krebs (1998); Red colobus: Von Oirschot (1999); Korstjens (2001); King colobus: Paukert (2002); Nijssen (1999); Korstjens (2001); Diana monkeys: Wolters (2001); Eckardt (2002); Campbell's monkeys: Wolters (2001); putty-nosed monkeys: Eckardt (2002); guenons: Buzzard (2004)
Conservation	Refisch (2001); Kone (2004)
Positional behavior	Cercopithecids: McGraw (1996); putty-nosed monkeys: Bitty, E. A. (in prep)
Non-primate studies	Crowned eagle: Shultz (2003); mongooses: Dunham (2003); bats: Gordon (2001); hornbills: Rainey (2004)

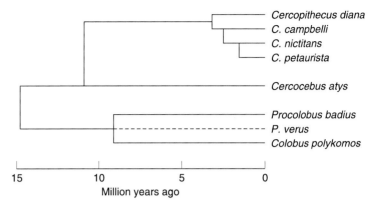

Figure 1.6. Approximate divergence dates of the eight cercopithecid species in the Taï forest (after Kingdon 1997, Disotell & Rauum 2002, Tosi *et al.* 2005).

explain the recent evolution of at least several cercopithecid groups. In any case, the colobus monkeys (*Procolobus* and *Colobus* spp.) and mangabeys (*Cercocebus* sp.) may have been the first monkeys to radiate into the Upper Guinea forest (Kingdon 1989). Early members of these groups may have met significant ecological competition by westerly radiating guenons which could explain why the colobines at Taï today (*Procolobus badius*, *P. verus*, and *Colobus polykomos*) are specialist in terms of their niches, diets, and ecological strategies (Kingdon 1989). The three common guenon species are descendants of distinct radiations, each of a different age. For example, the Diana monkey has no East African equivalent and may be descendants of the first arboreal lineage that migrated into the narrow coastal forests of Upper Guinea (see Disotell & Raaum 2002, Tosi *et al.* 2002). *C. campbelli* represents the most conservative member of the mona super-species, while *C. petaurista* may be the contemporary descendant of ancestors of the moustached or red-tail group (Kingdon 1989).

The diversity of sympatric species at Taï and their ecological profiles seem typical for an African forest. Several recent studies have compared the ecological characteristics and taxonomic makeup of primate communities globally (Fleagle & Reed 1996, 1999, Chapman *et al.* 1999) and these analyses have demonstrated that primates in African forested areas typically are characterized by, "a fairly high number of arboreal frugivores, 2–3 arboreal folivores, terrestrial cryptic foragers of the Papionin tribe, and 2–5 nocturnal gumivores/insectivores" (Reed & Bidner 2004:23). This accurately describes the Taï primate community and is similar to that at Kibale, Uganda (Struhsaker 1997) and Lope, Gabon (Tutin *et al.* 1997). The concordance of taxa occupying specific

Table 1.2. *Estimated annual percentage of food items consumed by Taï monkeys*

	DIA[a]	CAM[b]	PET[b]	NIC[a]	BAD[d]	POL[e]	VER[e]	ATY[c]
Fruit	70.9	46.3	33.6	58.9	28.8	48.0	9.0	68.4
Foliage	1.6	8.4	39.7	4.2	50.0	48.0	91.0	2.4
Invertebrate	26.5	33.1	12.3	31.3	–	–	–	26.4
Flower	0.4	1.0	6.2	2.5	19.5	3.0	–	1.3
Fungi	0.2	1.6	1.1	1.3	–	–	–	0.9
Other	0.5	9.3	6.4	1.8	2.1	1.0	–	0.6
N samples	1,828	953	924	1,424	6,480	4,090	991	406

DIA (*Cercopithecus diana*); CAM (*Cercopithecus campbelli*); PET (*Cercopithecus petaurista*); NIC (*Cercopithecus nictitans*); BAD (*Procolobus badius*); POL (*Colobus polykomos*); VER (*Procolobus verus*); ATY (*Cercocebus atys*)
Source: [a]Buzzard (2004); [b]Eckardt (2002); [c]Bergmüller (1998); Bergmüller *et al.* submitted; [d]Korstjens unpublished data; [e]Korstjens & Galat-Luong in press; Oates *et al.* in press

niches defined by body size, diet, locomotion, and activity patterns, strongly suggests that the structure of the primate community at Taï and elsewhere in Africa is the product of a complex but common series of evolutionary events and limiting agents including forest productivity, availability of keystone resources, predation pressure and, perhaps, historical anthropogenic factors (Fleagle & Reed 1996, Tutin & White 1999, Struhsaker 1999, Reed & Bidner 2004). These factors are discussed throughout this volume. Table 1.2 summarizes general feeding data on each species while Table 1.3 provides species means for group size, home range size, number of adult males per group, canopy use, and group density.

Procolobus badius badius *(Kerr 1792) Western red colobus*
The Western red colobus monkey *Procolobus badius badius* is the most abundant monkey in the study area and prior to reductions by human poachers, the most common primate in the park. Red colobus are medium-sized, slender monkeys that exhibit little sexual dimorphism: mean male body weight is 8.3 kg and mean female body weight is 8.2 kg (Oates *et al.* 1990). Individuals are predominantly red with a black band running the length of the dorsum and a rust-colored tail (see Figure 1.7). Adult females exhibit large sexual swellings when in estrus. The inter-membral index of *P. badius badius* is 87 (Fleagle 1999).

Red colobus monkeys have a wide distribution across equatorial Africa. Their taxonomy is in need of revision, however recent classifications recognize at least five species including 15 subspecies (Grubb *et al.* 2003).

Table 1.3. *Summary overview of some socio-ecological features of the Taï monkeys*

Species	Group size[a]	Home range	N males[a]	Habitat[b]	Density[a]
Procolobus badius	52.9	0.58	10.1	0.4	123.8
Procolobus verus	6.7	0.56	1.43	1.3	17.3
Colobus polykomos	15.4	0.78	1.42	13.2	35.5
Cercopithecus diana	20.2	0.63	1	6.1	48.2
Cercopithecus campbelli	10.8	0.60	1	36.8	24.4
Cercopithecus petaurista	17.5	0.69	1	9.9	29.3
Cercopithecus nictitans	10.5	0.96	1	0.7	2.1
Cercocebus atys	69.7	4.92	9.0	88.9	11.9

Density: estimated number of individuals per square kilometer
Group size: average number of individuals per group
N males: average number of adult males per group
Habitat: per cent time observed in lower forest strata
Home range: estimated size of annual range in km^2
[a]Data compiled by Zuberbühler and Jenny (2002);
[b]Data from McGraw (1998a, 2000); Eckardt (2002)
Home range data from the following studies: *C. diana*: Eckardt (2002): 0.66 km^2; Buzzard (2004): 0.59 km^2; *C. campbelli*: Buzzard (2004): 0.67 & 0.52 km^2; *C. petaurista*: 0.74 & 0.64 km^2; *C. nictitans*: Eckardt (2002): 0.96 km^2; *P. badius*: Korstjens (2001): 0.50 & 0.66 km^2 *P. verus*: Korstjens (2001): 0.54 & 0.58 km^2, *C. polykomos*: Korstjens (2001): 0.83 & 0.72 km^2; *Cercocebus atys*: Rutte (1998): 4.92 km^2

The Taï subspecies, *P. badius badius*, is found from the Ivory Coast's Bandama River in the east to Sierra Leone in the west and except for an outlying population on the Gambia River (*Procolobus badius temminckii*), represents the most western extent of the red colobus radiation (Kingdon 1989, Starin 1991, 1994, Galat-Luong & Galat 2005).

The red colobus is possibly the best-known forest monkey in Africa, having been the subject of many important studies in East Africa (Clutton-Brock 1973, 1974, 1975a, 1975b, Struhsaker 1974, 1975, 1978, 1980, Struhsaker & Oates 1975, Busse 1977, Marsh 1979a, 1979b, 1981, Baranga 1982, 1983, 1986, Isbell 1984, Struhsaker & Leland 1985, Decker 1994, Stanford *et al.* 1994, Stanford 1995, Chapman & Chapman 1996, 2002, Chapman *et al.* 2002a, 2002b), Central Africa (Maisels *et al.* 1994), and West Africa (Gatinot 1977, Galat & Galat-Luong 1985, Starin 1990, 1991, 1994, 2001, Fimbel 1992, Teichroeb *et al.* 2003). These studies have highlighted great variation in red colobus socio-ecology, diet preferences, group sizes, association tendencies and anti-predator adaptations across the continent.

At Taï, red colobus monkeys live in loud, large groups of between 40 and 90 individuals. These groups often divide into subgroups during periods

Figure 1.7. Western red colobus monkey *Procolobus badius badius* (Photo: Scott McGraw).

of low resource availability (Höner *et al.* 1997). Group composition is multi-male, multi-female and groups exhibit male philopatry. Solitary females are regularly observed in mono-specific and mixed-species groups. Red colobus are frequently found in association with other monkey species, primarily as a response to predation pressure (Holenweg *et al.* 1996, Honer *et al.* 1997, Wachter *et al.* 1997, Bshary & Noe 1997a, 1997b). They are the favored monkey prey of chimpanzees which appear to specifically target them during the rainy season. The response to monkey-hunting chimpanzees by Taï red colobus differs markedly from that of red colobus at sites in East Africa (e.g. Boesch 1994, Stanford 1998). Red colobus are frequent leapers who suffer a high incidence of injuries from falls (Hellmer & McGraw 2005) (see Figure 1.8). They use all layers of the forest but prefer the main canopy (McGraw 1996, 1998a). Taï red colobus

Figure 1.8. Red colobus monkeys are spectacular leapers. Here, an adult is passing between a large discontinuity at the top of the main canopy forest layer (Photo: Scott McGraw).

feed predominantly on leaves, fruit, and flowers (Korstjens 2001). The vocal behavior of red colobus has been described by Struhsaker (1975); to date, no thorough vocal studies have been conducted at Taï. The IUCN Red List of Threatened Species considers this taxon as Endangered, based on an estimation of rate of population decline (www.redlist.org). The species is now rare or absent in parts of Taï National Park.

Procolobus verus *(van Beneden 1838) Olive colobus*
The monotypic olive colobus is the smallest colobine monkey. The species is not particularly dimorphic: average body mass for males is 4.7 kg and for females is 4.2 kg (Oates *et al.* 1990). It is cryptically colored with a dull grayish underside and a greenish olive upper side. The face is hairless of a dark gray color and framed by a dull-white ruff (see Figure 1.9). The drab coat of *Procolobus verus* makes locating this species extremely difficult, particularly in the shadows of the forest understory where it spends the majority of its time feeding and resting in vine tangles and other areas of dense vegetation.

   Olive colobus are characterized by a unique combination of features. These small, mysterious monkeys are restricted to West Africa and most authors believe they are the most primitive African colobine having

Figure 1.9. An adult male olive colobus *Procolobus verus* (Photo: Scott McGraw).

retained many ancestral traits (Kingdon 1989, Davies & Oates 1994). Olive colobus have the most reduced thumb and the largest feet of any African colobine (Fleagle 1999). The inter-membral index is 80 (McGraw, this study) and they are the most frequent leapers at Taï, capable of propelling themselves over distances many times their body lengths. They prefer the forest understory for all activities where they frequent dense vine tangles (McGraw 1996, 1998a). Olive colobus are the only anthropoids known to carry their dependent offspring in their mouth for as yet unknown reasons (Booth 1957, 1960).

Olive colobus are extremely cryptic monkeys who vocalize infrequently and primarily when alarmed (Koffi, B. J.-C. in prep). They are nearly always found in association with other monkey species, especially Diana monkeys (*Cercopithecus diana*) (see Figure 1.10). Oates and Whitesides (1990) argue that this small colobine has evolved a specific strategy to associate with other monkey species – most likely for anti-predation benefits – without compromising its dietary strategy.

Group structure of olive colobus varies significantly, however a typical social unit consists of several adult males, three or more adult females and their infants (Korstjens & Schippers 2003, Korstjens & Noë 2004). Females display prominent sexual swellings during estrus. The diet of olive colobus has been well studied in an area of old secondary forest on Sierra Leone's

Figure 1.10. Olive colobus monkeys are found in association with Diana monkeys (*Cercopithecus diana*) over 95 per cent of the time. Here, a female Diana monkey is resting with an adult male olive colobus (Photo: Florian Möllers).

Tiwai Island, (Oates 1988a, Oates & Whitesides 1990). The most preferred food item is young leaves with fruit comprising between 10 and 20 per cent of the annual diet (Korstjens 2001). The IUCN Red List of Threatened Species considers this taxon as *Near Threatened*.

Colobus polykomos polykomos *(Zimmerman 1780) King Colobus*
Most authorities recognize five species of black and white colobus; the King colobus *Colobus polykomos* is the western most species and is separated from *Colobus vellerosus* to the east by the Sassandra River. *C. polykomos* has more conservative features than the well-known *C. guereza* of East Africa (Oates 1977a, 1977b, 1977c), perhaps because

Figure 1.11. An adult male king colobus *Colobus polykomos*
(Photo: Scott McGraw).

this species is the direct descendant of a lineage that has continued to
evolve further east (Kingdon 1989). Analyses of loud calls, cranial
morphology and pelage indicate that King colobus diverged early from
the ancestral black and white colobus while *C. vellerosus* and *C. guereza*
to the east are more recent, derived forms (Oates & Trocco 1983, Oates
et al. 2000a).

King colobus are the largest arboreal monkeys at Taï. Adults are
dimorphic in size: average male body weight is 9.9 kg and average female
body weight is 8.3 kg (Oates *et al.* 1990). The inter-membral index is
78 (McGraw, this study). Adults possess jet-black coats and a long,
rope-like tail with no tuft. The monkey's black face is fringed by tufts of
white or gray hair, which often extends to the shoulders in adults
(see Figure 1.11). Infants are born completely white and develop their
adult coat within a few months. Females show no evidence of sexual
swelling during estrus.

The typical social unit of *C. polykomos* consists of one or two adult
males, 3 to 7 adult females and between 6 and 12 infants, juveniles and
subadults (Korstjens *et al.* 2002). Neighboring groups have strongly
overlapping home ranges. The species is generally cryptic and tends
to actively avoid associating with sympatric species. In these respects,
it provides a striking contrast to the closely related red colobus sharing

the same forest. King colobus males emit roaring loud calls, which function in predator defense and may also serve in intergroup spacing (Walek 1978, Tranquilli 2003). Contact calls are soft and consist of snorts and grunts.

Species of black and white colobus are generally not as sensitive to habitat disturbance, nor as reliant on primary forest, as are red colobus (e.g. Saj & Sicotte 2004). Several studies have shown that densities of black and white colobus may actually be greater in areas of colonizing forest or secondary growth (Oates 1977a, 1977b, Struhsaker 1997). In the undisturbed forest at Taï, King colobus can be found exploiting all layers of the canopy, particularly lianas in the understory, although they most frequently use large supports of the main canopy. The diet of *C. polykomos* at Taï and elsewhere is characterized by a marked preference for seeds from fruit, particularly those of *Pentaclethera macrophylla*, as well as liana leaves (Dasilva 1992, 1994, Hayes *et al.* 1996, Davies *et al.* 1999, Daegling & McGraw 2001). The IUCN Red List of Threatened Species considers this taxon as *Near Threatened*.

Cercocebus atys atys (*Audebert 1797*) *sooty mangabey*

Mangabeys are diphyletic and consist of two groups: the arboreal members are placed in the genus *Lophocebus* with three species (*aterrimus*, *albigena* and *kipunji*) and the predominately terrestrial members are placed in the genus *Cercocebus* with upwards of six species (*atys*, *torquatus*, *agilis*, *galeritus*, *sanjei*, *chrysogaster* (Cronin & Sarich 1976, Groves 1978, Disotell 1994, Nakatsukasa 1996, Fleagle & McGraw 1999, 2002). *Cercocebus atys atys* is the western-most species and the Ivory Coast's Sassandra River serves as the approximate boundary separating it from *C. atys lunulatus* (the White-naped mangabey) further east (Booth 1956a, 1956b). It is the former that is found in the Taï forest.

Sooty mangabeys are large, long-limbed, predominantly terrestrial monkeys (McGraw 1998a, 1998b). Their coat color is charcoal gray and they have lighter, flesh-colored faces (see Figure 1.12). Sexual dimorphism is high: mean male body weight is 11 kg while mean female body weight is 6.2 kg (Oates *et al.* 1990). The inter-membral index is 84 (McGraw, this study).

Apart from some anecdotal observations (Booth 1956a, 1956b, Struhsaker 1971, Harding 1984, Galat & Galat-Luong 1985), the monkey had not been systematically studied under natural conditions prior to our project. A well-established colony at the Yerkes National Primate Research Center has been the subject of numerous studies and

Figure 1.12. The sooty mangabey *Cercocebus atys* is the only predominantly terrestrial monkey at Taï (Photo: Scott McGraw).

most early information on this species was based on this captive population (Bernstein 1971a, 1971b, 1976, Hadidian & Bernstein 1979, Aidara *et al.* 1981, Busse & Gordon 1984, Ehardt 1988a, 1988b, Fultz *et al.* 1986, Gordon *et al.* 1991, Gust & Gordon 1991, 1993, 1994, Gust 1994).

Studies on the dynamics of social behavior in Taï sooty mangabeys (Range & Noe 2002, 2005, Range & Fischer 2004, Range 2005) provide an interesting contrast to conclusions based on observations of captive individuals (Ehardt 1988a, 1988b). A typical sooty mangabey group at Taï numbers approximately 100 individuals. Groups frequently splinter into subgroups as an adaptation to seasonal fluctuations of preferred resources. Group structure is multi-male, multi-female with female philopatry. Solitary or groups of non-resident males are frequently observed and are known to invade resident groups during the breeding season. Females exhibit marked swellings during estrus.

Sooty mangabeys obtain most of their food from within the leaf litter on the forest floor (see Figure 1.13). Here, members search through the forest debris looking for insects and fallen hard object foods that resist decomposition and are generally too hard for other cercopithecines to process. It is the suite of adaptations for manual foraging and hard object

Figure 1.13. Sooty mangabeys spend a large portion of their foraging time searching for fallen fruits and nuts amid the leaf litter on the forest floor. Many of these hard object foods resist decomposition and are not available as resources to other Taï monkeys because they are too difficult to open (Photo: Ralph Bergmüller).

feeding – including powerful forelimbs, large teeth and strong jaws for crushing – that links *Cercocebus* mangabeys with their sister taxon, *Mandrillus* (Fleagle & McGraw 1999, 2002). The vocal behavior of the species has been described by Range & Fischer (2004).

In recent years, sooty mangabeys have received considerable attention from the biomedical community. The human immunodeficiency virus type 2 (HIV-2) is thought to have originated from simian immunodeficiency viruses, which occur naturally in sooty mangabeys (SIVsm). Chen *et al.* (1996) showed that viruses of eight feral sooty mangabeys from West Africa belonged to the SIVsm/HIV-2 family, although they were widely divergent from SIVs found earlier in captive monkeys at American primate centers. Their findings support the hypothesis that each HIV-2 subtype in West Africans originated from widely divergent SIVsm strains transmitted by independent cross-species events in the same geographic locations (see also Marx *et al.* 1991, Gao *et al.* 1992). One current research topic concerns the question why SIV does not induce acquired immuno-deficiency syndrome (AIDS) in sooty mangabeys (Silvestri 2005). Surprisingly, the mangabeys' immune system does not suffer any damage despite highly replicating viruses. It may thus be the case that sooty mangabeys have evolved a special mechanism for resisting AIDS development (Ling *et al.* 2004), suggesting that understanding this mechanism may provide clues to understanding the pathogenesis of immunodeficiency in HIV-infected humans (Santiago *et al.* 2005). Research into this and related questions is being carried out cooperatively by the TMP and Yerkes National Primate Research Center. The IUCN Red List of Threatened Species considers this taxon as *Near Threatened.*

Cercopithecus diana diana *(Linnaeus 1758) Diana monkey*

There are two subspecies of Diana monkeys: *Cercopithecus diana roloway* (Roloway monkeys) east of Ivory Coast's Sassandra River and *C. diana diana* (Diana monkey, proper) to the west. All reports indicate that both subspecies require high, primary rainforest and that they do not fare well in disturbed areas and secondary forest (Booth 1958a, 1958b, Oates 1988a, 1988b, Whitesides 1989). The Roloway monkey is one of the world's 25 most endangered primates and the Diana monkey appears to be not far behind (McGraw 1998b, McGraw & Oates in press).

Diana monkeys are the most active, acrobatic and conspicuous monkeys at Taï. They are beautifully adorned, with black faces, short white beards, white chests, black/auburn coats, bright reddish orange hair on the rump and inner thighs and black tails (see Figure 1.14). The monkey is sexually dimorphic; mean male body weight is 5.2 kg and mean female body weight is 3.9 kg (Oates *et al.* 1990). The inter-membral index for *C. diana* is 79.

The first intensive studies of Diana monkeys were conducted at Tiwai Island, Sierra Leone where it was revealed that the typical Diana monkey social unit consisted of a single adult male, 6 or 7 females and their offspring (Whitesides 1989, Oates *et al.* 1990, Hill 1994). At Tiwai, Diana monkeys are almost always found in association with other monkey species (Oates & Whitesides 1990, Whitesides 1991). The same is true for Diana monkeys at Taï (Bshary & Noe 1997b, Wachter *et al.* 1997, Korstjens *et al.* 2002). Previous dietary information is available from Tiwai Island (Oates & Whitesides 1990) and from Taï (Galat & Galat-Luong 1985). These studies, as well as those at Taï, indicate that Diana monkeys eat large amounts of fruit and insects with smaller amounts of flowers and leaves.

Diana monkeys are noisy, active, fast and agile primates that appear to be in constant motion throughout the forest, from the top of the emergent layer to the ground (see Figure 1.15). Their foraging regime, alert nature and constant vigilance make them excellent early warning signallers for predators. Evidence from Taï indicates that at least one monkey – red

Figure 1.14. The Diana monkey *Cercopithecus diana* (Photo: Scott McGraw).

colobus – preferentially associates with Diana monkeys because of its ability to detect monkey-hunting chimpanzees (Bshary & Noe 1997a, 1997b, Noe & Bshary 1997) (see Figure 1.16). Detailed studies of the acoustic and semantic properties of Diana monkey vocalizations, particularly their loud calls, have revealed that males and females produce acoustically distinct alarm calls for two of their predators, leopards and crowned eagles (see Chapter 8). For several reasons, Diana monkeys can be regarded as the central species in the Taï monkey community: two colobine species frequently associate with Diana monkeys (one – *Procolobus verus*, permanently), Diana monkey groups dominate those of the two other guenon species (*C. campbelli* and *C. petaurista*) that share a common home range, and Diana monkeys have competitively displaced a third guenon at Taï (*C. nictitans*). Thus, an understanding of the Taï monkey community depends to a large extent on exploring the relationship of all cercopithecids to *C. diana*. The 2000 IUCN Red List of Threatened Species considers this taxon as *Endangered,*

Figure 1.15. Diana monkeys are agile, active foragers that are found at all layers of the forest canopy (Photo: Scott McGraw).

Figure 1.16. Red colobus monkeys associate with Diana monkeys because of their ability to detect monkey-hunting chimpanzees. The active foraging strategy, extensive use of all forest layers and alert nature of Diana monkeys make them excellent sentinels (Photo: Scott McGraw).

Figure 1.17. An adult male Campbell's monkey *Cercopithecus campbelli* (Photo: Florian Möllers).

facing a very high risk of extinction in the wild in the near future (Hilton-Taylor 2000).

Cercopithecus campbelli campbelli *(Waterhouse 1838)*
Campbell's monkey

Campbell's monkey – *Cercopithecus campbelli campbelli* – is one of several species in the Mona super-species of guenons (Booth 1955, 1956a, 1956b, 1958a, 1958b). It is found from Senegal to parts east of the Taï Forest meaning that the Cavallay River forming the Ivory Coast-Liberia border is not an effective barrier to dispersion. The species is capable of exploiting many habitats including highly disturbed areas and low, secondary forest (Booth 1955, 1956a, 1956b, 1958a, 1958b, Bourliere *et al.* 1970, Hunkeler *et al.* 1972, Galat-Luong & Galat 1979, Harding 1984, Oates 1988b). For this reason, Campbell's monkey is one of the most abundant monkeys in West Africa (McGraw 1998b, Oates *et al.* 2000b).

The coat of Campbell's monkey is drab olive-gray with darker hair towards the distal ends of the limbs (see Figure 1.17). A yellow brow band above the darkened blue shading around the orbits marks the face. Campbell's monkeys are among the most sexually dimorphic of all guenons with mean male body weights of 4.5 kg and mean female body weights of 2.7 kg (Oates *et al.* 1990). The inter-membral index for *C. campbelli* is 85 (McGraw, this study).

C. campbelli has not been studied intensively outside of Taï, however available reports suggest that the behavior of Campbell's monkey at Taï is representative of this species throughout its range. *C. campbelli* is a cryptic monkey that, like the olive colobus, is adapted to the shadows of the dark understory. Group sizes average approximately 11 individuals and consist of single adult male, 3–4 adult females and their offspring. In contrast to the raucous Diana monkey, Campbell's monkeys produce much softer contact calls, travel and forage primarily at levels below the main canopy (including the ground) and are generally inconspicuous. They can be found in association with all other cercopithecid species, but show a marked tendency to move and feed with another cryptic guenon at Taï – the lesser spot-nosed monkey *Cercopithecus petaurista*. Indeed, the niches of these two species are quite similar and adult males of associated groups perform an unusual three-unit loud call duet, initiated by the male Campbell's monkey's two booms, which are immediately answered by the male lesser spot-nosed monkey's loud calls, completed by the Campbell's monkey's hacks. The function of this intricate behavior is currently under investigation but preliminary data suggest that it may serve as a means of joint territorial defense. The IUCN Red List of Threatened Species considers Campbell's monkeys as *Not Threatened*.

Cercopithecus petaurista buettikoferi *(Jentink 1886) Western lesser spot-nosed monkey*

There are two subspecies of lesser spot-nosed monkeys: *C. petaurista petaurista* is found east of the Cavally River and *C. petaurista buettikoferi* to the west, including Taï. Intermediate forms have been reported between Sassandra-Cavally-N'zo river systems. The lesser spot-nosed monkey is part of the *cephus* group of guenons whose members are characterized by small body size, frugivorous diet, cryptic behavior, and habitat flexibility (Lernould 1988, Oates 1988b). These shy, cryptic monkeys are quite difficult to follow in the forest because of their ability to quickly and quietly leave an area (all members of our team have – at one time or another – been frustrated by the ability of this monkey to seemingly vanish!). This quality, combined with the ability to exploit multiple habitat types including degraded forest, explains why this small and adaptable guenon is one of the most common primates in West Africa (McGraw 1998a, 1998b, Oates *et al.* 2000b).

Lesser spot-nosed monkeys are the smallest monkeys at Taï; mean body weight for males is 4.4 kg, and females average 2.9 kg (Oates *et al.* 1990).

Figure 1.18. An adult female lesser spot-nosed monkey *Cercopithecus petaurista* (Photo: Scott McGraw).

Their coat is agouti-brown and the underside of the limbs and chest are light colored (see Figure 1.18). A white stripe marks the side of the dark face but it is the white or pinkish heart-shaped spot on the nose that is the most recognizable feature of this monkey. The inter-membral index is 80 (McGraw, this study). The Western lesser spot-nosed monkey has not been intensively studied outside of Taï. Data culled from various reports indicate that like *Cercopithecus campbelli*, *C. petaurista* is generally found in small, single male groups that exploit the lower levels of forests. Lesser spot-nosed monkeys associate frequently with all other cercopithecids in the forest (see Figure 1.19). Their diet consists of large percentages of fruit, flowers and insects. Their vocal behavior has not yet been studied intensively. The IUCN Red List of Threatened Species considers lesser spot-nosed monkey as *Not Threatened*.

Cercopithecus nictitans stampflii *(Jentinck 1888) Stampfli's putty-nosed monkey*

Putty-nosed monkeys are large, long-tailed, arboreal primates with dark, grizzled olive fur on the back, crown, cheeks, and base of the tail. The limbs and distal half of the tail are black or dark gray. The brilliant white nose spot in a dark face is striking (see Figure 1.20).

Figure 1.19. Like all Taï guenons, lesser spot-nosed monkeys frequently associate and interact with other cercopithecids in the forest. This photograph shows a red colobus monkey grooming a female lesser spot-nosed monkey (Photo: Scott McGraw).

Putty-nosed populations east of the River Cross, Nigeria, are generally numerous (e.g. Mitani 1991, Garcia & Mba 1997). In contrast, population densities of Stampfli's putty-nosed monkeys in Côte d'Ivoire and Liberia are very low, suggesting that they may have been recent and non-competitive colonists to habitats already occupied by other primates, particularly the lesser spot-nosed monkeys (*C. petaurista*) and Diana monkeys (*C. diana*). For example, in the Odzala National Park, Republic of the Congo, density was estimated to be about 1.4 groups/km^2 (Bermejo 1999). Similarly, in the Campo-Ma'an area, Southwestern Cameroon, group density was 1.43 groups/km^2 (Mathews & Matthews 2002), while in Ipassa-Makou, north-east Gabon, density was estimated at 56.43 ind./km^2 (Okouyi *et al.* 2002).

Putty-nosed monkeys live in groups of 12 containing a single adult male and four adult females (Eckardt & Zuberbühler 2004). The feeding ecology of *C. nictitans* has been documented at Lope, Gabon (Tutin *et al.* 1997) and Taï (Eckardt 2002, Eckardt & Zuberbühler 2004). Other ecological work has been conducted at Bioko Island, Republic of Equatorial Guinea (Gonzalez-Kirchner 1996). The vocal behavior of *C. nictitans* has been studied at Taï where males exhibit an interesting pattern of combining two

Figure 1.20. Stampfli's putty-nosed monkey, *Cercopithecus nictitans stampflii*. This monkey is probably a recent arrival to the Taï Forest having migrated from northern savannah regions. It lives at low densities and is dominated by groups of Diana monkeys (Photo: Winnie Eckardt).

basic types of alarm calls – the pyows and the hacks – into structurally more complex sequences with novel meanings (Eckardt & Zuberbühler 2004). The positional behavior of putty-nosed monkeys is currently being investigated (Bitty & McGraw 2006, in press).

The taxonomic position of Stampfli's putty-nosed monkeys is debated (Groves 2001, Grubb *et al.* 2003). If afforded full species status, the species is considered *Critically Endangered* and faces a high risk of extinction due to its very patchy distribution in areas of heavy hunting pressure.

Format of the book

The book contains 11 chapters authored by 22 researchers. The chapters are grouped into four parts dealing with questions of social behavior, anti-predator strategies, habitat use, and conservation.

Part I – Social behavior

The first section consists of three chapters, each covering key elements of social behavior in the three radiations of Taï monkeys: guenons, mangabeys, and colobines. Buzzard and Eckardt describe the social systems of the four guenon species, reviewing a set of previously unpublished material. Obtaining social data from forest guenon species has proven to be challenging, which makes Buzzard and Eckardt's contribution particularly important. The authors find dramatic differences in group-densities with Diana monkeys, Campbell's monkeys, and lesser spot-nosed monkeys being relatively equally common, while putty-nosed monkeys are much rarer. The typical group structure was single male/ multi-female, although occasionally two adult males were observed for certain periods. Stable all male groups were only observed in Campbell's monkeys. Buzzard and Eckardt then provide data on these species' social behavior, noting that levels of social interactions are substantially lower than those of macaques or baboons. Across species, grooming rates were comparable, although Diana monkeys were characterized by higher rates of agonistic behavior to other group members (e.g. McGraw *et al.* 2002).

There are few forests in Africa that boast three sympatric species of colobus monkeys. Because the red colobus, King colobus, and olive colobus differ in so many aspects of their behavior, Taï provides a superb opportunity to explore the determinants of group size, dispersal patterns, diet and anti-predation defenses in a comparative context while controlling for phylogeny. Perhaps not surprisingly, the three colobus species have been the most studied monkeys at Taï and over a dozen students have carried out long-term projects on the socio-ecology of one or more social groups. The chapter by Korstjens *et al.* represents the work of nine students whose combined data address the relationship between food competition, contestability and social structure within the framework of current socio-ecological theory. Among other things, these authors demonstrate that the King colobus is the most frugivorous and has the highest intra-group feeding competition of the three colobine species present. Females of all three species disperse at least occasionally suggesting that affiliative relationships among females are weak.

Range *et al.*'s chapter deals with social relationships among sooty mangabeys. These authors demonstrate that females are the philopatric sex and form linear, stable dominance hierarchies. Range *et al.* show how high rank is beneficial in terms of feeding competition and particularly predator avoidance. Among males, access to females was rank-dependent, suggesting that social dominance is closely linked with reproductive success in both sexes. Finally, Range *et al.* show that individuals of both sexes form well-differentiated relationships with preferred partners, whom they groomed, associated with and supported during agonistic interactions. These authors demonstrate that high-ranking females tended to form close relationships with high-ranking males. In several interesting ways, these results contrast with what is known about this species in captivity.

Part II – Anti-predation strategies

Understanding the evolution and mechanisms underlying the anti-predator behavior of the Taï monkeys has been a major focus of the project, particularly in its early stages. This interest was prompted by earlier work by Boesch and Boesch (1989), showing the extraordinary high hunting pressure exerted by the Taï chimpanzees on the sympatric cercopithecids. Recognizing predation as a key selective factor, an overarching hypothesis was that the Taï monkeys have responded to chimpanzees by evolving species-specific anti-predator strategies. It quickly became apparent that chimpanzees were part of a larger, more complex system in which three other monkey predators played a major role: crowned eagles (*Stephanoaetus coronatus*), leopards (*Panthera pardus*) and, more recently, human poachers. Over the years, it has been possible to collect a large set of behavioral data on these monkey predators and we are now able to provide a near complete picture on the impact of these predators both individually and collectively.

Zuberbühler and Jenny review their work on the impact of leopard predation on the primate community. These authors studied the content of leopard feces as well as the ranging behavior of several radio-tagged individuals in relation to the ecological characteristics of each monkey species. A number of unexpected findings have emerged. First, Taï leopards appear to be selective hunters that develop individual prey preferences. Monkeys and duikers are particularly targeted. Second, it became apparent that the various species differed in how vulnerable each was to leopard predation, in ways that often contradicted current theory. For instance, larger species living in larger groups suffered disproportionately high losses suggesting that body size, group size, and the number of males per group are not adaptive traits to avoid predation by leopards. Zuberbühler and Jenny show how the various species use their vocal behavior in response to different predators and demonstrate how the hunting techniques of each predator have resulted in the evolution of species-specific anti-predator strategies.

The next chapter by Bshary concerns the interactions between red colobus monkeys and chimpanzees. Bshary carried out experiments using recordings of chimpanzee vocalizations and drumming on tree buttresses to simulate the sounds produced with chimpanzees prior to a hunt. He was primarily interested in the responses of red colobus monkeys, who routinely fall victim to chimpanzees, as well as those of Diana monkeys, who rarely do. Bshary's data strongly indicate that a major "anti-chimpanzee" strategy of red colobus is to seek out associations with the vigilant Diana monkeys. Bshary shows that red colobus monkeys respond to the sounds of chimpanzees by moving towards and remaining in association with Diana monkeys and that chimpanzees tend to avoid groups of red colobus when they determine that Diana monkeys are in the vicinity. The red colobus-chimpanzee relationship at Taï differs strikingly from that at sites in East Africa and can be explained by differences in forest structure, the size of red colobus groups and the hunting tactics required in each system (Boesch 1994, Stanford 1998). Bshary concludes by discussing the adaptations and counter adaptations in this evolving arms race.

The sudden and fleeting nature of attacks by raptors on forest dwelling primates does not lend itself to systematic scrutiny and long-term studies on the relationship between arboreal cercopithecids and monkey-eating crowned eagles are few. Nevertheless, Shultz and Thomsett use a combination of techniques to provide a detailed look at the interactions between crowned eagles and their jungle prey. Data from radio-transmitters affixed to two adult and two juvenile eagles provide information on ranging behavior and activity budgets. Feeding remains in the form of bones collected from within and beneath 12 eagle nests are used to compile prey profiles that are then examined in light of specific ecological characteristics of each prey species. In so doing, Shultz and Thomsett highlight the biases in eagle diets and discuss the criteria eagles use to preferentially hunt different taxa. These data are complemented by the author's observations of eagle attacks as well as the responses of monkeys to eagle alarm calls to present a complex portrait of eagle-monkey interactions (Shultz 2001 and Shultz *et al.* 2004).

The section on anti-predator strategies concludes with a review of one of the most prominent aspects of monkey anti-predator behavior: alarm calls. All Taï monkeys produce these vocalizations in response to danger, but there are substantial inter-species differences. A chapter by Zuberbühler provides a survey of the experimental work conducted on the alarm call behavior of the various monkey species. Although still incomplete, the emerging picture is one of remarkable species-specific differences in alarm calling behavior. In the case of guenons the idiosyncratic solutions are especially apparent: although Diana monkeys, Campbell's monkeys, and spot-nosed monkeys are all hunted by the same predators, the natural selection pressure exerted by them has led to widely different vocal behavior. The conclusion from these observations is that the vocal repertoire of a species has evolved in response to the

Figure 1.21. A Diana monkey twines its tail around a vertical trunk as it scans the under story for fruit, insects and, perhaps, predators (Photo: Scott McGraw).

particularities of their species-specific ecological niche and social organization. How exactly the monkeys use their various calls, however, is a rather flexible affair. Compliant call usage allows individuals to respond adaptively to differences in type and degree of predation pressure, which are likely to occur over evolutionary times. It also leads to dramatic species difference in calling behavior.

Part III – Habitat use

The single chapter in this section is a comparative examination of the locomotion, posture and habitat use of the seven species found in the study area's central grid. McGraw provides positional behavior, canopy use, and support use profiles for all species and discusses how these can be used to understand additional aspects of each monkey's natural history (see Figure 1.21). He then evaluates the extent that variables known to co-vary in other primate groups are similarly related in the Taï monkeys. The results show that canopy partitioning is critical for permitting monkey sympatry and that species with similar canopy preferences tend to avoid one another. Maintenance activities are not strongly associated with particular canopy levels, but rather with support size. There is a strong relationship between body size and large supports, but not with medium and small supports. Body size is only weakly associated with particular locomotor behaviors and inter-membral index is not a consistent predictor of leaping frequencies. McGraw discusses the significant differences in the postural behavior between cercopithecid subfamilies and argues that these are almost certainly related to the spatial configuration and digestive requirements of colobines versus cercopithecine foods.

Part IV – Conservation

The future of the Taï monkeys is uncertain. Hunting for bushmeat is a daily event, and those who have worked in Taï are unlikely to forget the sound of gunshots that shatter the forest calm. Due to the continuous presence of researchers, poaching is low in some areas of the park, but the vast majority of the forest is used as a hunting ground for both commercial and local subsistence hunters. Some success has been made in curbing illegal hunting activity, but we have yet to detect a widespread political will to aggressively confront the problem. Bushmeat is very popular in large segments of West African society suggesting that until habits change, demand will remain high (see Figure 1.22). The monkeys have little means to protect themselves against these activities. Nevertheless, some adaptations have occurred and Koné and Refisch examine the impact of the most dangerous monkey predator – humans – on two Taï cercopithecids. These authors compare the behavior of Diana monkeys and red colobus monkeys in an area of high poaching with that in a region not frequented by human hunters. They conclude the behavior of Diana monkeys differs significantly between localities but that of red colobus monkeys does not. The inability of red colobus to alter its behavior in response to poaching pressure significantly increases its vulnerability, an idea explored in more detail in the next chapter.

In the final paper, McGraw uses data from preceding chapters, including those on habitat sensitivity, diet, and response to predators to assess the extirpation risk of the seven main monkey species at Taï. He discusses the synergistic factors responsible for putting several species at great risk and concludes that naturally high densities do not necessarily

Figure 1.22. An Ivorian soldier holds a red colobus monkey being sold immediately adjacent to Taï National Park. Poaching is widespread in the forest (Photo: Scott McGraw).

provide protection if naturally abundant primates are conspicuous and ill-equipped to avoid and flee from poachers. The chapter concludes with a discussion of the threats to the Taï fauna as well as recommendations for the conservation of the park and its inhabitants.

References

Adachi-Kanazawa, K. (2004). *Polyspecific associations of Cercopithecus monkeys in the Taï National Park, Ivory Coast*. Ph.D. Thesis, Kyoto University, Japan.

Aidara, D., Tahirizagret, C. and Robyn, C. (1981). Serum prolactin concentrations in Mangabey (*Cercocebus atys lunulatus*) and Patas (*Erythrocebus patas*) monkeys in response to stress, ketamine, Trh, sulpiride and levodopa. *Journal of Reproduction and Fertility*, **62**, 165–72.

Aldrich-Blake, F. P. G. (1968). A fertile hybrid between two *Cercopithecus* species in the Budongo Forest, Uganda. *Folia Primatologica*, **9**, 15–21.

Aldrich-Blake, F. P. G. (1970). *The ecology and behavior of the blue monkey Cercopithecus mitis stuhlmanni*. Ph.D. Thesis, University of Bristol, UK.

Alexandre, D. Y. (1978). Le role disseminateur des éléphants en forêt de Taï, Côte d'Ivoire. *La Terre et la Vie*, **32**, 47–72.

Allen, C. and Bekoff, M. (1997). *Species of Mind: the Philosophy and Biology of Cognitive Ethology*. Cambridge, MA: MIT Press.

Altmann, S. A. and Altmann, J. (1970). *Baboon Ecology*. Chicago: University of Chicago Press.

Balchin, C. S. (1988). Recent observations of birds from the Ivory Coast. *Malimbus*, **10**, 201–6.

Baranga, D. (1982). Nutrient composition and food preferences of colobus monkeys in Kibale Forest, Uganda. *African Journal of Ecology*, **20**, 113–21.

Baranga, D. (1983). Changes in chemical composition of food plants in the diet of colobus monkeys. *Ecology*, **64**, 668–73.

Baranga, D. (1986). Phenological observation on two food-tree species of colobus monkeys. *African Journal of Ecology*, **24**, 209–14.

Benneton, C. (2002). *The reproductive tactics of adult male sooty mangabeys in the Taï National Park, Ivory Coast*. Dipl. University Louis Pasteur of Strasbourg, France.

Bergmann, K. (1998). *Niche differentiation between three sympatric colobine monkeys in Taï National Park Ivory Coast*. Dipl. Johann Wolfgang Goethe-University, Frankfurt am Main, Germany.

Bergmüller, R. (1998). Feeding ecology of the sooty mangabey (*Cercocebus torquatus atys*). A key for its social organisation? Dipl. Friedrich-Alexander-Universität Erlangen-Nürnberg, Germany.

Bergmüller, R., Rutte, C. and Noë, R. submitted. Sooty mangabeys (*Cercocebus torquatus atys*) adjust group size to cope with fluctuating food availability. *Behavioural Ecology and Sociobiology*.

Bermejo, M. (1999). Status and conservation of primates in Odzala National Park, Republic of the Congo. *Oryx*, **33**, 323–31.

Bernstein, I. S. (1971a). Agonistic behavior during group formation and inter-group interactions in sooty mangabeys (*Cercocebus atys*). *Proceedings 3rd International Congress Primatology*, Zurich 1970, **3**, pp. 66–70.

Bernstein, I. S. (1971b). The influence of introductory techniques on the formation of captive mangabey groups. *Primates*, **12**, 33–44.

Bernstein, I. S. (1976). Activity patterns in a sooty mangabey group. *Folia Primatologica*, **26**, 185–206.

Bitty, E. A. (2001). What determines the ranging pattern of the black and white colobine, *Colobus polykomos* in the Taï National Park (Ivory Coast). Dipl. University of Cocody, (Abidjan, Ivory Coast).

Bitty, E. A. and McGraw, W. S. (2006). Locomotor behavior of *Cercopithecus nictitans stampflii* in the Taï National Park, Ivory Coast. *American Journal of Physical Anthropology*, **42**, 65.

Blank, C. (1997). *A poly-specific association between western red colobus (Colobus badius) and Diana monkeys (Cercopithecus diana) of the Taï National Park, Ivory Coast (West Africa): the influence of the association on the food choice of both species*. Dipl. Institute of Ecology and Evolutionary Biology: Bremen, Germany.

Boesch, C. (1994). Hunting strategies of Gombe and Taï chimpanzees. In *Chimpanzee Cultures*, ed. R. Wrangham, W. McGrew, F. D. Waal and P. Heltne. Cambridge, MA: Harvard University Press, pp. 77–92.

Boesch, C. and Boesch, H. (1989). Hunting behavior of wild chimpanzees in the Taï National Park. *American Journal of Physical Anthropology*, **78**, 547–73.

Boesch, C. and Boesch-Achermann, H. (2000). *The Chimpanzees of the Taï Forest. Behavioural ecology and evolution*. Oxford: Oxford University Press.

Booth, A. H. (1954). A note on the colobus monkeys of the Gold and Ivory Coasts. *Annals and Magazine of Natural History*, **12**, 857–60.

Booth, A. H. (1955). Speciation in the mona monkeys. *Journal of Mammalogy*, **36**, 434–49.

Booth, A. H. (1956a). The Cercopithecidae of the Gold and Ivory Coasts: geographic and systematic observations. *Annals and Magazine of Natural History, 9th Ser.*, **9**, 476–80.

Booth, A. H. (1956b). The distribution of primates in the Gold Coast. *Journal of the West African Science Association*, **2**, 122–33.

Booth, A. H. (1957). Observations of the natural history of the olive colobus monkey, *Procolobus verus* (van Beneden). *Proceedings of the Zoological Society of London*, **129**, 421–30.

Booth, A. H. (1958a). The Niger, the Volta and the Dahomey Gaps as geographic barriers. *Evolution*, **12**, 48–62.

Booth, A. H. (1958b). The zoogeography of West African primates: a review. *Bulletin de l'Institut Français d'Afrique Noire*, **20**, 587–622.

Booth, A. H. (1960). *Small Mammals of West Africa*. London: Longmans.

Bourliere, F., Hunkeler, C. and Bertrand, M. (1970). Ecology and behavior of Lowe's guenon (*Cercopithecus campbelli lowei*) in Ivory Coast. In *Old World Monkeys: Evolution, Systematics and Behaviour*, ed. J. R. Napier and P. H. Napier. New York: Academic.

Bshary, R. (1995). Red colobus and Diana monkeys in the Taï National Park, Ivory Coast: Why do they associate? Ph.D. Thesis, Ludwig-Maximilian-Universität, Munich, Germany.

Bshary, R. and Noe, R. (1997a). Anti-predation behavior of red colobus monkeys in the presence of chimpanzees. *Behavioral Ecology and Sociobiology*, **41**, 321–33.

Bshary, R. and Noe, R. (1997b). Red colobus and Diana monkeys provide mutual protection against predators. *Animal Behaviour*, **54**, 1461–74.

Busse, C. D. (1977). Chimpanzee predation as a possible factor in the evolution of red colobus monkey social organisation. *Evolution*, **31**, 907–11.

Busse, C. D. and Gordon, T. P. (1984). Infant carrying by adult male mangabeys (*Cercocebus atys*). *American Journal of Primatology*, **6**, 133–41.

Buzzard, P. (2004). Interspecific competition among *Cercopithecus campbelli*, *C. petaurista*, and *C. diana* at the Taï Forest, Côte d'Ivoire. Ph.D. Thesis, Columbia University, USA.

Chalmers, N. R. (1968a). Group composition, ecology and daily activities of free living mangabeys in Uganda. *Folia Primatologica*, **8**, 247–62.

Chalmers, N. R. (1968b). The social behaviour of free living mangabeys in Uganda. *Folia Primatologica*, **8**, 263–81.

Chapman, C. A. and Chapman, L. J. (1996). Mixed-species primate groups in the Kibale Forest: ecological constraints on association. *International Journal of Primatology*, **17**, 31–50.

Chapman, C. A. and Chapman, L. J. (2002). Foraging challenges of red colobus monkeys: influence of nutrients and secondary compounds. *Comparative Biochemistry and Physiology Part A*, **133**, 861–75.

Chapman, C. A., Chapman, L. J., Bjorndal, K. A. and Onderdonk, D. A. (2002a). Application of protein-to-fiber ratios to predict colobine abundance on different spatial scales. *International Journal of Primatology*, **23**, 283–310.

Chapman, C. A., Chapman, L. J. and Gillespie, T. R. (2002b). Scale issues in the study of primate foraging: red colobus of Kibale National Park. *American Journal of Physical Anthropology*, **117**, 349–63.

Chapman, C. A., Gautier-Hion, A., Oates, J. F. and Onderdonk, D. A. (1999). African primate communities: determinants of structure and threats to survival. In *Primate Communities*, ed. J. Fleagle, C. Janson and K. Reed. Cambridge: Cambridge University Press, pp. 1–37.

Chen, Z. W., Telfer, P., Gettie, A. *et al.* (1996). Genetic characterization of new West African simian immunodeficiency virus SIVsm: geographic clustering of household-derived SIV strains with human immunodeficiency virus type 2 subtypes and genetically diverse viruses from a single feral sooty mangabey troop. *Journal of Virology*, **70**, 3617–27.

Clutton-Brock, T. H. (1973). Feeding levels and feeding sites of red colobus (*Colobus badius tephrosceles*) in the Gombe National Park. *Folia Primatologica*, **19**, 368–79.

Clutton-Brock, T. H. (1974). Activity pattern of red colobus (*Colobus badius tephrosceles*). *Folia Primatologica*, **21**, 161–87.

Clutton-Brock, T. H. (1975a). Feeding behavior of red colobus and black and white colobus in East Africa. *Folia Primatologica*, **23**, 165–207.

Clutton-Brock, T. H. (1975b). Ranging behavior of red colobus (*Colobus badius tephrosceles*) in Gombe National Park. *Animal Behaviour*, **23**, 706–22.

Cords, M. (1984). Mixed species groups of blue and redtail monkeys in the Kakamega Forest, Kenya. *International Journal of Primatology*, **5**, 329.

Cronin, J. E. and Sarich, V. M. (1976). Molecular evidence for the dual origin of mangabeys among Old World monkeys. *Nature*, **260**, 700–2.

Crook, J. H. (1966). Gelada baboon herd structure and movement. A comparative report. *Symposium of the Zoological Society of London*, **18**, 237–58.

Crook, J. H. and Aldrich-Blake, P. (1968). Ecological and behavioral contrasts between sympatric ground-dwelling primates in Ethiopia. *Folia Primatologica*, **8**, 192–227.

Daegling, D. J. and McGraw, W. S. (2001). Feeding, diet, and jaw form in West African *Colobus* and *Procolobus*. *International Journal of Primatology*, **22**, 1033–55.

Dasilva, G. L. (1992). The western black-and-white colobus as a low-energy strategist: activity budgets, energy expenditure and energy intake. *Journal of Animal Ecology*, **61**, 79–91.

Dasilva, G. L. (1994). Diet of *Colobus polykomos* on Tiwai Island – selection of food in relation to its seasonal abundance and nutritional quality. *International Journal of Primatology*, **15**, 655–80.

Davies, A. G. and Oates, J. F. (1994). *Colobine Monkeys: Their Ecology, Behaviour and Evolution*. Cambridge, UK: Cambridge University Press.

Davies, A. G., Oates, J. F. and Dasilva, G. L. (1999). Patterns of frugivory in three West African colobine monkeys. *International Journal of Primatology*, **20**, 327–57.

Decker, B. S. (1994). Effects of habitat disturbance on the behavioral ecology and demographics of the Tana River red colobus (*Procolobus badius rufomitratus*). *International Journal of Primatology*, **15**, 703−37.

Dekeyser, P. L. (1955). *Les mammiferes de l'Afrique noire française*. Dakar: Institut Fondamontale d'Afrique Noire.

Deschner, T. (1996). Aspects of social behavior of the olive colubus, *Colobus verus* (van Beneden 1838) in the Taï National park, Ivory Coast. M.A. Thesis, University of Hamburg, Germany.

Disotell, T. R. (1994). Generic level relationships of the Papionini (Cercopithecoidea). *American Journal of Primatology*, **94**, 47−57.

Disotell, T. R. and Raaum, R. L. (2002). Molecular timescale and gene tree congruence in the guenons. In *The Guenons: Diversity and Adaptation in African Monkey*, ed. M. Glenn and M. Cords. New York: Kluwer Academic/Plenum, pp. 27−36.

Dunbar, R. I. M. and Dunbar, E. P. (1974). The reproductive cycle of the gelada baboon. *Animal Behaviour*, **22**, 203−10.

Dunham, A. E. (2003). Effects of understory insectivores on community dynamics and ecosystem processes in a tropical forest. Ph.D. Thesis, Stony Brook University, USA.

Eckardt, W. (2002). Niche comparison between putty-nosed monkeys and Diana monkeys in the Taï National Park, Ivory Coast. Unpublished Masters Thesis, Universitaet Leipzig, Dept. Biowissenschaften, Leipzig.

Eckardt, W. and Zuberbühler, K. (2004). Cooperation and competition in two forest monkeys. *Behavioral Ecology*, **15**, 400−11.

Ehardt, C. L. (1988a). Absence of strongly kin-preferential behavior by adult female sooty mangabeys (*Cercocebus atys*). *American Journal of Physical Anthropology*, **76**, 233−43.

Ehardt, C. L. (1988b). Affiliative behavior of adult female sooty mangabeys (*Cercocebus atys*). *American Journal of Primatology*, **15**, 115−27.

Elliot, D. (1913). *A Review of the Primates*. New York: American Museum of Natural History.

Fimbel, C. C. (1992). Cross-species handling of colobine infants. *Primates*, **33**, 545−649.

Fleagle, J. G. (1999). *Primate Adaptation and Evolution*. New York: Academic Press.

Fleagle, J. G., Janson, C. H. and Reed, K. E. (1999). *Primate Communities*. Cambridge: Cambridge University Press.

Fleagle, J. G. and McGraw, W. S. (1999). Skeletal and dental morphology supports diphyletic origin of baboons and mandrills. *Proceedings of the National Academy of Sciences*, **96**, 1157−61.

Fleagle, J. G. and McGraw, W. S. (2002). Skeletal and dental morphology of African papionins: unmasking a cryptic clade. *Journal of Human Evolution*, **42**, 267−92.

Fleagle, J. G. and Reed, K. E. (1996). Comparing primate communities: A multivariate approach. *Journal of Human Evolution*, **30**, 489−510.

Fleagle, J. G. and Reed, K. E. (1999). Phylogenetic and temporal perspectives on primate ecology. In *Primate Communities*, ed. J. G. Fleagle, C. H. Janson and K. E. Reed. Cambridge: Cambridge University Press.

Förderer, T. (2001). Orientation in free ranging sooty mangabeys at Taï National Park/Ivory Coast. M.A. Thesis, Westfälische Wilhelms-Universität Münster, Germany.

Fultz, P. N., McClure, H. M., Anderson, D. C., Swenson, R. B., Anand, R. and Srinivasan, A. (1986). Isolation of a T-lymphotropic retrovirus from naturally infected sooty mangabey monkeys (*Cercocebus atys*). *Proceedings of the National Academy of Sciences*, **83**, 5286–90.

Galat, G. (1978). Données ecologiques sur les singes de la région de Bozo. Adiopodoume, Côte d'Ivoire. ORSTROM: Rapport de Mission a Bozo, Empire Centrafricain.

Galat, G. and Galat-Luong, A. (1985). La communauté de primates diurnes de la fôret de Taï, Côte d'Ivoire. *Revve d'Ecologie (la Terre et la Vie)*, **40**, 3–32.

Galat-Luong, A. (1981). Some observations on a newborn pigmy hippopotamus in the Taï Forest, Ivory-Coast. *Mammalia*, **45**, 39.

Galat-Luong, A. and Galat, G. (1979). Consequences comportementales de perturbations sociales répétées sur une troupe de mones de lowe *Cercopithecus campbelli lowei* de Côte d'Ivoire. *Revue d'Ecologie (la Terre et la Vie)*, **33**, 49–58.

Galat-Luong, A. and Galat, G. (2005). Conservation and survival adaptations of Temminck's red colobus (*Procolobus badius temmincki*), in Senegal. *International Journal of Primatology*, **26**, 585–603.

Gao, F., Yue, L., White, A. T. *et al.* (1992). Human Infection by genetically diverse Sivsm-related Hiv-2 in West Africa. *Nature*, **358**, 495–9.

Garcia, J. E. and Mba, J. (1997). Distribution, status and conservation of primates in Monte Alen National Park, Equatorial Guinea. *Oryx*, **31**, 67–76.

Gartlan, J. S. and Struhsaker, T. T. (1972). Polyspecific associations and niche separation of rain-forest anthropoids in Cameroon, West Africa. *Journal of Zoology* (London, England), **168**, 221–66.

Gartshore, M. E. (1989). An avifaunal survey of Taï National Park, Ivory Coast. In *International Council for Bird Preservation Report No. 39*.

Gatinot, B. L. (1977). Diet of West-African red colobus in Senegal. *Mammalia*, **41**, 373–402.

Gautier, J. P. and Gautier-Hion, A. (1969). Les associations polyspecifiques chez les Cercopithecidas du Gabon. *La Terre et la Vie*, **2**, 164–201.

Gautier-Hion, A., Bourliere, F., Gautier, J.-P. and Kingdon, J. (1988). *A Primate Radiation: Evolutionary Biology of the African Guenons*. Cambridge: Cambridge University Press.

Gautier-Hion, A., Colyn, M. and Gautier, J. P. (1999). *Histoire Naturelle des Primates d'Afrique Centrale*. Libreville, Gabon: Ecofac.

Gleitman, H. (1999). *Psychology*. New York: WW Norton Company.

Glenn, M. E. and Cords, M. (eds.) (2002). *The Guenons: Diversity and Adaptation in African Monkeys*. New York: Kluwer Academic/Plenum.

Gonzalez-Kirchner, J. (1996). Habitat preference of two lowland sympatric guenons on Bioko island, equatorial Guinea. *Folia Zoologica*, **45**, 201–8.

Goodall, J. (1965). Chimpanzees of the Gombe Stream Reserve. In *Primate Behavior*, ed. I. Devore. New York: Holt, Rhinehart and Winston, pp. 425–73.

Goodall, J. (1968). The behavior of free-living chimpanzees of the Gombe Stream Reserve. *Animal Behavior Monographs*, **1**, 161–311.

Gordon, T. T. (2001). The calling behavior and mating system of a non-lekking population of *Hypsignathus monstrosus*. Ph.D. Thesis, SUNY Stony Brook, USA.

Gordon, T. P., Gust, D. A., Busse, C. D. and Wilson, M. E. (1991). Behavior associated with postconception perineal swelling in sooty mangabeys, *Cercocebus torquatus atys*. *International Journal of Primatology*, **12**, 585–97.

Groves, C. P. (1978). Phylogenetic and population systematics of the mangabeys (Primates: Cercopithecoidea). *Primates*, **19**, 1–34.

Groves, C. P. (2001). *Primate Taxonomy*. Washington: Smithsonian Institution Press.

Grubb, P. (1978). Patterns of speciation in African mammals. *Bulletin Carnegie Museum of Natural History*, **6**, 152–67.

Grubb, P. (1982). Refuges and dispersal in the speciation of African forest mammals. In *Biological Diversification in the Tropics*, ed. G. T. Prance. New York: Columbia University Press, pp. 537–53.

Grubb, P., Butynski, T. M., Oates, J. F. *et al.* (2003). Assessment of the diversity of African primates. *International Journal of Primatology*, **24**, 1301–57.

Gust, D. A. (1994). Alpha-male sooty mangabeys differentiate between females' fertile and their post-conception maximal swellings. *International Journal of Primatology*, **15**, 289–301.

Gust, D. A. and Gordon, T. P. (1991). Male age and reproductive behaviour in sooty mangabeys *Cercocebus torquatus atys*. *Animal Behaviour*, **41**, 277–84.

Gust, D. A. and Gordon, T. P. (1993). Conflict resolution in sooty mangabeys. *Animal Behaviour*, **46**, 685–94.

Gust, D. A. and Gordon, T. P. (1994). The absence of matrilineally based dominance system in sooty mangabeys, *Cercocebus torquatus atys*. *Animal Behaviour*, **47**, 589–94.

Haddow, A. J. (1952). Field and laboratory studies on an African monkey, *Cercopithecus ascanius schmidti* Matschie. *Proceedings of the Zoological Society of London*, **122**, 297–394.

Hadidian, J. and Bernstein, I. S. (1979). Female reproductive cycles and birth data from an Old World monkey colony. *Primates*, **20**, 429–42.

Hall, K. R. L. (1962). Numerical data, maintenance activities, and locomotion of the wild chacma baboon, *Papio ursinus*. *Proceedings of the Zoological Society of London*, **10**, 1–28.

Hall, K. R. L. (1965). Behavior and ecology of wild Patas monkeys, *Erythrocebus patas*, in Uganda. *Journal of Zoology*, **148**, 15–87.

Hamilton, A. C. (1988). Guenon evolution and forest history. In *A Primate Radiation: Evolutionary Biology of the African Guenons*, ed. A. Gautier-Hion, F. Bourliere, J.-P. Gautier and J. Kingdon. Cambridge: Cambridge University Press, pp. 13–34.

Hansen, F. (1996). *The relation between niche and predation in three guenons*. Diploma, Kiel University: Kiel, Germany.

Harding, R. S. O. (1984). Primates of the Kilimi Area, Northwest Sierra-Leone. *Folia Primatologica*, **42**, 96–114.

Hayes, V. J., Freedman, L. and Oxnard, C. E. (1996). Dental sexual dimorphism and morphology in African colobus monkeys as related to diet. *International Journal of Primatology*, **17**, 725–57.

Hellmer, E. and McGraw, W. S. (2005). Patterns of skeletal trauma in seven species of cercopithecoid monkeys. *American Journal of Primatology*, **66**, 93.

Hill, C. M. (1994). The role of female Diana monkeys, *Cercopithecus diana*, in territorial defense. *Animal Behaviour*, **47**, 425–31.

Hilton-Taylor, C. (2000). *2000 IUCN Red List of Threatened Animals*. Gland, Switzerland: IUCN.

van der Hoeven, C. (1996). Hiding in the crowd. The anti-predation strategy of *Colobus verus* in comparison to two sympatric colobines: *Colobus badius* and *Colobus polykomos*. Diploma, University of Utrecht, Netherlands.

Holenweg, K. (1992). Associations between *Procolobus badius* and *Cercopithecus diana* groups in Taï National Park. Diploma, University of Zurich, Switzerland.

Holenweg, A. K., Noë, R. and Schabel, M. (1996). Waser's gas model applied to associations between red colobus and diana monkeys in the Taï National Park, Ivory Coast. *Folia Primatologica*, **67**, 125–36.

Höner, O. (1993). Associations of red colobus and diana monkeys in the Taï National Park, Ivory Coast. Diploma, Swiss Federal Institute of Technology (ETH) Zurich, Switzerland.

Höner, O. P., Leumann, L. and Noë, R. (1997). Dyadic associations of red colobus and diana monkey groups in the Taï National Park, Ivory Coast. *Primates*, **38**, 281–91.

Hoppe-Dominik, B. (1984). Etude du spectre des proies de la panthere, *Panthera pardus*, dans le Parc National de Taï en Côte d'Ivoire. *Mammalia*, **48**, 477–87.

Hunkeler, C., Bertrand, M. and Bourliere, F. (1972). Le comportement social de la Mone de Lowe (*Cercopithecus campbelli lowei*). *Folia Primatologica*, **17**, 218–36.

Isbell, L. A. (1984). Daily ranging behaviour of red colobus (*Colobus badius tephrosceles*) in Kibale Forest, Uganda. *Folia Primatologica*, **39**, 145–59.

Janmaat, K. R. L. (2006). Fruits of enlightenment – Fruit localisation skills in wild mangabeys. Diploma, University of St Andrews, Scotland, UK.

Jenny, D. (1996). Spatial organization of leopards (*Panthera pardus*) in Taï National Park, Ivory Coast: is rain forest habitat a tropical haven? *Journal of Zoology*, **240**, 427–40.

Kingdon, J. (1989). *Island Africa: The Evolution of Africa's Rare Animals and Plants*. Princeton, NJ: Princeton University Press.

Kingdon, J. (1997). *The Kingdon Field Guide to African Mammals*. New York: Academic Press

Kone, I. (2004). The impact of poaching on behavioural aspects of the Diana guenon and the West African red colobus in the Taï National Park, Côte d'Ivoire. Ph.D. Thesis, University of Cocody (Abidjan, Ivory Coast).

Korstjens, A. H. (2001). The mob, the secret sorority, and the phantoms: an analysis of the socio-ecological strategies of the three colobines of Taï. Ph.D. Thesis, University of Utrecht, Netherland.

Korstjens, A. H. and Galat-Loung, A. in press. *Colobus polykomos* (Zimmerman). In *Mammals of Africa*, ed. J. Kingdon, D. C. D. Happold and T. M. Butynski. Oxford: Oxford University Press.

Korstjens, A. H. and Noë, R. (2004). Mating system of an exceptional primate, the olive colobus (*Procolobus verus*). *American Journal of Primatology*, **62**, 261–73.

Korstjens, A. H. and Schippers, E. P. (2003). Dispersal patterns among olive colobus in Taï National Park. *International Journal of Primatology*, **24**, 515–39.

Korstjens, A. H., Sterck, E. H. M. and Noë, R. (2002). How adaptive or phylogenetically inert is primate social behaviour? A test with two sympatric colobines. *Behaviour*, **139**, 203–25.

Krebs, M. (1998). The role of males in family groups of *Colobus verus*, Taï National Park, Ivory Coast. Diploma, University of Osnabrück, Germany.

Kukla, G. G. H. (1977). Pleistocene land-sea correlations. 1: Europe. *Earth Science Review*, **13**, 307–74.

Kummer, H. (1968). *Social Organization of Hamadryas Baboons*. Chicago: University of Chicago Press.

Lernould, J.-M. (1988). Classification and geographical distribution of guenons: a review. In *A Primate Radiation: Evolutionary Biology of the African Guenons*, ed. A. Gautier-Hion, F. Bourliere, J.-P. Gautier and J. Kingdon. Cambridge, UK: Cambridge University Press, pp. 54–78.

Leumann, L. (1994). Mixed species association of red colobus (*Procolobus badius*) and Diana monkeys (*Cercopithecus diana*) with particular emphasis on the context of predation. Diploma, University of Zurich, Switzerland.

Ling, B. H., Apetrei, C., Pandrea, I. *et al.* (2004). Classic AIDS in a sooty mangabey after an 18-year natural infection. *Journal of Virology*, **78**, 8902–8.

Livingstone, D. A. (1975). Late Quaterary climatic change in Africa. *Annual Review of Ecology & Systematics*, **6**, 249–80.

Livingstone, D. A. (1982). Quarternary geography of Africa and the refuge theory. In *Biological Diversification in the Tropics*, ed. G. T. Prance. New York: Columbia University Press.

Lonnberg, E. (1929). The development and distribution of the African fauna in connection with and depending upon climatic changes. *Arkiv For Zoologie (Stockholm)*, **21A**, 1–33.

Maisels, F., Gautier-Hion, A. and Gautier, J. P. (1994). Diets of two sympatric Colobines in Zaire – more evidence on seed-eating in forests on poor soils. *International Journal of Primatology*, **15**, 681–701.

Marsh, C. W. (1979a). Comparative aspects of social organization in the Tana River red colobus, *Colobus badius rufomitratus*. *Zeitschrift für Tierpsychologie*, **51**, 337–62.

Marsh, C. W. (1979b). Female transfer and mate choice among Tana River Red Colobus. *Nature*, **281**, 568–9.

Marsh, C. W. (1981). Time budget of Tana River Red Colobus. *Folia Primatologica*, **35**, 30–50.

Martin, C. (1991). *The Rainforests of West Africa: Ecology, Threats, Conservation*. Basel: Birkhäuser.

Marx, P. A., Li, Y., Lerche, N. W. *et al.* (1991). Isolation of a simian immuno-deficiency virus related to human-immunodeficiency-virus Type-2 from a West African pet sooty mangabey. *Journal of Virology*, **65**, 4480–5.

Mathews, A. and Matthews, A. (2002). Distribution, population density, and status of sympatric cercopithecids in the Campo-Ma'an area, Southwestern Cameroon. *Primates*, **43**, 155–68.

McGraw, W. S. (1996). Cercopithecid locomotion, support use, and support availability in the Taï Forest, Ivory Coast. *American Journal of Physical Anthropology*, **100**, 507–22.

McGraw, W. S. (1998a). Comparative locomotion and habitat use of six monkeys in the Taï Forest, Ivory Coast. *American Journal of Physical Anthropology*, **105**, 493–510.

McGraw, W. S. (1998b). Three monkeys nearing extinction in the forest reserves of eastern Côte d'Ivoire. *Oryx*, **32**, 233–6.

McGraw, W. S. (2000). Positional behavior of *Cercopithecus petaurista*. *International Journal of Primatology*, **21**, 157–82.

McGraw, W. S. and Oates, J. F. (in press). Roloway monkey *(Cercopithecus diana roloway)*. In *Primates in Peril: The World's Most Endangered Primates 2006–2008*, eds. R. A. Mittermeier *et al.* Primate Conservation.

McGraw, W. S., Plavcan, J. M. and Adachi-Kanazawa, K. (2002). Adult female *Cercopithecus diana* employ canine teeth to kill another adult female *C. diana*. *International Journal of Primatology*, **23**, 1301–8.

Merz, G. (1986). Counting elephants (*Loxodonta africana cyclotis*) in tropical rain-forests with particular reference to the Taï National Park, Ivory-Coast. *African Journal of Ecology*, **24**, 61–8.

Meystre-Storrer, Y. (2002). Dominance hierarchy and relationships between male mangabeys, *Cercocebus torquatus atys*. M.A. Thesis, Université de Neuchâtel, Switzerland.

Mitani, M. (1991). Niche overlap and polyspecific associations among sympatric cercopithecids in the Campo Animal Reserve, Southwestern Cameroon. *Primates*, **32**, 137–51.

Nakatsukasa, M. (1996). Locomotor differentiation and different skeletal morpho-logies in mangabeys (*Lophocebus* and *Cercocebus*). *Folia Primatologica*, **66**, 15–24.

Napier, J. and Napier, P. (1970). *Old World Monkeys: Evolution, Systematics and Behaviour*. New York: Academic Press.

Newing, H. (2001). Bushmeat hunting and management: implications of duiker ecology and interspecific competition. *Biodiversity and Conservation*, **10**, 99–118.

Nijssen, E. C. (1999). Female philopatry in *Colobus polykomos polykomos*: Fact or fiction? Diploma, University of Utrecht, Netherlands.

Noë, R. and Bshary, R. (1997). The formation of red colobus-Diana monkey associations under predation pressure from chimpanzees. *Proceedings of the Royal Society of London Series B-Biological Sciences*, **264**, 253–9.

Oates, J. F. (1977a). The guereza and its food. In *Primate Ecology*, ed. T. H. Clutton-Brock. London: Academic Press, pp. 275–321.

Oates, J. F. (1977b). The guereza and man. In *Primate Conservation*, ed. G. H. Bourne and Prince Rainier, III of Monaco. New York: Academic Press.

Oates, J. F. (1977c). The social life of a black-and-white colobus monkey, *Colobus guereza. Zeitschrift fur Tierpsychologie*, **45**, 1–60.

Oates, J. F. (1988a). The diet of the olive colobus monkey, *Procolobus verus*, in Sierra Leone. *International Journal of Primatology*, **9**, 457–78.

Oates, J. F. (1988b). The distribution of cercopithecus monkeys in West African forests. In *A Primate Radition*, ed. A. Gautier-Hion, F. Bourliere, J. P. Gautier and J. Kingdon. Cambridge: Cambridge University Press, pp. 79–103.

Oates, J. F., Abedi-Lartey, M., McGraw, W. S., Struhsaker, T. T. and Whitesides, G. H. (2000b). Extinction of a West African red colobus monkey. *Conservation Biology*, **14**, 1526–32.

Oates, J. F., Bocian, C. M. and Terranova, C. J. (2000a). The loud calls of black and white colobus monkeys: their adaptive and taxonomic significance in light of new data. In *Old World Monkeys*, ed. P. W. Whitehead and C. J. Jolly. Cambridge: Cambridge University Press, pp. 431–57.

Oates, J. F., Korstjens, A. H. and Galat-Loung, A. in press. The olive colobus, *Procolobus verus*. In *Mammals of Africa*, ed. J. Kingdon, D. C. D. Happold and T. M. Butynski. Oxford: Oxford University Press.

Oates, J. F. and Trocco, T. F. (1983). Taxonomy and phylogeny of black-and-white colobus monkeys. Inferences from an analysis of loud call variation. *Folia Primatologica*, **40**, 83–113.

Oates, J. F. and Whitesides, G. H. (1990). Association between olive colobus (*Procolobus verus*), Diana Guenons (*Cercopithecus diana*), and other forest monkeys in Sierra-Leone. *American Journal of Primatology*, **21**, 129–46.

Oates, J. F., Whitesides, G. H., Davies, A. G. *et al.* (1990). Determinants of variation in tropical forest primate biomass – new evidence from West-Africa. *Ecology*, **71**, 328–43.

van Oirschot, B. (1999). Group fission systems of red colobus monkeys, Taï National Park, Ivory Coast. In *Institute of Evolutionary and Ecological Sciences*. Leiden: Leiden University.

Okouyi, J. O., Posso, P., Lepretre, A. and Scaps, P. (2002). Estimation of large mammal densities in the Ipassa-Makokou reserve (Gabon). *Bulletin De La Société Zoologique De France*, **127**, 121–35.

Paukert, C. (2002). Aspects of range use and inter-group relations of three groups of black-and-white colobus monkeys (*Colobus polykomos*) in Taï National Park, Ivory Coast. Diploma, Johann Wolfgang Goethe-Universität Frankfurt am Main, Germany.

Pocock, R. I. (1907). A monographic revision of the genus *Cercopithecus. Proceedings of the Zoological Society of London*, **1907**, 677–746.

Rainey, H. J. (2004). The responses of birds to alarm calls and predators. Ph.D. Thesis, University of St Andrews, UK.

Range, F. (1998). *Social system and competition in female sooty mangabeys*. Diploma, University of Bayreuth, Germany.

Range, F. (2004). *The strategies employed by sooty mangabeys in a socially complex world*. Ph.D. Thesis, University of Pennsylvania, USA.

Range, F. (2005). Female sooty mangabeys (*Cercocebus torquatus atys*) respond differently to males depending on the male's residence status – preliminary data. *American Journal of Primatology*, **65**, 327–33.

Range, F. and Fischer, J. (2004). Vocal repertoire of sooty mangabeys (*Cercocebus torquatus atys*) in the Taï National Park. *Ethology*, **110**, 301−21.

Range, F. and Noë, R. (2002). Familiarity and dominance relations among female sooty mangabeys in the Taï National Park. *American Journal of Primatology*, **56**, 137−53.

Range, F. and Noë, R. (2005). Can simple rules account for the pattern of triadic interactions in juvenile and adult female sooty mangabeys? *Animal Behaviour*, **69**, 445−52.

Reed, K. E. and Bidner, L. R. (2004). Primate communities: past, present, and possible future. *American Journal of Physical Anthropology*, **4**, 2−39.

Refisch, J. (2001). Impact of poaching on primates and secondary effects on the vegetation in Taï National Park, Côte d'Ivoire. Ph.D. Thesis, University of Bayreuth, Germany.

Rode, P. (1937). *Les Primates de l'Afrique*. Paris: Librairie Larose.

Roth, H. H. and Hoppe-Dominik, B. (1987). Repartition et statut des grandes espèces de mammiferes en Côte d'Ivoire. IV. Buffles. *Mammalia*, **51**, 89−109.

Roth, H. H., Merz, G. and Steinhauer, B. (1984). Repartition et statut des grandes especes de mammiferes en Côte d'Ivoire, I. Introduction, II. Les elephants. *Mammalia*, **48**, 207−26.

Rowell, T. E. (1966). Forest living baboons in Africa. *Journal of Zoology*, **149**, 344−64.

Rutte, C. (1998). Foraging strategies of sooty mangabeys (*Cercocebus torquatus atys*). Diploma, Friedrich-Alexander-Universität Erlangen-Nürnberg, Germany.

Saj, T. L. and Sicotte, P. (2004). The population status of *Colobus vellerosus* at Boabeng-Fiema sacred grove, Ghana. In *Commensalism and Conflict: Human-Primate Interface*, ed. J. Patterson. American Society of Primatology, pp. 264−87.

Sanderson, I. T. (1940). The mammals of the North Cameroons forest area. *Transactions of the Zoological Society of London*, **24**, 623−725.

Santiago, M. L., Range, F. Keele, B. F. *et al.* (2005). Simian immunodeficiency virus infection in free-ranging sooty mangabeys (*Cercocebus atys atys*) from the Taï Forest, Côte d'Ivoire: implications for the origin of epidemic human immunodeficiency virus type 2. *Journal of Virology*, **79**, 12515−27.

Schaaff, M. (1995). Differences in anti-predation strategies, food choice and social structure between sympatric *Colobus polykomos* and *Procolobus badius*: testing a model. Diploma, University of Utrecht, Netherlands.

Schabel, M. (1993). Die Rolle des Futters in gemischte Assoziationen Zwischen Roten Coloben und Dianameerkatzen in Taï National Park. Diploma, University of Zurich, Switzerland.

Schaller, G. B. (1963). *The Mountain Gorilla*. Chicago: University of Chicago Press.

Schippers, E. (1999). Dispersal of the olive colobus (*Colobus verus*) in Taï National Park. Diploma, University of Utrecht, Netherlands.

Schwarz, E. (1928). Notes on the classification of the African monkeys in the genus *Cercopithecus* Erxleben. *Annals and Magazine of Natural History*, **10**, 649−63.

Schwarz, E. (1929). On the local races and distribution of the black and white colobus monkeys. *Proceedings of the Zoological Society of London*, 585−98.

Shultz, S. (2001). Notes on interactions between monkeys and African crowned eagles in Taï National Park, Ivory Coast. *Folia Primatology*, **72**, 248–50.

Shultz, S. (2003). Of monkeys and eagles: predator-prey interactions in the Taï National Park, Côte d'Ivoire. Ph.D. Thesis, University of Liverpool, UK.

Shultz, S., Noë, R., McGraw, W. S. and Dunbar, R. I. M. (2004). A community-level evaluation of the impact of prey behavioural and ecological character-istics on predator diet composition. *Proceedings of the Royal Society of London Series B-Biological Sciences*, **271**, 725–32.

Silvestri, G. (2005). Naturally SIV-infected sooty mangabeys: are we closer to understanding why they do not develop AIDS? *Journal of Medical Primatology*, **34**, 243–52.

Stanford, C. B. (1995). The influence of chimpanzee predation on group size and antipredator behavior in red colobus monkeys. *Animal Behaviour*, **49**, 577–87.

Stanford, C. B. (1998). *Chimpanzee and Red Colobus. The Ecology of Predator and Prey*. Cambridge, MA: Harvard University Press.

Stanford, C. B., Wallis, J., Matama, H. and Goodall, J. (1994). Patterns of predation by chimpanzees on red colobus monkeys in Gombe National Park, 1982–1991. *American Journal of Physical Anthropology*, **94**, 213–28.

Starin, E. D. (1990). Object manipulation by wild red colobus monkeys living in the Abuko Nature reserve, Gambia. *Primates*, **31**, 385–92.

Starin, E. D. (1991). *Socioecology of the Red Colobus Monkey in the Gambia Particular Reference to Female-Male Differences and Transfer Pattern*. New York: CUNY.

Starin, E. D. (1994). Philopatry and affiliation among red colobus. *Behaviour*, **130**, 253–70.

Starin, E. D. (2001). Patterns of inbreeding avoidance in Temminck's red colobus. *Behaviour*, **138**, 453–65.

Struhsaker, T. T. (1967). *Behavior of Vervet Monkeys (Cercopithecus aethiops)*. Berkeley: University of California Press.

Struhsaker, T. T. (1969). Correlates of ecology and social organisation among African cercopithecines. *Folia Primatologica*, **11**, 80–118.

Struhsaker, T. T. (1971). Notes on *Cercocebus a. atys* in Senegal, West Africa. *Mammalia*, **35**, 343–4.

Struhsaker, T. T. (1974). Correlates of ranging behavior in a group of red colobus monkeys (*Colobus badius tephrosceles*). *American Zoologist*, **14**, 177–84.

Struhsaker, T. T. (1975). *The Red Colobus Monkey*. Chicago: University of Chicago Press.

Struhsaker, T. T. (1978). Food habits of five monkey species in the Kibale Forest, Uganda. In *Recent Advances in Primatology*, ed. D. J. Chivers and J. Herbert. New York: Academic Press, pp. 225–48.

Struhsaker, T. T. (1980). Comparison of the behavior and ecology of red colobus and redtail monkeys in the Kibale forest. *African Journal of Ecology*, **18**, 33–51.

Struhsaker, T. T. (1997). *Ecology of an African Rain Forest: Logging in Kibale and the Conflict Between Conservation and Exploitation*. Gainsville: University of Florida Press.

Struhsaker, T. T. (1999). Primate communities in Africa: The consequence of long term evolution or the artifact of recent hunting? In *Primate Communities*, ed. J. G. Fleagle, C. H. Janson and K. E. Reed. Cambridge: Cambridge University Press, pp. 289–94.

Struhsaker, T. T. and Hunkeler, P. (1971). Evidence of tool using by chimpanzees in the Ivory Coast. *Folia Primatologica*, **15**, 212–19.

Struhsaker, T. T. and Leland, L. (1985). Infanticide in a patrilineal society of red colobus monkeys. *Zeitschrift für Tierpsychologie*, **69**, 89–132.

Struhsaker, T. T. and Oates, J. F. (1975). Comparison of the behaviour and ecology of red colobus and black and white colobus monkeys in Uganda: a summary. In *Socioecology and Psychology of Primates*, ed. R. H. Tuttle. The Hague: Mouton, pp. 103–23.

Tallerman, M. (2005). *Language Origins: Perspectives on Evolution*. Oxford: Oxford University Press.

Teichroeb, J. A., Saj, T. L., Paterson, J. D. and Sicotte, P. (2003). Effect of group size on activity budgets of *Colobus vellerosus* in Ghana. *International Journal of Primatology*, **24**, 743–58.

Terborgh, J. (1983). *Five New World Primates*. New Jersey: Princeton University Press.

Thiollay, J.-M. (1985). Birds of Ivory Coast: status and distribution. *Malimbus*, **7**, 1–59.

Tosi, A. J., Buzzard, P. J., Morales, J. C. and Melnick, D. J. (2002). Y-chromosomal window onto the history of terrestrial adaptation in the cercopithecini. In *The Guenons: Diversity and Adaptation in African Monkeys*, ed. M. E. Glenn and M. Cords. New York: Kluwer Academic, pp. 15–26.

Tosi, A. J., Detwiler, K. M. and Disotell, T. R. (2005). X-chromosomal window into the evolutionary history of the guenons (Primates: Cercopithecini). *Molecular and Phylogenetic Evolution*, **36**, 58–66.

Tranquilli, S. (2003). Vocalizations and referential function in the alarm calls of the black and white colobus (*Colobus polykomos*) in Taï Forest, Ivory Coast. Diploma, Universita' degli studi di Roma "La Sapienza," Italy.

Tutin, C. E. G., Ham, R. M., White, L. J. T. and Harrison, M. J. S. (1997). The primate community of the Lope Reserve, Gabon: diets, responses to fruit scarcity, and effects on biomass. *American Journal of Primatology*, **42**, 1–24.

Tutin, C. E. G. and White, L. J. T. (1999). The recent evolutionary past of primate communities: likely environmental impacts during the past three millennia. In *Primate Communities*, ed. J. Fleagle, C. Janson and K. Reed. Cambridge: Cambridge University Press, pp. 220–36.

Uster, D. (2000). The functional significance of diana monkey contact calls. Diploma, Braunschweig: Technische Universität Braunschweig, Germany.

Wachter, B. (1993). The role of food in polyspecific associations of red colobus and Diana monkeys in the Taï National Park, Ivory Coast. M.A. Thesis, Swiss Federal Institute of Technology (ETH) Zurich, Switzerland.

Wachter, B., Schabel, M. and Noë, R. (1997). Diet overlap and polyspecific associations of red colobus and Diana monkeys in the Taï park, Ivory Coast. *Ethology*, **103**, 514–26.

Waitkuwait, W. E. (1981). Untersuchungen zur Brutbiologie des Panzerkrokodils (*Crocodilis cataphractus*) im Taï Nationalpark in der Republik Elfenbeinkuste. Thesis, Ruprecht-Karls Universitat: Heidelberg, Germany.

Walek, M. L. (1978). Vocalizations of black and white colobus-monkey (*Colobus polykomos* Zimmerman 1780). *American Journal of Physical Anthropology*, **49**, 227–40.

Washburn, S. L. and Devore, I. (1961a). Social behaviour of baboons and early man. In *Social life of Early Man*, ed. S. L. Washburn. Chicago: Aldine Publishing Company, pp. 91–105.

Washburn, S. L. and Devore, I. (1961b). The social life of baboons. *Scientific American*, **204**, 62–71.

Whitehead, P. F. and Jolly, C. J. (eds.) (2000). *Old World Monkeys*. Cambridge: Cambridge University Press.

Whitesides, G. H. (1989). Interspecific associations of Diana monkeys, *Cercopithecus diana*, in Sierra Leone, West Africa: Biological significance or chance? *Animal Behaviour*, **37**, 760–76.

Whitesides, G. H. (1991). Patterns of foraging, ranging, and interspecific associations of Diana monkeys (*Cercopithecus diana*) in Sierra Leone, West Africa. Ph.D. Thesis, University of Miami, FL, USA.

Wolters, S. (2001). The effect of mixed-species associations on the behaviour of Campbell's and Diana monkeys. Diploma, Westfälische Wilhelms-Universität, Germany.

Zuberbühler, K. M. (1998). Natural semantic communication in wild Diana monkeys. Proximate mechanisms and evolutionary function. Ph.D. Thesis, University of Pennsylvania, USA.

Zuberbühler, K. (1993). Akustische Kommunikation im Feindkontext: die Bedeutung der Vokalisationen als Antiprädator-Strategien bei zwei waldlebenden Cercopitheciden. Diploma, Zurich: University of Zurich, Switzerland.

Zuberbühler, K. and Jenny, D. (2002). Leopard predation and primate evolution. *Journal of Human Evolution*, **43**, 873–86.

I *Social behavior*

2 The social systems of the guenons

P. Buzzard and W. Eckardt

Introduction

The social system of a species includes the nature of the interactions of individuals between and within social units and the spatial distribution of different age/sex classes. For primates, theories concerning the evolution of social systems are typically based on field data from a restricted number of species, with a clear bias towards species living in more open habitats (Sterck *et al.* 1997). This is problematic because the forest is a major primate habitat, housing a large number of primate species. The social behavior of most forest-living primates is not well described, primarily due to the difficulties in accessibility and observation conditions. In this respect the forest guenons (*Cercopithecus* spp.) are of particular interest for evolutionary theories because they represent a major group of Old World primates.

It is theorized that female primates live in social groups because of anti-predation benefits (van Schaik 1983, van Schaik & van Hoof 1983) and because group-living improves their capacity to defend resources against other groups of conspecifics (Wrangham 1980). Across species, female primates differ in the types of social relations they maintain with one another to achieve these goals. It has been proposed that the relative strengths of inter- and intra-group competition are the two main factors that determine the nature of the females' social relationships and their social system (Sterck *et al.* 1997, Table 2.1).

Across primate species, inter-group encounters can vary from friendly intermingling to hostile fights. Home ranges may overlap completely with neighboring groups ignoring or avoiding each other or they may be defended vigorously against neighbors. In most baboons and in some colobines, for example, home ranges overlap largely and adjacent groups normally do not fight but avoid each other (e.g. Rowell 1988, Korstjens 2001). In contrast, guenon groups usually maintain territories by defending a particular geographical area (Rowell 1988, Butynski 1990, Rowell *et al.* 1991, Cords 2000a, Payne *et al.* 2003). For example, blue monkey (*C. mitis*)

Monkeys of the Taï Forest, ed. W. Scott McGraw, Klaus Zuberbühler and Ronald Noë.
Published by Cambridge University Press. © Cambridge University Press 2007.

Table 2.1. *A classification scheme of primate social systems*
(see Sterck et al. *1997)*

In dispersal-egalitarian populations, females disperse from their natal group and do not show clear hierarchical relationships. In the remaining types, females remain in their natal group showing either unclear or clear hierarchical relationships in resident-egalitarian and resident-nepotistic populations respectively. In resident-nepotistic-tolerant populations, females show clear hierarchical relationships, but the benefits of cooperating against other female groups in territorial defense prevents highly despotic behavior of higher-ranking individuals.

	Competition			
Type	Within group	Between group	Female philopatry	Social dominance
Dispersal-Egalitarian	Weak	Weak	No	Weak
Resident-Egalitarian	Weak	Strong	Yes	Weak
Resident-Nepotistic	Strong	Weak	Yes	Strong
Resident-Nepotistic-Tolerant	Strong	Strong	Yes	Weak

females form alliances almost daily against neighboring females and as much as half of a group's home range is used exclusively (Cords 2000a).

A prevalent belief about guenons is that these are animals with low rates of social interactions. Several authors have stated that individuals only engage in low frequencies of agonistic and affiliative behavior and that individual relationships are difficult to discern (Bourlière *et al.* 1970, Glenn 1996, Cords 2000a, Pazol 2001). For example, female baboons and macaques may interact agonistically 1.5 to 20 times more often than female blue monkeys (Seyfarth 1976, Barton 1993, Barton *et al.* 1996, Sterck & Steenbeek 1997, Cords 2000a, Pazol 2001). Consequently, forest guenons were categorized as "resident-egalitarian" and baboons as "resident-nepotistic," using terminology listed in Table 2.1 (Sterck *et al.* 1997).

However, it is possible that guenon behavior has been misrepresented due to the difficult observation conditions and because interactions may be more subtle than in other primate species. For example, long-term data on blue monkey social behavior at Kakamega forest (Kenya) showed that these primates form highly linear and matrilineally-based dominance hierarchies where individuals maintain stable rank positions over several years (Cords 2002a). Thus, the low rates of social interactions may simply be a consequence of high social stability in these groups. Analogous results have recently been obtained from captive Campbell's monkeys (*Cercopithecus campbelli campbelli*) (Lemasson *et al.* 2003,

Lemasson & Hausberger 2004), questioning whether the guenons as a whole should be classified as "resident-egalitarian" (Sterck *et al.* 1997, Cords 2002a). More field data on guenon social systems are needed to resolve this controversy.

Unlike the baboons and some colobines, forest guenons typically live in single-male groups for most of the year (Rowell 1988, Cords 2000b). During the breeding season, single-male groups are sometimes invaded by non-resident males who may stay from several days to months and occasionally manage to expel the resident male (Cords 1988, 2000b, 2002b, Macleod *et al.* 2002). It is possible that male forest guenons follow two alternative reproductive strategies. While resident males try to remain with a female group throughout the year non-resident males may try to associate with a female group during the breeding season only. Alternatively, non-resident males may simply try to make the best of a bad situation by joining groups at times when the cost-benefit ratio is most in their favor (Macleod *et al.* 2002). During the non-breeding season, non-resident males have been observed to range solitarily, to form a polyspecific association with other primate species, or to associate in all-male groups. The latter two strategies are thought to be advantageous in providing anti-predation and foraging benefits (Struhsaker 1969, Wrangham 1980, van Schaik & van Horstermann 1994, Noë & Bshary 1997). All-male groups may additionally provide opportunities to form coalitions, which may improve the chances of successfully invading a group of females (Pusey & Packer 1987). In sum, one of the main reasons for guenon males to live in social groups is to gain access to females during the breeding season, and the relative costs and benefits at any one time appear to be the main determinant of male sociality.

In this chapter, we present several sets of data on the four guenon species present in the Taï forest, the Diana monkeys (*Cercopithecus diana*), Campbell's monkeys (*C. campbelli*), lesser spot-nosed monkeys (*C. petaurista*), and putty-nosed monkeys (*C. nictitans*). The Taï primates are of particular interest for evolutionary theories because their habitat, the Taï forest, could be representative of a typical primate forest habitat, containing natural densities of the major primate predators and relatively low levels of human activity. We will first provide data on the population density for each species because of the potential impact of this variable on social systems (Smuts *et al.* 1987). Next, we will describe the typical group compositions for each species and assess the nature of the social relations of individuals between and within social groups. Finally, we will utilize these sets of information to contribute to the debate about the proper classification of guenon social systems.

Methods
Study animals

Data were collected from seven habituated study groups, two each of Diana monkeys, Campbell's monkeys and lesser spot-nosed monkeys, as well as one putty-nosed monkey group. We conducted additional group counts and some behavioral observations on a third Diana monkey group, on two all-male Campbell's monkey groups, and on several partly habituated neighboring groups of Campbell's monkeys, lesser spot-nosed monkeys, and one group of putty-nosed monkeys.

Sampling procedure

On most observation days we followed one group from about 7:30 to 18:00 hrs until the monkeys stopped moving in sleeping trees. At other times, we followed the groups for half-days from about 7:30 to 12:30 hrs or from 12:00 to 18:00 hrs. Groups were followed for 2−4 consecutive days. The observation regime for the putty-nosed monkey group was slightly different because the group's home range was much further away, outside the main study area. This group was followed from 8:00 to 17:00 hrs (8:00 to 12:30 hrs or 12:30 to 17:00 hrs). Paul Buzzard (PB) and a trained field assistant followed the Diana monkeys, Campbell's monkeys, and lesser spot-nosed monkeys from August 2000 until November 2001. Winnie Eckardt (WE) followed the putty-nosed monkey group from June to December 2001. Data collection was based on scan sampling (Altmann 1974). Scans were conducted twice per hour throughout the day, and individual scans lasted for up to 15 min. During each scan we attempted to sample as many different individuals as possible once per scan by walking through the group.

Group composition

We classified individuals into five different age/sex classes: adult males, adult females, sub-adults, juveniles, and infants. Adult males were slightly bigger than other adult individuals and they produced loud calls (Gautier & Gautier 1977, Zuberbühler *et al.* 1997, Zuberbühler 2001, Eckardt & Zuberbühler 2004). Individuals were classified as adult females if they appeared to be pregnant or had pendulous nipples, suggesting that they had given birth before. Sub-adult males were the same size or slightly smaller than adult males but did not give loud calls. Sub-adult females were the size of adult females but were not pregnant and had no pendulous nipples. Juveniles were smaller than adult females but were no longer carried by them and, if at all, only suckled infrequently. Infants were frequently carried by their mothers and were still suckling regularly.

Home range size, home range overlap, and population density

At the end of each scan, we determined the group's center of mass (Cords 1987) and marked it on a grid system that was made up of 0.25 ha quadrants. Since the putty-nosed monkey group was outside this grid system, their position was determined with a GPS receiver (GARMIN XL16). We estimated home range overlap by noting the position of neighboring groups on an all-occurrence basis while we followed a study group or during days when we collected other data (2−4 days/month).

Home range sizes and overlap with neighboring groups were determined and used to calculate group density, using the "block method" (Struhsaker 1981, Whitesides *et al.* 1988, Fashing and Cords 2000). Blocks (i.e. grid quadrants) used by one group only contributed with a value of 1.0; quadrants used by two conspecific groups contributed with 0.5, quadrants used by three groups contributed with 0.33, and so forth. Each group's value was then summed up to produce an "adjusted home range size." Group density (groups/km^2) was determined by using the inverse of the average adjusted home range sizes for each species. Individual density (individuals/km^2) was determined by multiplying group density with the average group size for each species. Putty-nosed monkeys are extremely rare in the Taï forest and adjacent groups did not overlap in home range use (see Eckardt & Zuberbühler 2004). We calculated this group's home range by assigning GPS readings to an imaginary grid system equivalent to one used for the other study groups. We located a total of four different putty-nosed monkey groups in a 73 km^2 area surrounding the main study grid, which provided the basis for our population density estimates.

Inter-group encounters

An inter-group (or between-group) encounter was recorded when a study group was less than 50 m apart from a neighboring group of conspecifics. Since our study groups were often in mixed-species associations (60−90 per cent of monthly scans: Wolters & Zuberbühler 2003, Buzzard 2004, Eckardt & Zuberbühler 2004), we were able to recognize some inter-group encounters while following other study groups. Only data from PB were included for inter-group encounters since PB followed Diana, spot-nosed and Campbell's, and only these species were involved in inter-group encounters. If another group was associated with the study group for more than eight hours, we were able to enlarge our sample size and counted such days as possible days of encounters for both groups. While sampling a study group, PB walked not only through that study group but also through most of the other associated groups to look for inter-group encounters. We distinguished between passive and aggressive types of inter-group encounters. An aggressive encounter was scored if individuals gave acoustically distinct vocalizations at high amplitude, or if individuals chased or were chased by members of the neighboring group. During aggressive encounters the males frequently exchanged loud calls. Passive encounters did not involve any of the behavioral patterns described. Instead, the two groups simply stayed in each other's vicinity for a while and then travelled on to other areas of their home range.

Intra-group interactions

To assess the nature of relationships between individuals within the same group, we recorded all affiliative and agonistic behaviors during the scans. Grooming and mounting were scored as affiliative behaviors; attacks, threats, and avoidances were scored as agonistic behaviors. Grooming behavior was defined as the manual inspection of another individual's fur with one or both hands or with the mouth. Mounts, which also occurred between females, could take place with or without pelvic thrusts. An attack was scored if one individual engaged in a physical confrontation with another individual involving biting and/or hitting. A threat was scored if one individual lunged at another individual with bared teeth. Finally, an avoidance occurred if one individual moved away in response to another individual's approaching to a distance of two meters or less. We calculated relative rates of affiliative and agonistic behaviors for each species by dividing the number of behaviors by the number of individuals scanned.

To further describe the social dynamics in these groups we determined the age/sex-specific interaction patterns of affiliative and agonistic behaviors. In particular, we compared the expected and the observed number of agonistic and affiliative behaviors for each age/sex class combination. To calculate expected frequencies, we multiplied the total number of grooming or agonistic bouts by the random interaction frequency ratios for each age/sex class combination. To obtain the random interaction frequency ratio, the marginal totals of the actor and recipient were each divided by the total number of grooming or agonistic bouts and then multiplied together. The observed number was then tested to see if it was significantly greater or less than the expected number by using the binomial (or z-test approximation to the binomial) wherever the expected number exceeded 10 (Siegal & Castellan 1988, Bernstein 1991).

Results

Group composition

The average group size for Diana monkeys was 23.5 individuals, more than twice the number of an average Campbell's monkey (9.3 individuals),

Table 2.2. *Typical group compositions of the Taï forest guenons (November 2001)*

Group	Adult male	Adult female	Sub-adult	Juvenile	Infant	Total
Diana monkeys						
DIA1[a]	1	11	2	7	0	21
DIA2[a]	1	13	2	10	0	26
DIA3	1	11	?	?	?	>15
Campbell's monkeys						
CAM1[a]	1	6	0	2	1(?)	9 (10?)
CAM2[a]	1	5	1	2	0	9
CAM3	1	5	0	4	0	10
Spot-nosed monkeys						
PET1[a]	1, (2)[b]	8	1	6	0	16
PET2[a]	1	4	0	2	0	7
PET3	1	5	1	4	0	11
Putty-nosed monkeys						
NIC1[a]	1	4	5	2	0	12
NIC2[c]	1	2(?)	0(?)	2	1(?)	6(?)

[a]Study groups; [b]PET 1 frequently contained 2 males from February 2001 to December 2001; [c]Census in 2000

lesser spot-nosed monkey (11.3 individuals) and putty-nosed monkey group (11.3 individuals). All study groups contained one resident male through most of the study period (see Table 2.2). However, in February 2001, a sub-adult male in a lesser spot-nosed monkey group (PET1) became adult but remained in the group resulting in two adult males. About five months later, the original male was involved in several fights and chases with the male from a neighboring group (PET2). Soon thereafter, both males of PET1 disappeared and the PET2 male took over the PET1 group. In addition, a new adult male was often seen with the PET1 group until the end of the study period. After the PET2 male successfully transferred to the PET1 group another new male took over the PET2 group quickly.

In addition, we observed two all-male groups of Campbell's monkeys in the study area. One of them (CAMAM1) contained two to four adult males and one juvenile male and was often (68.7 per cent, N = 115) observed in association with one Diana monkey (DIA1) and one lesser spot-nosed monkey group (PET2). The other all-male group (CAMAM2) contained two adult males and was observed five times in the home range of a Campbell's monkey group (CAM2). Individuals in both all-male groups were observed to give loud calls in response to Diana monkey loud calls, falling trees, or other disturbances.

Table 2.3. *Home range size, home range overlap, and population density of the Taï guenons*

Group	Home range (ha)	Overlap (%)	Adjusted range (ha)	Density Groups (/km²)	Individuals (/km²)
DIA1	59.3	67	37.0	2.7	63
DIA2	58.5	65	38.8	2.6	61
CAM1	67	54	42.0	2.4	24
CAM2	52	56	37.6	2.7	25
PET1	73.5	60	49.3	2	23
PET2	64	69	38.5	2.6	29
NIC1	93	–	93.0	0.05	0.45

Diana monkeys (DIA); Campbell's monkeys (CAM); Spot-nosed monkeys (PET); Putty-nosed monkeys (NIC)

Home range size, home range overlap, and population density

Diana monkeys were the most common guenons in the Taï forest (62 ind./km²) followed by lesser spot-nosed monkeys (26 ind./km²), Campbell's monkeys (24 ind./km²) and putty-nosed monkeys (0.6 ind./km²). Group densities (groups/km²) were similar for Diana monkeys (2.6 groups per km²), Campbell's monkeys (2.5 groups per km²) and lesser spot-nosed monkeys (2.3 groups per km²), but lower for putty-nosed monkeys (0.05 groups per km²; Table 2.3). Diana monkeys used an average home range of 56.8 ha. About 66 per cent of it was shared with neighboring groups, resulting in an average adjusted home range of 37.9 ha. Campbell's monkeys used an average home range of 56.0 ha. About 55 per cent was shared with neighboring groups, resulting in an average adjusted home range of 39.8 ha. Lesser spot-nosed monkeys used an average home range of 65.3 ha. About 64 per cent was shared with neighboring groups, resulting in an average adjusted home range of 43.9 ha. The putty-nosed monkey group used a home range of 93.0 ha, and there was no overlap with neighboring groups.

Inter-group encounters

Although the group densities were roughly equal, the three species in the study grid differed greatly in how frequently they encountered neighboring groups and in the consequences of these encounters (see Table 2.4). Campbell's monkeys were least likely to encounter neighboring groups, although the few times this happened it was usually of an aggressive nature. Additionally, the all-male group CAMAM1 was observed in two

Table 2.4. *Inter-group encounters in the Taï guenons*

Species (observation days)	Encounter rate (N/day)	Aggressive encounters	Male calling
Diana monkeys (N = 95)	0.358	35%	24% (N = 34)
Campbell's monkeys (N = 91)	0.033	67%	67% (N = 3)
Spot-nosed monkeys (N = 104)	0.125	31%	31% (N = 13)

inter-group encounters with a neighboring group. All males from the all-male group exchanged repeated loud calls and threats with the resident male from the neighboring group. Inter-group encounters were four times more common in spot-nosed monkeys compared to Campbell's monkeys, although during encounters groups typically ignored each other and aggressive interactions were uncommon. Three of the four observed aggressive encounters were over access to a feeding tree. In all cases, the males took an active role, engaging in loud calling and threatening neighboring group members. Finally, Diana monkeys had the highest rates of inter-group encounters, about ten times more frequent than Campbell's monkeys. However, only a minority of them resulted in aggressive interactions, often about access to a feeding tree. Males called in 8 out of 12 aggressive encounters (66.7 per cent). The home ranges of putty-nosed monkey groups were over one kilometer apart and inter-group encounters were never observed.

Intra-group interactions

We conducted a total of 7,258 scans on the seven different study groups (see Table 2.5). Grooming rates were relatively similar for all species and highest in putty-nosed monkeys, followed by Campbell's monkeys, spot-nosed monkeys, and Diana monkeys (see Table 2.5). Only Diana monkey adult females, however, mounted each other (N = 5 bouts). The order was different for agonistic interactions. Diana monkeys showed the highest rates of agonistic behaviors, followed by putty-nosed monkeys, spot-nosed monkeys, and Campbell's monkeys (see Table 2.5). Diana monkey females were also the only ones observed to form coalitions with one another. In six instances, two or more individuals chased or threatened one or more other group members. Coalitions occurred frequently during conflicts with members of another species (Buzzard 2004).

In each species, the adult females were involved in the majority of grooming bouts (see Table 2.6). Diana and lesser spot-nosed adult females groomed other adult females more than expected, but the differences were

Table 2.5. *Intra-group affiliative and agonistic interactions*

			Relative frequency (bouts/ind. $\times 10^3$)	
Species (N observation days)	N scans	N individual samples	Affiliative	Agonistic
Diana monkeys (N=75)[a]	1,320	9,243	9.6	4.9
Campbell's monkeys (N=145)	2,262	9,093	12.5	0.8
Spot-nosed monkeys (N=156)	2,362	9,425	10.3	1.0
Putty-nosed monkeys (N=90)	1,314	5,577	17.6	2.7

[a]Includes the 5 mounts

not significant ($z = -0.86$, -0.11; $p = 0.19$, 0.46, respectively; Table 2.6). Males groomed rarely, in some cases never (see Table 2.6). In Diana monkeys and Campbell's monkeys the majority of agonistic interactions involved adult females, but in lesser spot-nosed monkeys most agonistic interactions involved juveniles (see Table 2.7). Adult female Diana monkeys were involved in more agonistic bouts than expected with other females, but the difference was not significant (see Table 2.7, $z = -0.62$, $p = 0.27$). In addition, juvenile Diana monkeys were involved in fewer agonistic bouts than expected with other juveniles, but this difference was not significant either (see Table 2.7, $p = 0.24$).

Discussion
Group composition

All study groups lived in single-male groups for the majority of the time, as documented in other forest guenons (Rowell 1988, Cords 2000b). The numbers of adult females per group was comparable to other published records (e.g. Cords 2000b), and group sizes for Diana monkey, Campbell's monkeys and spot-nosed monkeys were comparable to data reported from another West African study site, Tiwai Island, Sierra Leone (Oates *et al.* 1990, Hill 1994). Putty-nosed monkey groups were smaller at Taï than at other field sites, probably due to higher inter-specific competition with Diana monkeys (Gautier & Gautier-Hion 1969, Struhsaker 1969, Gautier-Hion & Gautier 1974, Whitesides 1981, Mitani 1991, White 1994, Eckardt & Zuberbühler 2004).

Resident forest guenon males defend their group against intruding males, but extra-group male influxes have still been reported in some forest guenons, particularly during the breeding season (Cords 2000b). Our documentation of a male take-over, the likely eviction of the resident PET1 male by his neighbor, suggests that this may also be part of the

Table 2.6. *Number of observed (expected) grooming bouts between each age/sex class in (a)* C. diana, *(b)* C. campbelli, *(c)* C. petaurista, *and (d)* C. nictitans

Marginal totals are for observed bouts; for each species the total number of bouts is in bold. To calculate the expected values, the marginal totals of the actor and recipient were each divided by the total number of bouts and then multiplied together; this product was then multiplied by the total number of bouts (see Methods). The differences between observed and expected bouts were insignificant in all cases (see Results).

Groomer	Recipient					Marginal totals
	Adult male	Adult female	Sub-adult	Juvenile	Unknown	
(a) *C. diana*						
Adult male	—	0 (0)	0 (0)	0 (0)	—	0
Adult female	0 (0)	28 (21)	2 (5)	10 (14)	—	40
Sub-adult	0 (0)	6 (8)	2 (2)	6 (5)	—	14
Juvenile	0 (0)	11 (16)	6 (3)	13 (10)	—	30
Marginal totals	0 (0)	45	10	29	—	**84**
(b) *C. campbelli*						
Adult male	—	1 (1)	0 (0)	0 (0)	—	1
Adult female	10 (9)	63 (65)	2 (1)	15 (15)	—	90
Sub-adult	0 (0)	2 (3)	0 (0)	2 (1)	—	4
Juvenile	1 (2)	16 (14)	0 (0)	2 (3)	—	19
Marginal totals	11	82	2	19	—	**114**
(c) *C. petaurista*						
Adult male	0 (0)[a]	0 (0)	0 (0)	0 (0)	—	0
Adult female	13 (14)	45 (43)	3 (3)	12 (14)	—	73
Sub-adult	1 (1)	1 (2)	0 (0)	1 (1)	—	3
Juvenile	4 (4)	11 (12)	1 (1)	5 (4)	—	21
Marginal totals	18	57	4	18	—	**97**
(d) *C. nictitans*						
Adult male	—	0 (0)	0 (0)	0 (0)	0 (0)	0
Adult female	6 (4)	23 (24)	0 (0)	7 (11)	11 (9)	47
Sub-adult	0 (0)	4 (2)	0 (0)	0 (1)	0 (1)	4
Juvenile	2 (4)	23 (22)	0 (0)	15 (10)	3 (8)	43
Unknown	0 (0)	0 (2)	0 (0)	0 (1)	4 (1)	4
Marginal totals	8	50	0	22	18	**98**

[a]In *C. petaurista*, adult male interaction with other adult males in the group was possible since PET1 frequently contained 2 males from February 2001 to December 2001

Taï guenon social system. An additional male was frequently seen in the group for another five months. Similarly, Hill (1994) recorded a second male associating with a group of Diana monkeys for five months.

The stable all-male groups found in Campbell's monkeys are also found in Mona monkeys in Benin and Grenada (Glenn 1996, 1997,

Table 2.7. *Number of observed (expected) agonistic bouts between age/sex classes of (a)* C. diana, *(b)* C. campbelli, *(c)* C. petaurista, *and (d)* C. nictitans

Marginal totals are for observed bouts; for each species the total number of bouts is in bold. To calculate the expected values, the marginal totals of the actor and recipient were each divided by the total number of bouts and then multiplied together; this product was then multiplied by the total number of bouts (see Methods). The differences between observed and expected bouts were insignificant in all cases (see Results).

| | Victim | | | | | |
Aggressor	Adult male	Adult female	Sub-adult	Juvenile	Unknown	Marginal totals
(a) *C. diana*						
Adult male	—	0 (2)	2 (0)	1 (1)	—	3
Adult female	0 (0)	23 (18)	4 (5)	7 (11)	—	34
Sub-adult	0 (0)	1 (3)	1 (1)	4 (2)	—	6
Juvenile	0 (0)	0 (1)	0 (0)	2 (1)	—	2
Marginal totals	0	24	7	14	—	**45**
(b) *C. campbelli*						
Adult male	—	0 (0)	0 (0)	0 (0)	—	0
Adult female	0 (0)	3 (2)	0 (0)	2 (3)	—	5
Sub-adult	0 (0)	0 (0)	0 (0)	0 (0)	—	0
Juvenile	0 (0)	0 (1)	0 (0)	2 (1)	—	2
Marginal totals	0 (0)	3	0	4	—	**7**
(c) *C. petaurista*						
Adult male	0 (0)[a]	0 (0)	1 (0)	0 (1)	—	1
Adult female	0 (0)	3 (1)	0 (0)	0 (2)	—	3
Sub-adult	0 (0)	0 (0)	0 (0)	0 (0)	—	0
Juvenile	0 (0)	0 (2)	0 (1)	5 (3)	—	5
Marginal totals	0 (0)	3	1	5	—	**9**
(d) *C. nictitans*						
Adult male	—	0 (0)	0 (0)	0 (0)	0 (0)	0
Adult female	0 (0)	4 (2)	0 (1)	0 (2)	1 (1)	5
Sub-adult	0 (0)	0 (1)	2 (0)	0 (1)	0 (0)	2
Juvenile	0 (0)	2 (3)	0 (1)	5 (2)	0 (1)	7
Unknown	0 (0)	0 (0)	0 (0)	0 (0)	1 (0)	1
Marginal totals	0 (0)	6	2	5	2	**15**

[a]In *C. petaurista*, adult male interaction with other adult males in the group was possible since PET1 frequently contained 2 adult males from February 2001 to December 2001

Glenn *et al.* 2002) as well as the Lowe's subspecies of Campbell's monkeys (*C. campbelli lowei*, Bourlière *et al.* 1970). Because these monkeys are all members of the *mona* superspecies (Butynski 2002), the presence of all-male Campbell's monkey groups in Taï forest suggests that this form of grouping may be a hallmark of the *mona* superspecies social system

(Glenn *et al.* 2002). Among the other forest guenons, all-male groups have also been reported in blue monkeys (Tsingalia & Rowell 1984, Macleod 2000) and de Brazza's guenons (*C. neglectus*, Chism & Cords 1998), but these males tend to move in loose associations, and grooming is rare (Glenn *et al.* 2002, Cords, personal communication). Struhsaker (1969) suggested that the formation of all-male groups is an adaptation against ground predators, perhaps because terrestrial species are more vulnerable to predation (Clutton-Brock & Harvey 1977, van Schaik & van Noordwijk 1985 but see Zuberbühler & Jenny 2002). Consistent with Struhsaker's prediction Campbell's and Mona monkeys frequently forage on the ground and in lower forest strata (McGraw 1996, Glenn *et al.* 2002, Buzzard 2004). All-male groups are also common in other terrestrial species such as Japanese macaques (*Macaca fuscata*) and patas monkeys (*Erythrocebus patas*). Further research on the presence of all-male groups in crowned monkeys (*C. pogonias*) would be valuable since this species is a member of the *mona* superspecies but uses higher forest strata (Gautier-Hion 1988) than the other members.

Population densities

Diana monkeys, Campbell's monkeys, and lesser spot-nosed monkeys were found at similar group densities, whereas putty-nosed monkeys were much more rare (see Table 2.2, Höner *et al.* 1997, Eckardt & Zuberbühler 2004). At Tiwai Island, densities of Diana monkeys, Campbell's monkeys, and lesser spot-nosed monkeys were similar (Oates *et al.* 1990, Fimbel 1994). Putty-nosed monkeys (*C. n. nictitans* and *C. n. martini*) are normally much more common than what we found at Taï (*C. n. stampflii*), however. For example, in Gabon (White 1994) and Cameroon (Mitani 1991), putty-nosed monkeys can reach densities of 2.0 groups/km^2 and are over three times more populous than sympatric moustached guenons (*C. cephus*) and crowned guenons. The low density at Taï is most likely because of competition with the other guenons, especially the Diana monkeys. Putty-nosed monkeys most likely originated in central Africa (Kingdon 1997) and it has been suggested that putty-nosed monkeys are later arrivals to the Taï Forest and have not been able to establish themselves as well there (Oates 1988, Eckardt & Zuberbühler 2004).

Inter-group encounters

Jolly (1985 p. 152) defined a territory as "a geographical area, defended by the owners, or exclusively used by the owners, or both." According to this definition, Taï Diana monkeys, Campbell's monkeys and spot-nosed monkeys can be classified as territorial as are Diana monkeys at

Tiwai Island (Oates & Whitesides 1990, Hill 1994). Inter-group encounters were often aggressive, and we never observed affiliative interactions between neighbors as described, for example, by Lawes and Henzi (1995) for blue monkeys. Diana monkeys were involved in more inter-group encounters than the other species with similar group densities, perhaps the result of the greater population density of this species (see Table 2.3). In blue monkeys, higher population density was related to higher rates of aggressive inter-group interactions (Butynski 1990). At Kakamega Forest (Kenya) the density of blue monkey groups was almost twice as high as in Taï and aggressive encounters were much more common occurring every other day on average (Rowell *et al.* 1991, Cords 2002a). Diana monkeys rely on large fruit trees (> 50 cm DBH) more extensively than Campbell's monkeys and lesser spot-nosed monkeys (Buzzard 2006), and the more clumped distribution of these larger trees may be responsible for the 3 to 10 times higher inter-group encounter rates seen in Diana monkeys. However, our data also showed that the percentage of aggressive encounters in Diana monkeys was not exceptional (see Table 2.4). Alternatively, it is possible that part of the species difference could be attributed to observer presence. The home ranges of our Campbell's and spot-nosed monkey study groups bordered on those of neighboring groups that were less habituated to human presence than the neighbors of our Diana monkey study groups, suggesting that inter-group encounters might be much more common, particularly for Campbell's monkeys (Buzzard unpublished data).

Intra-group interactions: affiliative and agonistic behavior

Since we did not use focal animal sampling our data on rates of affiliative and agonistic interactions are not directly comparable with other studies. Agonistic interactions in these monkeys are usually very brief, indicating that the scan sampling method is likely to generate underestimated values concerning the frequency of agonistic behavior. Nevertheless, rates of affiliative and agonistic behaviors of forest guenons in Taï appeared to be considerably lower compared to what is normally reported from many other non-guenon primate species in the wild (Seyfarth 1976, Barton 1993, Barton *et al.* 1996, Sterck & Steenbeek 1997).

Adult females were mainly responsible for observed grooming bouts, and grooming rates were similar in most species (see Table 2.6). The higher grooming rate in putty-nosed monkeys could reflect tighter group cohesion in this species (Chism & Rogers 2002). Males of all species did not normally groom other individuals, but they were sometimes groomed by others, with the exception of the Diana monkey males. Male Diana monkeys appear to be much less integrated into the group than other guenon males, corroborating earlier findings on captive Diana monkeys (Byrne *et al.* 1983) and free-ranging Campbell's monkeys (Hunkeler *et al.* 1972). The lower number of male-female interactions in Diana monkeys compared to the other species may reflect the low willingness of the Diana male to play an active role during inter-group encounters. In blue monkeys, male-female interactions are also rare, and blue monkey males also rarely play an active role during inter-group encounters (Pazol 2001).

Diana monkeys and putty-nosed monkeys exhibited considerably more agonistic interactions (threats, chases, and avoids) than Campbell's monkeys and lesser spot-nosed monkeys. Diana monkeys live in larger groups than Campbell's monkeys and lesser spot-nosed monkeys, suggesting that increased feeding competition could explain higher rates of agonistic interactions (van Schaik & van Hoof 1983, Wrangham *et al.* 1993, Janson & Goldsmith 1995). The putty-nosed monkey group was often associated with Diana monkeys and the two groups frequently foraged on the same large fruit trees (Eckardt & Zuberbühler 2004). Perhaps individuals found it more difficult to maintain comfortable inter-individual distances on these spatially restricted patches, leading to higher rates of agonistic behavior (Clutton-Brock & Harvey 1977, van Schaik 1989). Campbell's monkeys and spot-nosed monkeys foraged more dispersed than Diana monkeys leading to lower rates of agonistic interactions, which supports previous data from Campbell's monkeys (Bourlière *et al.* 1970, Hunkeler *et al.* 1972).

Theoretical implications

Although the low density of putty-nosed monkeys in Taï precluded comparisons of inter- and intra-group competition, our data showed clear differences in the relative importance of both types of competition in the other Taï guenons. In Campbell's monkeys and spot-nosed monkeys, within-group competition appeared to be low, while inter-group encounters were often aggressive suggesting that their social systems can best be classified as "resident-egalitarian" (Sterck *et al.* 1997). For Diana monkeys, however, both intra- and inter-group competition were high, as evidenced by the high rates of agonistic interactions among group members and the high encounter rates with neighboring groups, suggesting that the Diana monkeys may deviate from other guenons and that their social system may be better classified as "resident-nepotistic-tolerant." Additionally, female Diana monkeys engaged in mountings and coalition formation, behaviors often considered aspects of nepotistic societies (e.g. Smuts *et al.* 1987, Chadwick-Jones 1989, Srivastava *et al.* 1991). Finally, McGraw *et al.* (2002) reported a fatal attack of several

Diana monkey females on another one, further suggesting that the label "resident-egalitarian" is not suitable for Diana monkeys. Our data and the results from work on blue monkeys (Cords 2002a) suggest that the resident-egalitarian social system is not a hallmark of guenon social systems.

Conclusions

1. Diana monkeys, Campbell's monkeys, and lesser spot-nosed monkeys were found at similar group densities in the Taï forest, although Diana monkeys lived in consistently larger groups. Putty-nosed monkey groups were much more rare.

2. Study groups consisted of one single adult male and several adult females with their offspring. One lesser spot-nosed monkey group had two adult males during a five-month period following a male take-over. Some extra-group Campbell's monkey males formed stable all-male groups.

3. All species had low levels of social interactions in comparison to macaques and baboons.

4. All species had relatively equal rates of grooming while Diana monkeys showed higher rates of inter-group aggression and intra-group agonistic behavior in relation to the other species. In addition, Diana females formed coalitions and mounted each other, suggesting that their social system was "resident-nepotistic-tolerant," using a recent classification scheme (Sterck *et al.* 1997). Campbell's monkeys and lesser spot-nosed monkeys showed high levels of inter-group aggression, but low levels of intra-group agonistic behavior, suggesting that their social system was a "resident-egalitarian" one. Putty-nosed monkeys could not be classified because they did not have neighboring groups to interact with. High rates of grooming and agonistic acts in putty-nosed monkeys compared to the other species suggest that putty-nosed relationships may be differentiated, however.

References

Altmann, J. (1974). Observational study of behavior: sampling methods. *Behaviour*, **49**, 227–67.

Barton, R. A. (1993). Sociospatial mechanisms of feeding competition in female olive baboons, *Papio anubis. Animal Behaviour*, **46**, 791–802.

Barton, R. A., Byrne, R. W. and Whiten, A. (1996). Ecology, feeding competition and social structure in baboons. *Behavioral Ecology and Sociobiology*, **38**, 321–9.

Bernstein, I. S. (1991). The correlation between kinship and behaviour in non-human primates. In *Kin Recognition*, ed. P. G. Hepper. Cambridge: Cambridge University Press, pp. 6–29.

Bourlière, F., Hunkeler, C. and Bertrand, M. (1970). Ecology and behaviour of Lowe's guenon (*Cercopithecus campbelli lowei*) in the Ivory Coast. In *Old World Monkeys: Evolution, Systematics and Behaviour*, ed. J. R. Napier and P. H. Napier. New York: Academic Press, pp. 367–405.

Butynski, T. (1990). Comparative ecology of blue monkeys (*Cercopithecus mitis*) in high and low density subpopulations. *Ecological Monographs*, **60**, 1–26.

Butynski, T. (2002). The guenons: an overview of diversity and taxonomy. In *The Guenons: Diversity and Adaptation in African Monkeys*, ed. M. E. Glenn and M. Cords. New York: Kluwer, pp. 3–13.

Buzzard, P. J. (2004). *Interspecific competition among Cercopithecus campbelli, C. petaurista, and C. diana*. Ph.D. Thesis. Columbia University, New York.

Buzzard, P. J. (2006). Ecological partitioning of *Cercopithecus campbelli, C. petaurista*, and *C. diana* in the Taï Forest. *International Journal of Primatology*, **27**(2), 529–58.

Byrne, R. W., Conning, A. M. and Young, J. (1983). Social relationships in a captive group of Diana monkeys (*Cercopithecus diana*). *Primates*, **24**, 360–70.

Chadwick-Jones, J. K. (1989). Presenting and mounting in non-human primates: theoretical developments. *Journal of Social and Biological Stress*, **12**, 319–33.

Chism, J. and Cords, M. (1998). De Brazza's monkeys in the Kisere National Reserve, Kenya. *African Primates*, **3**, 18–22.

Chism, J. and Rogers, W. (2002). Grooming and social cohesion in patas monkeys and other guenons. In *The Guenons: Diversity and Adaptation in African Monkeys*, ed. M. E. Glenn and M. Cords. New York: Kluwer, pp. 233–44.

Clutton-Brock, T. H. and Harvey, P. H. (1977). Primate ecology and social organization. *Journal of Zoology, London*, **183**, 1–39.

Cords, M. (1987). Mixed species associations of *Cercopithecus* monkeys in the Kakamega Forest. *University of California Publications in Zoology*, **117**, 1–109.

Cords, M. (1988). Mating systems of forest guenons: a preliminary review. In *A Primate Radiation: Evolutionary Biology of the African Guenons*, ed. A. Gautier-Hion, F. Bourlière, J.-P. Gautier and J. Kingdon. Cambridge: Cambridge University Press, pp. 323–39.

Cords, M. (2000a). The agonistic and affiliative relationships of adult females in a blue monkey group. In *Old World Monkeys*, ed. C. Jolly and P. Whitehead. Cambridge: Cambridge University Press, pp. 453–79.

Cords, M. (2000b). The number of males in guenon groups. In *Primate Males: Causes and Consequences of Variation in Group Composition*, ed. P. M. Kappeler. Cambridge: Cambridge University Press, pp. 84–96.

Cords, M. (2002a). Friendship among adult blue monkeys (*Cercopithecus mitis*). *Behaviour*, **139**, 291–314.

Cords, M. (2002b). When are there influxes in blue monkey groups? In *The Guenons: Diversity and Adaptation in African Monkeys*, ed. M. E. Glenn and M. Cords. New York: Kluwer, pp 189–201.

Eckardt, W. and Zuberbühler, K. (2004). Cooperation and competition in forest monkeys. *Behavioural Ecology*, **15**: 400–12.

Fashing, P. and Cords, M. (2000). Diurnal primate densities and biomass in the Kakamega Forest: an evaluation of census methods and a comparison with other forests. *American Journal of Primatology*, **50**, 139–52.

Fimbel, C. (1994). The relative use of abandoned farm clearings and old forest habitats by primates and a forest antelope at Tiwai, Sierra Leone, West Africa. *Biological Conservation*, **70**, 277–86.

Gautier, J.-P. and Gautier, A. (1977). Communication in Old World monkeys. In *How Animals Communicate*, ed. T. A. Seboek. Bloomington, IN: Indiana University Press, pp. 890–964.

Gautier, J.-P. and Gautier-Hion, A. (1969). Les associations chez les Cercopithecidae du Gabon. *Terre et Vie*, **23**, 164–201.

Gautier-Hion, A. (1988). Polyspecific associations among forest guenons: ecological, behavioural and evolutionary aspects. In *A Primate Radiation: Evolutionary Biology of the African Guenons*, ed. A. Gautier-Hion, F. Bourlière, J.-P. Gautier and J. Kingdon. Cambridge: Cambridge University Press, pp. 452–76.

Gautier-Hion, A. and Gautier, J.-P. (1974). Les associations polyspecifique des Cercopitheques du plateau de M'passa, Gabon. *Folia Primatologica*, **22**, 134–77.

Glenn, M. E. (1996). *The natural history and ecology of the mona monkey (Cercopithecus mona Schreber 1774) on the island of Grenada, West Indies*. Ph.D. Thesis, Northwestern University, Evansville, IL, USA.

Glenn, M. (1997). Group size and group composition of the mona monkey (*Cercopithecus mona*) on the island of Grenada, West Indies. *American Journal of Primatology*, **43**, 167–73.

Glenn, M., Matsuda, R. and Bensen, K. (2002). Unique behavior of the Mona Monkey (*Cercopithecus mona*): all-male groups and copulation calls. In *The Guenons: Diversity and Adaptation in African Monkeys*, ed. M. E. Glenn and M. Cords. New York: Kluwer Academic Publishers, pp. 133–45.

Hill, C. M. (1994). The role of female Diana monkeys (*Cercopithecus diana*) in territorial defence. *Animal Behaviour*, **47**, 425–31.

Höner, O. P., Leumann, L. and Noë, R. (1997). Dyadic associations of red colobus and Diana monkeys in the Taï National Park, Ivory Coast. *Primates*, **38**, 281–91.

Hunkeler, C., Bourlière, F. and Bertrand, M. (1972). Le comportement social de la mone de Lowe (*Cercopithecus campbelli lowei*). *Folia Primatologica*, **17**, 218–36.

Janson, C. H. and Goldsmith, M. L. (1995). Predicting group size in primates: foraging costs and predation risks. *Behavioral Ecology and Sociobiology*, **5**, 326–36.

Jolly, A. (1985). *The Evolution of Primate Behaviour*, 2nd ed. New York: Macmillan.

Kingdon, J. (1997). *The Kingdon Field Guide to African Mammals*. New York: Academic Press.

Korstjens, A. H. (2001). *The Mob, the Secret Sorority, and the Phantoms: an Analysis of the SocioEcological Strategies of the Three Colobines of Taï*. Ph.D. Thesis, Utrecht University, Utrecht, The Netherlands.

Lawes, M. J. and Henzi, S. P. (1995). Inter-group encounters in blue monkeys: how territorial must a territorial species be? *Animal Behaviour*, **49**, 240–3.

Lemasson, A., Gautier, J.-P. and Hausberger, M. (2003). Vocal similarities and social bonds in Campbell's monkeys (*Cercopithecus campbelli campbelli*). *Comptes Rendus Biologies*, **326**, 185–93.

Lemasson, A. and Hausberger, M. (2004). Patterns of vocal sharing and social dynamics in a captive group of Campbell's monkeys (*Cercopithecus campbelli campbelli*). *Journal of Comparative Psychology*, **118**, 347–59.

Macleod, M. C. (2000). *The Reproductive Strategies of Samango Monkeys* (Cercopithecus mitis erythrarchus). Ph.D. Thesis, University of Surrey, Roehampton, UK.

Macleod, M. C., Ross, C. and Lawes, M. J. (2002). Costs and benefits of alternative mating strategies in samango monkey males. In *The Guenons: Diversity and Adaptation in African Monkeys*, ed. M. E. Glenn and M. Cords. New York: Kluwer, pp. 203–16.

McGraw, W. S. (1996). *The Positional Behavior and Habitat Use of Six Monkeys in the Taï Forest, Ivory Coast*. Ph.D. Thesis, State University of New York, Stony Brook, USA.

McGraw, W. S., Plavcan, M. J. and Adachi-Kanawaza, K. (2002). Adult female *Cercopithecus diana* employ canine teeth to kill another *C. diana* female. *International Journal Primatology*, **23**, 1301–8.

Mitani, M. (1991). Niche overlap and polyspecific associations among sympatric cercopithecids in the Campo Animal Reserve, southwestern Cameroon. *Primates*, **32**, 137–51.

Noë, R. and Bshary, R. (1997). The formation of red colobus-Diana monkey associations under predation pressure from chimpanzees. *Proceedings Royal Society London B*, **264**, 253–9.

Oates, J. F. (1988). The distribution of *Cercopithecus* monkeys in West African forests. In *A Primate Radiation: Evolutionary Biology of the African Guenons*, ed. A. Gautier-Hion, F. Bourlière, J.-P. Gautier and J. Kingdon. Cambridge: Cambridge University Press, pp. 79–103.

Oates, J. F. and Whitesides, G. H. (1990). Association between olive colobus (*Procolobus verus*), Diana guenons (*Cercopithecus diana*), and other forest monkeys in Sierra Leone. *American Journal of Primatology*, **21**, 129–46.

Oates, J. F., Whitesides, G. H., Davies, A. G. *et al.* (1990). Determinants of variation in tropical forest primate biomass: new evidence from West Africa. *Ecology*, **71**, 328–43.

Payne, H. F. P., Lawes, M. J. and Henzi, S. P. (2003). Competition and the exchange of grooming among female samango monkeys (*Cercopithecus mitis erythrarchus*). *Behaviour*, **140**, 453–72.

Pazol, K. (2001). Social, ecological and endocrine influences on female relationships in blue monkeys (*Cercopithecus mitis stuhlmanni*). Ph.D. Thesis, University of Pennsylvania, Philadelphia, USA.

Pusey, A. E. and Packer, C. (1987). Dispersal and philopatry. In *Primate Societies*, ed. B. B. Smuts, D. L. Cheney, R. M. Seyfarth, R. W. Wrangham and T. T. Struhsaker. Chicago: University of Chicago Press, pp. 250–66.

Rowell, T. E. (1988). The social system of guenons compared with baboons, macaques, and mangabeys. In *A Primate Radiation: Evolutionary Biology of the African Guenons*, ed. A. Gautier-Hion, F. Bourlière, J.-P. Gautier and J. Kingdon. Cambridge: Cambridge University Press, pp. 439–51.

Rowell, T. E., Wilson, C. and Cords, M. (1991). Reciprocity and partner preference in grooming of female blue monkeys. *International Journal of Primatology*, **12**, 319–36.

Seyfarth, R. M. (1976). Social relationships among adult female baboons. *Animal Behaviour*, **24**, 917–38.

Siegel, S. and Castellan, N. J. Jr. (1988). *Nonparametric Statistics for the Behavioral Sciences*, 2nd ed. New York: McGraw-Hill.

Smuts, B. B., Cheney, D. L., Seyfarth, R. M., Wrangham, R. W. and Struhsaker, T. T. (1987). *Primate Societies*, Chicago: University of Chicago Press.

Srivastava, A., Borries, C. and Sommer, V. (1991). Homosexual mounting in free-ranging female Hanuman langurs (*Presbytis entellus*). *Archives Sexual Behavior*, **20**, 487–512.

Sterck, E. H. M. and Steenbeeck, R. (1997). Female dominance relationships and food competition in the sympatric Thomas langur and long-tailed macaques. *Behaviour*, **134**, 749–77.

Sterck, E. H. M., Watts, D. P. and van Schaik, C. P. (1997). The evolution of female social relationships in nonhuman primates. *Behavioral Ecology and Sociobiology*, **41**, 291–309.

Struhsaker, T. T. (1969). Correlates of ecology and social organization among African cercopithecines. *Folia Primatologica*, **11**, 80–118.

Struhsaker, T. T. (1981). Polyspecific associations among tropical rain forest primates. *Zietschrift für Tierpsychchologie*, **57**, 268–304.

van Schaik, C. P. (1983). Why are diurnal primates living in groups? *Behaviour*, **87**, 120–44.

van Schaik, C. P. (1989). The ecology of social relationships amongst female primates. In *Comparative Socioecology: The Behavioural Ecology of Humans and Other Mammals*, ed. V. Standen and R. Foley. Oxford: Blackwell, pp. 195–218.

van Schaik, C. P. and van Hoof, J. (1983). On the ultimate causes of primate social systems. *Behaviour*, **85**, 91–117.

van Schaik, C. P. and van Horstermann, M. (1994). Predation risk and the number of adult males in a primate group: a comparative test. *Behavioral Ecology and Sociobiology*, **35**, 273–81.

van Schaik, C. P. and Van Noordwijk, M. A. (1985). Evolutionary effect of the absence of felids on the social organisation of macaques on the island of Simeuleu (*Macaca fascicularis*). *Folia Primatologica*, **44**, 138–44.

Tsingalia, H. M. and Rowell, T. E. (1984). The behaviour of adult male blue monkeys. *Zietschrift für Tierpsychchologie*, **64**, 253–68.

White, L. J. T. (1994). Biomass of rain forest mammals in the Lope reserve, Gabon. *Journal of Animal Ecology*, **63**, 499–512.

Whitesides, G. H. (1981). Community and population ecology of non-human primates in the Douala-Edea forest reserve. M.Sc. Thesis, Johns Hopkins University, Baltimore, USA.

Whitesides, G. H., Oates, J. F., Green, S. M. and Kluberdanz, R. P. (1988). Estimating primate densities from transects in a West African rain forest: a comparison of techniques. *Journal of Animal Ecology*, **57**, 345–67.

Wolters, S. and Zuberbühler, K. (2003). Mixed-species associations of Diana and Campbell's monkeys: the costs and benefits of a forest phenomenon. *Behaviour*, **140**, 371–85.

Wrangham, R. W. (1980). An ecological model of female-bonded primate groups. *Behaviour*, **75**, 262–300.

Wrangham, R. W., Gittleman, J. L. and Chapman, C. A. (1993). Constraints on group size in primates and carnivores: population density and day range as assays of exploitation competition. *Behavioral Ecology and Sociobiology*, **32**, 199–209.

Zuberbühler, K. (2001). Predator-specific alarm calls in Campbell's guenons. *Behavioral Ecology and Sociobiology*, **50**, 414–22.

Zuberbühler, K. and Jenny, D. (2002). Leopard predation and primate evolution. *Journal of Human Evolution*, **43**, 873–86.

Zuberbühler, K., Noë, R. and Seyfarth, R. M. (1997). Diana monkey long-distance calls: messages for conspecifics and predators. *Animal Behaviour*, **53**, 589–604.

3 How small-scale differences in food competition lead to different social systems in three closely related sympatric colobines

A. H. Korstjens, K. Bergmann, C. Deffernez, M. Krebs,
E. C. Nijssen, B. A. M. van Oirschot, C. Paukert, and
E. Ph. Schippers

Introduction

As an essential aspect of life, food can evoke strong competition among individuals and shape a species' social system. Through a pathway of relationships we can link the competitive regime that food evokes in a population to such seemingly loosely related traits as social relationships within and between groups, ranging patterns and dispersal patterns. Food most strongly determines female relationships in many mammals because female reproductive success is mainly constrained by food acquisition. Male success, on the other hand, largely depends on access to mating partners (Trivers 1972, Emlen & Oring 1977). The competitive regime among females can be predicted based on the contestability or usurpability of their food (Wrangham 1980, van Schaik & van Hooff 1983, van Schaik 1989, 1996, Isbell 1991, Sterck *et al.* 1997). In most primates, food competition increases with group size (Clutton-Brock & Harvey 1977, Wrangham *et al.* 1993). However, folivores do not always fit into the general patterns found in such studies (Clutton-Brock & Harvey 1977, Isbell 1991, Janson & Goldsmith 1995). A comparative test of the effect of diet on social systems *within* a largely folivorous genus can solve some of these controversies. Since dietary category is only a proxy for a whole suite of traits that together determines the contestability of a species' food we need to look at each of these traits to investigate differences within a dietary specialization. Among these traits we find: caloric content and digestibility of food item(s), the size, abundance and distribution of

Monkeys of the Taï Forest, ed. W. Scott McGraw, Klaus Zuberbühler and Ronald Noë.
Published by Cambridge University Press. © Cambridge University Press 2007.

a food patch and how it is distributed in the larger space, as well as its abundance, the time it takes to get to the edible part of the food item, and the time that is spent at a food patch (Wrangham 1980, Whitten 1983, van Schaik 1989, Janson 1990, Isbell 1991, Isbell *et al.* 1998, Mathy & Isbell 2001). Considering the number of variables to be evaluated, it is not surprising that few studies have managed to plot the general diet of a species onto this continuum of contestability, as well as test the implications for specific levels of competition on the social organization of a species (van Schaik & van Noordwijk 1988, Borries 1993, Cords 2000, Koenig 2000). Furthermore, because the level of competition in one species or population is only meaningful when compared to that of another species or population, comparative studies produce the most convincing support for the proposed effects that food competition has on a species' social system (e.g. Mitchell *et al.* 1991, Barton *et al.* 1996, Sterck & Steenbeek 1997, Isbell & Pruetz 1998, Isbell *et al.* 1998, 1999, Thierry *et al.* 2000).

We use a unique natural experiment, the coexistence of three closely related colobines (western red colobus: *Pilio-/or Procolobus badius badius;* western black-and-white colobus: *Colobus polykomos polykomos*; olive colobus: *Procolobus verus*), to compare the impact of food on social systems. The enormous advantage of this set-up is that we control for the confounding effects of phylogeny, predation and seasonality of the forest simultaneously. We will discuss the suite of traits that determine the contestability of the foods chosen by these three folivorous species and test whether the differences in contestability can explain the differences in social relationships within and between groups as predicted by current socio-ecological theory.

Although the details and the categorization of social systems that different socio-ecological models produce vary (Isbell & Young 2002, Isbell 2004) the general basis of the theory behind them is very similar. The level of competition is determined by food available per individual at a certain food source (Clutton-Brock & Harvey 1979, Janson & Goldsmith 1995). The predominant mode of competition, either contest or scramble competition, depends on the distribution and size of food sources relative to the number of individuals using the source (e.g. Wrangham 1980, Janson 1988, van Schaik 1989, Milinski & Parker 1991, van Hooff & van Schaik 1992). Patchily distributed food sources that can be monopolized by one individual or a coalition evoke contest competition. Small and equally distributed food sources that are difficult to monopolize result in scramble competition. However, a group of primates does not exist in a vacuum and therefore, in order to understand a social system we have to investigate both competition within a group and competition between groups. The third level, competition between species will not be discussed here. The competitive regime between groups depends on the same variables as competition within groups, but it occurs at different levels: intra-group competition depends on distribution of food within a food source on which several group members feed, and inter-group food competition depends on distribution of food sources over the area that different groups use.

Based on general socio-ecological theory (Alexander 1974, Emlen & Oring 1977, Wrangham 1980, van Schaik 1983, 1989, van Schaik & van Hooff 1983, Dunbar 1988, Isbell 1991, 2004, Janson & Goldsmith 1995, Sterck *et al.* 1997, Isbell & Young 2002) we formulated a set of predictions for the social system of a species, as based on the characteristics of its food (see Table 3.1). The levels of within-group and between-group contest competition are expected to be positively correlated to the quality of food since only high quality food will make food defense worth the effort. Food quality is determined by the energy content, digestibility, and size of food items. Colobines tend to select foods especially on the basis of the levels of protein and fiber (Davies & Bennett 1988, Dasilva 1994, Waterman & Kool 1994, Brugière *et al.* 2002, Chapman *et al.* 2003, Wasserman & Chapman 2003). High quality items are young leaves, fruits, flowers, and seeds, while mature leaves are low quality. Contest competition within a group is also more likely to occur when foods are patchily distributed in a tree and relatively rare. High quality food, such as fruits, tends to be less common and more patchily distributed than low quality food, such as leaves (Clutton-Brock & Harvey 1979). Long processing time and a long time spent at a patch is also predicted to increase the levels of agonistic interactions an individual endures per unit time (Isbell 1991). A long handling time means that the quality of the food item increases with time until it has been processed completely. Frugivores and folivores do not generally spend much time processing their food but swallow most food items whole. Granivores, on the other hand, may take some time to get to the seeds, especially when these are encased in a wooden husk (e.g. Korstjens *et al.* 2002). The time spent at a patch increases the risk of being attacked by others. Several species have solved this problem by the use of cheek pouches: stuffing the cheek pouch reduces the time an individual spends at a food patch. It is noteworthy that cheek pouches are not found in colobines but they are in their closest frugivorous relatives the cercopithecines (Oates & Davies 1994) which supports the idea that fruits are more contestable than leaves. Finally, within a large tree, within-group

Table 3.1. *General predicted effect that various food traits and social traits have on each other based on socio-ecological theory and as investigated in this chapter*

	Effect	
Contestability of food at within-group (WG) level	Increases with:	Percentage fruit in diet Processing time Patchiness of food items within trees Food spot residence time
	Decreases with:	Percentage mature leaves in diet Density of items in tree Tree size relative to number of individuals
Contestability of food at between-group (BG) level	Increases with:	Percentage fruit in diet Processing time Patchiness of food trees Tree size relative to group size
	Decreases with:	Percentage mature leaves in diet Density of trees
Within-group F-F aggression	Increases with: Decreases with:	Contestability of food at WG-level Contestability of food at BG-level
Between-group F-F aggression	Increases with:	Contestability of food at BG-level
Within-group F-F cooperation	Strength increases with:	WG F-F aggression if food items are shareable
Group-level cooperation	Strength increases with:	BG F-F aggression
F-F affiliation	Increases with:	WG & BG F-F aggression
Home range size	Increases with:	Group size Percentage fruit in diet Spatio-temporal patchiness of food trees
	Decreases with:	Density of food trees
Day-Journey Length	Increases with:	Group size Percentage fruit in diet Spatial patchiness of food trees Percentage time in association with frugivores
	Decreases with:	Density of food trees Food tree size relative to group size Food digestion and processing time
Territoriality	Increases with:	BG competition Small home range relative to day-journey length
	Decreases with:	Home range size
Female philopatry	Increases with:	Cooperation Patchiness and rarity of food trees Predation risk during transfer
	Decreases with:	Inbreeding risk e.g. few males per group

competition will be less than in a small tree because there is enough food to share.

Contest competition between groups is expected to be high when food trees are patchily distributed and occur at low densities. Fruit bearing trees generally follow these patterns more than leaf-bearing trees due to seasonal availability of food items (Harvey & Clutton-Brock 1981). Food source density is negatively correlated with competition as it removes any incentive to compete over food. Finally, the larger the tree that animals feed in relative to the number of individuals in the group, the more likely it is that they will spend a long time at each tree as it depletes more slowly than a small tree (Janson 1990, Isbell *et al.* 1998). Inter-group competition will be especially strong if large, patchily distributed and rare food sources are regularly used.

Although highly contestable food sources should lead to high levels of agonistic interactions between groups (Wrangham 1980), the level of territoriality of a species depends on more variables than contestability of food sources alone. The size of the home range relative to the distance travelled per day, should determine whether territories or individual patches are defended (Mitani & Rodman 1979, Lowen & Dunbar 1994). When home ranges are largely relative to day journey length, not territories but individual patches will be defended depending on the contestability of the patch at which two groups meet. Day journey length and home range size themselves correlate to the biomass of a group and food character-istics, such as its density, quality, and temporal and spatial distribution (Clutton-Brock & Harvey 1977, Harvey & Clutton-Brock 1981, Janson & van Schaik 1988, Wrangham *et al.* 1993, Janson & Goldsmith 1995, Gillespie & Chapman 2001). Because a large patch depletes less fast, day journey length is expected to decrease with patch size. Furthermore, because the animals have more energy and require less time for digestion, the species with the highest quality food is expected to travel the farthest each day. Finally, patchily distributed rare food sources would result in relatively long distances travelled between food sources and a generally large home range.

The levels of contest competition within and between groups affect the affiliative relationships within groups. Cooperation among individuals increases each partner's resource holding potential when contest compe-tition prevails and resources can be shared (Wrangham 1980, Walters & Seyfarth 1987, Scheel & Packer 1991, Hawkes 1992, van Hooff & van Schaik 1992). Such cooperative relationships are often stronger and may require less maintenance if formed between kin than between non-kin.

This difference is reflected in more stable relationships over time and has been explained as a result of inclusive fitness benefits (Hamilton 1964a, 1964b, Trivers 1972). To maintain long-term cooperative relationships, partners often have regular affiliative interactions, such as allogrooming, embracing, and sitting close together (Dunbar 1991). With high levels of inter-group competition we expect strong affiliative relationships within the group to maintain a group-level coalition. Furthermore, if group-level alliances are important, within-group contest competition may be relaxed because the dominants depend on the subordinates to assist in inter-group conflicts (Sterck *et al.* 1997). Not only kinship, but also reciprocity or the pursuit of a common goal can form the basis for cooperation (Harcourt 1989, Noë *et al.* 1991, Noë 1992).

The importance of cooperation in a group may affect dispersal decisions of individuals. Female-biased philopatry is predicted when females depend heavily on cooperation in food competition (Wrangham 1980, van Schaik 1989, Sterck *et al.* 1997). As a result of the importance of cooperation, females may be especially reluctant to migrate when food is contestable. However, dispersal patterns are determined by a multitude of costs and benefits to leaving and staying in an area or social unit (Greenwood 1980, Moore 1988, Clutton-Brock 1989, Isbell & van Vuren 1996, Sterck & Korstjens 2000, Isbell 2004). It is generally assumed that any individual would rather stay in a familiar environment and thus not disperse away from its natal group and/or range. This is especially true if leaving the natal range means immigrating into an unknown area. Moving to or through an unknown area is especially costly if food sources are not easily detected, for example because they are patchily distributed and occur at low densities, and if predation risk depends on knowledge of safe havens. However, inbreeding avoidance or high levels of intra-group food competition may force a female to migrate.

We tested the suggested relationship between diet and social system of three folivores in the Taï National Park. Despite their close relatedness and the fact that they share the same forest areas, there is little overlap in dietary preferences (Bergmann 1998) and social systems (Korstjens 2001, Korstjens *et al.* 2002). We are not trying to fit the different species into different categories of the various models. Rather, we expect them to largely fit into the same general category: low levels of contest competition with an egalitarian type of dominance relationships and weakly defined affiliative relationships among females, leading to the potential for female dispersal. We are interested in testing if current socio-ecological theory is strong enough to predict subtle differences between the social systems of species that are using relatively similar food sources. We will address the following questions:

1.   Does the species with more contestable food items have higher levels of agonistic interactions and stronger affiliative or cooperative relationships than the species with less contestable food items?
2.   Does the species with more contestable food patches have more inter-group competition over food and stronger group-level cooperation?
3.   Does the species with the strongest need for female-female cooperation within the group have a higher degree of female philopatry than the others?

Methods
The study groups
All individuals in the black-and-white and olive colobus study groups were individually recognized. In red colobus, most adults were individually recognized. Data on social behavior was only collected on individuals that were individually recognized. Recognition of individuals was based on characteristics of face, tail, and coloring of skin or fur. Individuals in the study groups were classified into the following age-sex categories: "sub-adults" were slightly smaller than adults and presumably sexually mature. Sub-adult females were nulli-parous, had smaller nipples and, in the case of red colobus, smaller swellings than adult females. Red and olive colobus males were classified as sub-adult as soon as their testes started to descend and for red colobus when their faces started to broaden. Black-and-white colobus sub-adult males were as large as adult females and they produced an incomplete version of the male roar (loud-call). "Adults" were sexually active individuals with fully developed secondary sex characteristics. Throughout the chapter we used the following abbreviations for these age-sex classes: AM = adult male, AF = adult female, SM = sub-adult male, SF = sub-adult female.

Red colobus. The two study-groups of red colobus, Bad1 and Bad2, were followed from 1992−9. Each of these groups consisted of over 90 individuals in 1997. The groups each started to split up into two sister-fractions between 1994−8 (Bshary 1995, van Oirschot 1999, 2000). In June 1999 all groups were counted based on individual recognition of adults and sub-adults (see Table 3.2). The two sister fractions of each group shared the same home range for 2−3 years during and after the splitting process.

Table 3.2. Summary of social and ecological variables for red colobus (groups Bad1 and Bad2), black-and-white colobus (groups Pol1 and Pol3), and olive colobus (Ver1 and Ver3)

	Bad1[a]	Bad2[a]	Pol1	Pol3	Ver1	Ver3
Group size[a]	A 41[b] B 64[b]	A 60[b] B 44[b]	12–16	16–18	3–7	3–4
Adults: [a] males/females	A 6/14[b] B 15/22[b]	A 12/22[b] B 9/15[b]	1–2/4–6	1/6	1–2/1–3	1/1–2
% Fruit average[c]	29.7 ± 13.31	29.7 ± 19.1	45.1 ± 27.5	–	7.5 ± 7.2	–
Monthly averages	16.7, 36.9, 31.3	21.6, 40.1, 24.7	37.2, 48.4	8.1	7.5	17.0
Yearly averages	36.1, 32.5	21.6, 40.1, 24.7	46.6, 49.9	–	8.6	–
Tree size median	55	51	44	56	10	44
Tree biomass[d]	5197 ± 2771	6735 ± 2201	2530 ± 1262	5897 ± 2375	6440 ± 2344	5299 ± 2938
Yearly HR size[e] average ± SD	65.7 ± 8.5	50.3 ± 5.8	83.3 ± 10.5	71.5 ± 4.9	53.5 ± 4.9	57.7 ± 7.1
DJL average ± SD (# days)[f]	922 ± 214 (16)	822 ± 235 (54)	677 ± 216 (54)	637 ± 2.5 (25)	1202 ± 297 (12)	1222 ± 589 (7)
Rodman & Mitani Index D[g]	1.01	1.03	0.71	0.65	1.46	1.43
HR overlap ± SD[h] (% of HR size)	5.7 ± 2.1	7.0 ± 3.6	20.5 ± 12.0	21.5 ± 7.8	14.0 ± 5.7	14.0 ± 8.5

[a] For Bad1 and Bad2 we provide the values for the two two sister groups (A = group A and B = group B) separately if data were collected for the groups separately.

[b] In June 1999 we individually distinguished all adult and sub-adult individuals per group, we added an estimate of the number of immatures per group based on the average number of females with infant or juvenile in each group.

[c] Average percentage of fruit in the diet was calculated in multiple ways since we had several years of data. The top values give the annual monthly average. It was calculated by first taking the average per month of the values from 1996, 1997, and 1998. Then the average and standard deviation of these average months was calculated and presented. The second value provides the monthly average of each year in which data were collected during every month. The third value provides the total percentage of fruits consumed during a particular year for years (1996–8 only) in which data were collected during every month.

[d] Monthly average, see Table 3.3 for comments.

[e] HR = Annual home range size was averaged over two years, ± SD.

[f] DJL = average distance travelled from one night's sleeping spot to the next night's sleeping place.

[g] Defensibility index D (Rodman & Mitani index): territory defense is expected when D > 1.

[h] HR overlap: overlap in yearly home ranges of two neighboring groups measured as average ± SD per cent of the home range of study group for two consecutive years.

The smaller sister fractions slowly moved out of the main home-range area. Kin relationships within the study groups were not known.

Black-and-white colobus. We followed two black-and-white colobus study groups and all individuals in the groups were individually recognized (see Table 3.2). Pol1 was followed from 1992−9 and the mothers of one of the sub-adult females, all juveniles and all infants were known (Korstjens *et al.* 2002). Pol3 was followed from 1998−9 so no kin relationships were known in Pol3 except for mother-infant relationships.

Olive colobus. Four groups, Ver1 to Ver4, were studied: Ver1 from 1994−7; Ver2 (4−7 individuals: one AM and 1−4 AFs) from 1997−9; Ver3 from 1997−9; and Ver4 (4−10 individuals: 2−3 AMs and 2−5 AFs) from 1997−8 (details on group dynamics in Korstjens & Schippers 2003, Korstjens & Noë 2004). Kin relationships were known for mother-infant pairs only.

Group compositions and group sizes of study groups were representative for the population (Galat & Galat-Luong 1985, Korstjens 2001). Study-groups Bad2, Pol1, and Ver1 formed a cluster in the sense that their home ranges overlapped greatly or entirely. Bad1, Pol3, and Ver3 formed a similar cluster in an adjacent area of the study-site. This meant that ecological parameters were completely identical for the members of such clusters and, therefore, any differences in food choice were not a result of differences in availability of food but in species-specific preferences.

Data collection and analyses

Most data were collected during day-follows from 7:00−17:30 hrs. The same methods were used for all species (following Altmann 1974, Martin & Bateson 1993) unless mentioned otherwise. Data that were collected by various observers are used only when inter-observer reliability was at least 90 per cent (for details on number of observers for different data sets see Korstjens 2001 and references cited with results). Dietary and behavioral data were collected from adult and sub-adult individuals using scan sampling, focal sampling and ad libitum sampling. Throughout the day we collected ad libitum data on social interactions within and between groups.

Scan samples were taken every hour and lasted for a maximum of 25 minutes. The observer assured that no single individual was sampled twice using individual recognition and location in the group. For each scanned individual observers noted: age-sex class, identity, activity (lasting for at least 5 seconds), item consumed, diameter at breast height (DBH) if not feeding in a liana, and species of the tree or liana in which the animal fed (if applicable), and number, sex, and distance to neighbors.

Scan samples were used to determine diets, food tree size, and proximity to neighbors.

The diet is expressed as the percentage of scans in which a certain food item was consumed. Food items were categorized as: fruits, leaves, flowers, insects, or termite matter. Termite matter consisted of termite earth and/or termites. The category "fruits" contains all fruits, irrespective of whether fruit pulp, whole fruits or only seeds were consumed. Food tree size is obtained from the median of the DBH of trees used during scan samples. A "neighbor" was an adult or sub-adult individual located within two meters of the scanned individual. The time spent with neighbors was defined as the percentage of scans of a certain individual in which it had at least one neighbor. Wilcoxon matched pairs signed ranks tests and Mann Whitney U tests were performed using scans from individually recognized animals only, with the individual animal as the sampling unit. The neighbor data are presented in two ways: (1) as the percentage of the total number of scan samples of an individual in which it had a neighbor (the uncorrected percentage); and (2) as this percentage corrected for the sex ratio in the group (the corrected percentage). The corrected percentage was calculated by multiplying the uncorrected percentage of scans spent with a female neighbor by the sex ratio in the study-group: 0.5 for Bad2A, 0.17 for Pol1 and 0.5 for Ver1. Only data of A. H. Korstjens (AHK) on red and black-and-white colobus were used for statistical tests (AHK did not collect these data on olive colobus). The neighbor scans presented for Ver1 were collected by F. Bélé, who had been trained by AHK for this type of data collection.

Focal animal samples were collected on individually recognized females and lasted from three to ten minutes. Consecutive focal samples of the same individual were separated by at least one hour. The number of focals was distributed approximately equally among individuals (Korstjens 2001). During a focal animal sample all social interactions involving the focal individual were recorded continuously. We collected 21.3 hours of focal observations on 16 females of Bad2A and 129.9 hours of focal observation on 9 females of Pol1 between 1997 and 1998. No focal data of olive colobus were used because there were many different observers who did not always use the same protocols. Focals were used for calculating rates of social interactions as well as handling time and food patch residence time.

For calculating handling time of food items, we recorded the number of times per minute that an individual moved its arm in order to pick a food item during each focal sample. We used the average value for each focal sample as one sampling unit. Mann Whitney U tests were used to compare arm movements between red and black-and-white colobus. These tests were also used to compare arm movements per minute for different food items within the black-and-white colobus sample.

The "food spot residence time" was defined as the number of full minutes with no movements or only movements in which an animal moved with a food item in hand. Some of the food spot residence periods were truncated ("censored" data) when the focal ended but the individual still fed at the same spot. We used Kaplan Meijer tests of survival, in the loglinear analysis setting (SPSS 7.5 for Windows) to compare food spot residence time between the species.

Rates of agonistic or affiliative interactions were calculated from focal animal samples. During a focal sample, we counted the number of interactions of the focal animal with any other individual that was within two meters distance. Only very few agonistic interactions (threats) occurred at a greater distance. Agonistic interactions included all instances of submissive and/or aggressive behavior. Submission was recorded when one individual yielded to, fled from, or crouched in front of another individual. Aggressive actions included threatening, pushing, biting, hitting, chasing, and stealing food from an individual. Nearly all of the aggressive acts produced a submissive response. When several agonistic or submissive acts occurred in the same context within three minutes of each other, all of the events were considered to be part of a single interaction. When the individual that was approached or attacked was feeding or manipulating a food item, the context was labelled as "food."

For calculating the rate of grooming interactions, we counted the number of grooming bouts per minute focal sampling. A single grooming bout was defined as a series of grooming episodes between the same individuals with interruptions of less than three minutes. When two individuals groomed alternately this was recorded as one interaction for each individual, irrespective of how often the grooming direction switched during the grooming bout. Cooperation among individuals was recorded when two individuals simultaneously threatened or chased a third individual.

Ranging data were derived from hourly samples of the location of the center of mass of the group in relation to a painted grid system with 100 by 100 m cells. The day journey length (DJL) was the summed distances travelled between the hourly data points during the day, plus the distance from the last observation of the day to the first observation on the consecutive day. The home range was the sum of all one hectare cells entered by the groups during the sampling period. The Mitani and Rodman index (Mitani & Rodman 1979) was used to determine whether home range size and day journey length allowed for territory defense. The similar Lowen and Dunbar index (Lowen & Dunbar 1994) requires an estimate of visibility that would differ greatly between the species and between situations and was, therefore, not used here (Korstjens 2001). Ranging data for red colobus were collected in the years before observers noted the identity of the sister groups they followed. As a result data collection was biased towards the larger and more conspicuous fractions.

Inter-group interactions were recorded ad libitum every time that individuals of different conspecific groups approached each other to within 50 meters (the distance over which an observer was likely to notice the presence of neighboring groups, following Oates 1977, Stanford 1991, Steenbeek 1999, Fashing 2001b). Agonistic inter-group encounters were defined as those encounters in which threat displays, aggressive physical contact or chasing occurred between individuals of different groups. This category combines display and aggressive inter-group encounters as defined in Korstjens *et al.* 2005. Non-aggressive inter-group encounters were those in which no obvious interactions occurred between individuals of the different groups.

Food tree density and biomass

Three line transects (north−south) of 1 km long and 25 m wide were laid out through the middle of the home ranges of the two study group clusters. Each transect was divided into quadrants of 25 by 25 m. For every tree (girth > 20 cm) or liana (girth of the largest stem > 10 cm) in each quadrant the girth at breast height (translated into diameters at breast height, DBH) and species name was recorded. In total 98.5 per cent of the trees and lianas on the two transects were identified at the species level.

A tree or liana was defined as a food-tree when the species made up at least 5 per cent of the feeding scans recorded in a specific calendar month (diet was taken from all scan samples collected between 1996 and 1997). Korstjens (2001) presents more details and slightly different analyses of food biomass.

Spatial distribution of food trees was measured with the coefficient of distribution: $CD = variance/mean$ (following Chapman *et al.* 1995). The CD was calculated per transect for the number of trees per unit. Two consecutive quadrants (i.e. 25 by 50 meters) were taken as one unit. We used two quadrants because the average spread of the colobine groups was > 25 m (Korstjens unpubl data). The DBH's of all food-trees were summed as a measurement of food biomass (cm DBH/ha) (see Chapman *et al.* 1995). In the Kruskal-Wallis tests (Siegel & Castellan 1988) on monthly collected variables the months were taken as independent units.

Although monthly values cannot be completely independent this method allowed us to incorporate the great differences between months due to changes in the diets. Furthermore, comparing dietary details between species during matched months allowed us to identify the real differences in food selection between the species because they all had the same food to choose from.

General data analyses

Statistical comparisons between the three colobus species were limited to comparisons within the cluster of Bad2, Pol1, and Ver1. Data on Pol3 and Ver3 were insufficient for statistical comparisons in the Bad1, Pol3, and Ver3 cluster. Non-parametric statistics were used (Siegel & Castellan 1988). All tests were two-tailed and α was set at 0.05. An α'-value was calculated, using a sharper Bonferroni correction (Hochberg 1988) when multiple tests were performed with the same dataset. Statistical tests were conducted with SPSS for Windows.

Results

Contestability of food

All three colobus species consumed mainly unripe fruits and young leaves (diets from January 1996 − April 1998; see Figures 3.1a, 3.1b, and 3.1c; see Table 3.2). Of the three species, olive colobus females consumed the least fruits (N = 14 months; Ver1-Bad2: $Z = -3.04$, p = 0.002; Ver1-Pol1: $Z = -2.86$, p = 0.0043) and red colobus females fed less on fruits than did black-and-white colobus females (N = 28 months: $Z = -2.53$, p = 0.011). Similar results were found for adult males (Pol1-Bad2: N = 28 months, $Z = -2.32$, p = 0.020; Pol1-Ver1: N = 14 months, $Z = -2.35$, p = 0.019; Bad2-Ver1: N = 14 months, $Z = -1.60$, p = 0.109; see Figure 3.1). Fruit consumed by red and olive colobus consisted mainly of fleshy fruits that were eaten as a whole, while black-and-white colobus more often ate seeds of woody fruits without fruit flesh (tree species lists in Korstjens 2001). Handling time and food spot residence time were significantly longer for black-and white colobus than for red colobus (Korstjens *et al.* 2002). Such data were not available for olive colobus.

Red colobus selected the largest trees, while olive colobus selected the smallest trees (Mann Whitney U tests: Bad2-Pol1: $Z = -14.6$, p < 0.0001; Bad2-Ver1: $Z = -28.9$, p < 0.0001; Bad2-Ver1: $Z = -20.6$, p < 0.0001; $\alpha' = 0.05$; see Table 3.3). We compared the total biomass per ha (see Table 3.3) and the distribution (measured as coefficient of distribution) of the trees from which each species selected their food using Wilcoxon Signed Ranks Matched Pairs tests that paired matching months

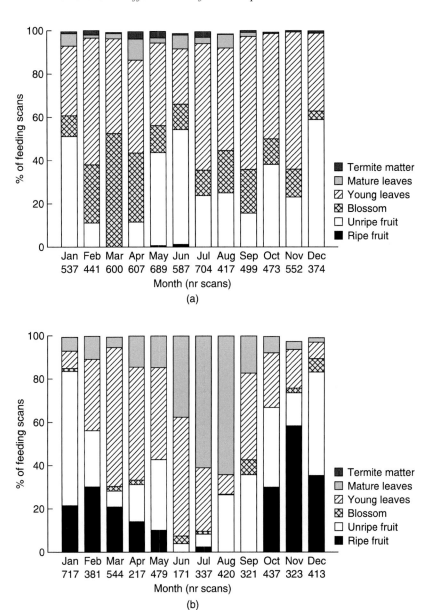

Figure 3.1. Monthly diet of: (a) red colobus; (b) black-and-white colobus.

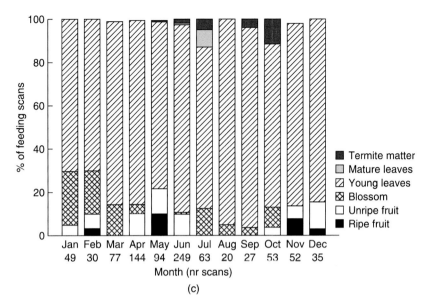

Figure 3.1. The monthly diet, separated into the major food categories, (measured as a percentage of feeding-scans of adults and sub-adults) of red colobus (for Jan 1996–Apr 1998; Figure 3.1a), black-and-white colobus (for Jan 1996–Apr 1998; Figure 3.1b), and olive colobus (for Jun 1996–May 1997; Figure 3.1c); "Termite matter" = material obtained from termite hills and from termite earth attached to tree branches.

together for the three transects combined and dietary data from 1996 to 1997 (maximum 24 months). Olive colobus and red colobus selected food from more abundant food sources (i.e. number of trees per ha) than black-and-white colobus (Bad2-Pol1: $N = 23$ months, $Z = -3.7$, $p = 0.0002$; Pol1-Ver1: $N = 17$ months, $Z = -3.6$, $p = 0.0004$; Bad2-Ver1: $N = 17$ months, $Z = -2.4$, $p = 0.017$; $\alpha' = 0.025$). Similarly, the biomass of food sources selected by red and olive colobus differed little but was higher than that for the black-and-white colobus (Bad2-Pol1: $N = 23$ months, $Z = -4.2$, $p < 0.0001$; Pol1-Ver1: $N = 17$ months, $Z = -3.6$, $p = 0.0004$; Bad2-Ver1: $N = 17$ months, $Z = -0.02$, $p = 0.98$; $\alpha' = 0.025$; see Figure 3.2). These results are the same if we use the trees that represented at least 10 per cent of the diet except that the biomass did not differ significantly between red and black-and-white colobus ($p = 0.09$). Note that the biomass of 5 and 10 per cent trees in the annual diet is higher for black-and-white colobus groups than for red colobus groups while this is not true for the monthly averages. This is the result of a more diverse diet for red colobus, e.g. red colobus consumed food from 77 tree/liana species while black-and-white colobus consumed food from 51 species

Table 3.3. *The diameter (DBH in cm) of food trees and the biomass of food trees used by the study groups (cm DBH/ha)*

| | | | | | Food biomass (cm DBH/ha) | | | |
| | Food tree size | | | | Monthly average | | Annual diet | |
Group	N	25%	Median	75%	5%	10%	5% trees	10% trees
Bad1	3535	44	55	71	5197	3469	4838; 4942	2798; 2798[a]
Bad2	2997	41	51	71	6735	3375	3266; 4063	2798; 2798[a]
Pol1	2390	23	44	69	2530	1858	5403; 6921	1521; 1748
(Pol3	274	45	56	117	5897	2866	4475	2798)
Ver1	430	6	10	21	6440	4227	6570; 5611	6570; 4098
(Ver3	59	36	44	58	5299	2314	5589; 2226	4311; 798)

The median and quartiles of the size of food trees selected by the three species were obtained from all scan samples (N = number of scans) in which the DBH of the tree was measured. Data stem from scan data collected in 1996 and 1997 for Bad1, Bad2, Pol1, and Ver1, in some months of 1997 and 1998 for Ver3 and March–December 1998 for Pol3. Tree biomass is given for two consecutive years of dietary data. Tree biomass was measured as the summed cm DBH per ha for all trees that represented either ≥ 5 per cent or ≥10 per cent of the annual diet. Data for Pol3, Ver3, and Ver1 are to be considered preliminary due to the small sample of dietary data. Bad = red colobus, Pol = black-and-white colobus, Ver = olive colobus, the numbers refer to the different study groups
[a]The biomass for the 10 per cent trees are the same for both red colobus groups in both years because only the tree species *Scytopetaleum tieghemii* occurred more than 10 per cent of the time in all situations in the annual diet

in the 1996 annual diet. This means that red colobus had fewer tree species that represented > 5 or > 10 per cent of their annual diet than did black-and-white colobus. The differences between the primates in the index of dispersion were less straightforward (see Figure 3.3). Further studies should be performed using a larger set of transects to conclude anything about the differences in the patchiness of food. The variation between months in the abundance of food trees tended to be higher for black-and-white colobus than for red colobus (Kremer 1999, Korstjens 2001 pp. 86–118). The results for olive colobus need to be considered with caution due to the small sample sizes for monthly diet.

Thus, based on the percentage of fruits, handling time, food spot residence time, and food density, black-and-white colobus food was more contestable than that of red and olive colobus at both the within- and between-group level. Considering the higher percentage of fruit in the red colobus diet their food may be slightly more contestable than olive colobus food. Both the percentage of fruits and the abundance and size of food

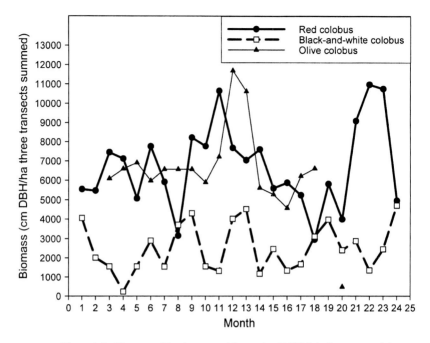

Figure 3.2. Biomass of food trees and lianas (cm DBH/ha) that occurred for at least 5 per cent of the monthly diet for the three primate species, values represent the sum of the three transects plotted against the months between January 1996 (1) through December 1997 (24) for the study groups Bad2, Pol1, and Ver1.

sources suggest that at the between-group level olive colobus' food might be less contestable than red colobus food.

Intra-group food competition

Indirect competition can be measured from the relationship between group size and the distance a group travels per day (DJL) or the size of the area (home range = HR) that the species uses. The black-and-white colobus study groups had the largest HR and shortest DJL of the three species. Red colobus had relatively small HR and intermediately long DJL. Olive colobus HR were slightly larger than those of the red colobus and DJL was the longest of the three species (see Table 3.2). The DJL and the HR of the olive colobus were almost entirely determined by those of the Diana monkey group with which the olive colobus travel. Therefore, DJL differences suggest that indirect competition is slightly more influential in the large red colobus group than in the black-and-white colobus group. Black-and-white colobus, however, required a larger HR per individual than red colobus. The dependence of black-and-white colobus on less

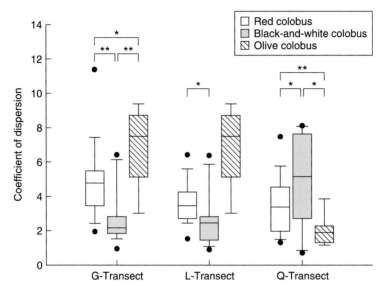

Figure 3.3. Summary of the coefficient of dispersion of the number of trees per 25 by 50 m section of the three transect lines for monthly values (taken from the 1996–1997 diets) of trees and lianas that represented at least 5 per cent of the monthly diet; a CD of 1 indicates a random distribution, CD > 1 indicates a clumped distribution, and a CD < 1 indicates an equal distribution (line through the boxes is the median, boxes represent the 25 and 75 percentiles and the dots depict the 5 per cent and 95 per cent percentiles) for the study groups from cluster 1 (uncorrected p-values from Wilcoxon matched pairs tests are depicted; *p < 0.05 level, **p < 0.01).

abundant and seasonally more variable food sources than those of red colobus, explains this discrepancy.

As predicted, agonistic interactions among red colobus females, rate 0.19 interactions/focal observation hour (21.3 hours), and among olive colobus females, n = 5 in 250 ad libitum observation hours (Deschner 1996) and N = 2 in 162 ad libitum observation hours (Krebs 1998) were less common than those among black-and-white colobus females, rate 0.60 interactions/focal observation hour (data from AHK, 21.7 hours; for combined E. C. Nijssen [ECN] and AHK dataset the value is 0.84 over 129.9 focal hours). In each species agonistic interactions among females were especially frequent during feeding (focal and ad libitum observations combined): in red colobus 5 of the 8 agonistic interactions and in black-and-white colobus 107 of 176 interactions concerned food (data of ECN & AHK). In olive colobus for all agonistic interactions by males and females combined: 14 out of 18 and 5 out of 8 agonistic interactions concerned food (Deschner 1996, Krebs 1998).

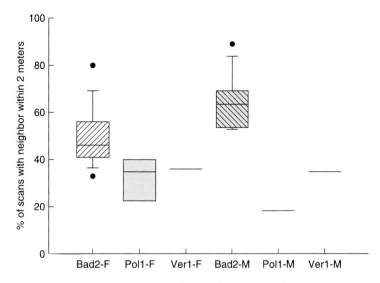

Figure 3.4. The percentage of scans that individuals spent with an adult or subadult neighbor of any sex; boxplots represent the quartiles (25 & 75 per cent) and the dots are outliers; red colobus data from 15 females (Bad2-F) and 14 males (Bad2-M), black-and-white colobus data from 6 females (Pol1-F) and 1 male (Pol1-M), olive colobus data from 2 females (Ver1-F) and 1 male (Ver1-M).

In black-and-white colobus agonistic interactions occurred more often when females fed on items that required long handling times and for which food spot residence time was long, namely seeds, than when they fed on soft fruits or on leaves for which handling time and food spot residence time was short (Korstjens *et al.* 2002). Thus, the differences in contest competition can be explained by differences in food choice: the more patchily distributed food items (fruits) with longer handling time (seeds) evoked the strongest contest competition.

Intra-group affiliation
The higher levels of food competition within and between groups in black-and-white colobus compared to red and olive colobus leads to the prediction that there are higher levels of within-group cooperation and affiliation among black-and-white colobus females than red and olive colobus females. Compared to black-and-white colobus females ($N = 6$), red colobus females ($N = 15$) spent a higher percentage of scans with a neighbor (adult or sub-adult) but an equal percentage with a female neighbor (Mann Whitney U tests: unsexed neighbor: $U = 10$, $p = 0.006$; male neighbor: $U = 4$, $p = 0.001$; female neighbor: $U = 38$, $p = 0.56$; see Figures 3.4 and 3.5). Red colobus females spent as much absolute

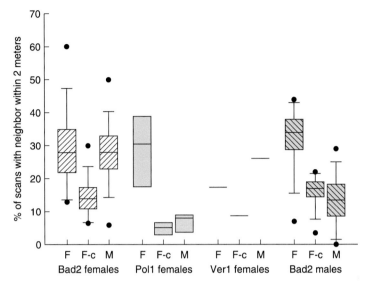

Figure 3.5. The percentage of scans that individuals spent with an adult or sub-adult female (F = uncorrected per cent, F-c = corrected per cent) or male (M) neighbor; boxplots represent the quartiles (25 and 75 per cent) and the dots are outliers; red colobus data from 15 females (Bad2 females) and 14 males (Bad2 males), black-and-white colobus data from 6 females (Pol1 females), olive colobus data from 2 females (Ver1 females), black-and-white and olive colobus males are not depicted because there was only 1 per group, thus, the time spent with a neighbor of any sex (Figure 3.4) corresponds to the time spent with females.

time with males as with females (Wilcoxon matched pair signed ranks test, $Z = -0.126$, $p = 0.90$) but less time with females if we corrected for group composition ($Z = -2.954$, $p = 0.003$; $\alpha' = 0.025$; Figure 3.5). Red colobus males spent more time near others than did females ($N = 14$; unsexed neighbor: $U = 31$, $p = 0.0013$; male neighbor: $U = 21$, $p = 0.0002$; female neighbor: $U = 66$, $p = 0.092$; see Figures 3.4 and 3.5). Red colobus males spent slightly more time with females than with males if we corrected for group composition (uncorrected: $Z = -1.014$, $p = 0.310$, corrected: $Z = -2.971$, $p = 0.003$; see Figure 3.5). Black-and-white colobus females spent more time with females than with the male if we did not correct for group composition ($N = 6$; uncorrected percentages: $Z = -2.00$, $p = 0.046$; corrected percentages: $Z = -0.54$, $p = 0.6$; Figures 3.4 and 3.5). Similar results were found when we investigated grooming interactions (Korstjens 2001, Korstjens *et al.* 2002). In olive colobus groups the male spent more time with others (Figure 3.4) and he groomed other adults more than he did the females (Deschner 1996, Deffernez 1999, Schippers 1999, Krebs 1998). The number of females in Ver1 ($N = 2$)

was too low for statistical comparisons of olive colobus with the other two species.

A coalition was formed in one of the eight agonistic interactions observed among red colobus females. A male supported another male in 13 of 100 agonistic interactions in which a male had a dispute with a female or another male of the group. We never observed a male to support an adult that was under attack. Rather, they always supported the attacker in an intra-group dispute. In 176 agonistic interactions among black-and-white colobus females we observed no coalitions. The only supportive behavior we saw was when a daughter aided her mother by carrying away an infant sibling during a sexual dispute between the mother and the adult male. In nine agonistic intra-group interactions among olive colobus, individuals were never seen to form a coalition.

Affiliative relationships among females and the tendency to form coalitions appeared weak in all three species. This was predicted for red and olive colobus. Black-and-white colobus females may be relatively unsocial despite high levels of contest competition because the most contestable food item (seeds) was not shareable. An individual could take over 15 minutes to open one pod, which it did not then share voluntarily with another individual.

Cooperation against intruders

Several types of intruders were recorded: conspecifics of another group, individuals of other primates species, human observers and predators. Due to the descriptive quality of the data on intruder defense we lumped data for different intruder types together in the descriptions below.

Red colobus males cooperatively attacked, mobbed or threatened intruders. Cooperation among males was often preceded by an embracing ritual that resembled a mount. It never occurred among females or between males and females. A similar ritualistic embrace or mount was described for Eastern red colobus (Struhsaker 1975). Red colobus females gave alarm calls but did not attack or form coalitions against intruders. R. Bshary and B. Beerlage performed two experiments in which the study groups (each group was tried once) were presented with an eagle dummy. Males attacked the dummy in twos or threes after elaborate embracing. During the experiments the females rarely entered the tree that contained the dummy. In black-and-white colobus females and males jointly threatened, mobbed, and alarm called when threatened by humans or predators but only males attacked such intruders. In olive colobus groups, males would face intruders and females would hide away (generally out of sight in large tall trees). In multi-male groups the males jointly attacked, mobbed or threatened (Schippers 1999, Korstjens & Noë 2004).

Inter-group food competition

Based on the contestability of food sources, black-and-white colobus females were predicted to compete most strongly and olive colobus least strongly. The Mitani and Rodman indices (Mitani & Rodman 1979) for the three colobus species suggest that olive and red colobus groups, in contrast to black-and-white colobus groups, could defend territories (see Table 3.2). Thus, the combination of food contestability and defensibility of the home range would predict that red and olive colobus *females* do not get involved much in inter-group disputes but red and olive colobus *groups* might defend territories. Black-and-white colobus females, on the other hand, are predicted to compete with females from other groups over individual food sources.

Indeed, concerning territory defense, we found that overlap between two neighboring groups was largest for black-and-white colobus, intermediate for olive colobus and smallest for red colobus (see Table 3.2). Furthermore, red colobus non-sister groups rarely encountered each other: once every 30 days (group Bad1, 122 observation days) and once every 21 days (group Bad2, 63 observation days). Agonistic interactions occurred in 50 per cent (Bad1) and 33 per cent (Bad2) of these encounters. Female red colobus were not observed to threaten members of non-sister groups (Korstjens *et al.* 2002). In black-and-white colobus, inter-group interactions occurred once every five days (99 observation days in 1997–8). Agonistic acts occurred in 76 per cent of encounters between entire groups (N = 83 in 1994–9) and females actively attacked members of conspecific groups in 68 per cent of the 62 agonistic inter-group encounters (Korstjens *et al.* 2002, Korstjens *et al.* 2005). However, when a single male attacked the group only the adult male would chase the intruder away and females ran from this male. Olive colobus inter-group encounters occurred when their partner Diana monkey groups met. During a well-monitored period this entailed an inter-group encounter once every 3 days (Ver3, 36.5 observation days). Females were not observed to attack members of other groups (Korstjens & Schippers 2003, Korstjens & Noë 2004).

Thus, in support of the predictions, red and olive colobus had relatively clearly demarcated territories, which were defended by males but not females. On the other hand, female and male black-and-white colobus defended locations but not territories. Korstjens *et al.* (2005) showed that female aggression was more common during the time that

Pentaclethra macrophylla fruits (that had long handling time) constituted a major portion of the animals' diet. However, clear proof that black-and-white colobus females fought over individual food sources is lacking.

Dispersal

Based on competitive relationships in the groups, we would only predict the strongest reluctance of females to disperse in black-and-white colobus. Furthermore, our black-and-white colobus study groups used the most widely distributed and unpredictable food sources, which suggests that knowledge of the home range was important. The number of males in the groups leads to the prediction that inbreeding avoidance is only a problem for maturing olive and black-and-white colobus females. However, red colobus are known for male philopatry (Struhsaker 1975, Stanford 1998) and potentially, some level of inbreeding could still occur. We predict, therefore, that females of all three species could disperse but that black-and-white colobus females may be least likely to migrate due to a stronger need for cooperation and knowledge of food sources.

Dispersal in red colobus

Dispersal patterns in red colobus were difficult to observe because we could not distinguish between male and female juveniles and we recognized only a few, older juveniles. We observed one female immigrating into Bad1 and one nulliparous female transferring from Bad2A to Bad2B (after the group composition of the sister-groups had become stable). Group membership was relatively stable for breeding females (0 disappearances for 20 well-recognized parous red colobus females between February 1997 and August 1999). The three nulliparous females that were individually recognized disappeared before they reproduced but after they had experienced their first few receptive cycles with sexual swelling (during which they mated with many males). Adult male group membership was stable over long periods of time in red colobus groups, normally only changing due to deaths and new recruitment from the sub-adult age class. One adult red colobus male transferred permanently from Bad3 into Bad1A, at a time when Bad1A had three males left (van Oirschot 1999). Another male of Bad3 joined him in Bad1A during his first two months there before disappearing again. Thus, we could confirm only two immigrations of females and one immigration of a male and suspected three emigrations of nulliparous females. More female immigrations were suspected on the basis of the shyness of several females in the study groups.

In addition, we regularly observed extra-group red colobus individuals either alone or in pairs of flexible composition, but always mingled with black-and-white colobus. Due to the difficulties in determining the sex of these individuals, we followed 12 of them for several days. Of these 12 individuals, three were most likely nulliparous females, eight were sub-adult males (testes were observed), and one was an adult male (December 1996–January 1999). In addition, one solitary juvenile male (recognizable because of a deformity) returned as a sub-adult to his natal group (Bad1) after an exile of about nine months. We saw the extra-group males generally over a period of several months, whereas, extra-group females were encountered only over periods of less than a week. Although this suggests that males migrated more than females, we suspect that these were males that were temporary or permanent exiles from their natal group who were not able to immigrate into a new group. This is supported by the fact that they remained in the area for longer periods of time than females. This scenario is based on the one well-documented case in Taï and observations on other red colobus populations (Struhsaker 1975, Starin 1994, Starin 2001). Not all sub-adult males spent time in exile: at least five natal red colobus males matured in Bad2A (1997–9).

Dispersal in black-and-white colobus

In Pol1, between 1992 and 1999, two parous females immigrated and three of four sub-adult females disappeared in healthy condition simultaneously from the group (Nijssen 1999, Korstjens *et al.* 2002). Support for the possibility that they emigrated rather than fell victim to predators or poachers comes from the intra-group relationships in this group. Aggression within the group was more common during the three months before their disappearance, 1.1 interactions/hour (in 58.6 focal hours), than during the 6 months before that, 0.62 interactions/hour (in 62.5 focal hours), and after their disappearance, 0.34 interactions/hour (in 8.85 focal hours). The three females that disappeared were harassed disproportionately during the period before their disappearance (Nijssen 1999). Two males disappeared simultaneously at the beginning of our study; one was a sub-adult and the other an adult. At least one female bred in the group in which she matured. Coincidentally, this was the female that had immigrated into the group with her mother when she was a juvenile. No males bred in their natal group. Three adult females and the adult male remained in Pol1 throughout the study.

Dispersal in olive colobus

In olive colobus, group composition changed regularly (Krebs 1998, Schippers 1999, Korstjens & Schippers 2003). Olive colobus individuals of all age-sex classes dispersed in groups or alone. Of the 12 observed juveniles at least 11 disappeared or dispersed before reproduction while the twelfth individual was still a juvenile at the end of this study. Olive colobus females seemed to spend only a short period of their breeding life with a particular group. Olive colobus males, on the other hand, seemed to spend most of their reproductive life in one particular group (Korstjens & Schippers 2003, Korstjens & Noë 2004).

Discussion

We investigated the links between the contestability of food and the social organization of three closely related folivorous primate species. The comparison of the species confirmed the idea that the contestability of food can be measured only if we incorporate a complete set of characteristics of the food. Based on differences in the contestability of food we could explain differences in social systems following the general logic of socio-ecological theory. In support of our predictions, contest competition was stronger in black-and-white colobus, who had the most contestable food items, than in red and olive colobus. Indeed, food competition among black-and-white colobus females was highest when they ate their most contestable food item: *Pentaclethra macrophylla* seeds. The seeds of this legume are contestable because they have a high protein to fiber ratio and high oil content (Dasilva 1994, Sicotte & MacIntosh 2004), they require a long processing time (Korstjens *et al.* 2002), and they are relatively few and unevenly distributed within the tree. Furthermore, black-and-white colobus food sources were more contestable than red and olive colobus food sources (based on temporal distribution, abundance, and relative size of trees). In addition, based on the Mitani and Rodman index we predicted that black-and-white colobus would not be able to defend territories. In support of our predictions, females more often interacted aggressively in inter-group encounters in black-and-white colobus than in the other two species but they did not defend territories. Although, red and olive colobus could be considered to be territorial, this territory defense was a male business.

Affiliative relationships among females were weak in all three species although, in support of the predictions, only black-and-white colobus females appeared to not favor males over females for affiliation. Female cooperation in defending food items from other group members did not occur regularly in any of the species. Thus, despite the contestability of black-and-white colobus food items, they did not cooperate to defend them. This is not surprising considering that the most contestable food of black-and-white colobus was not shareable. Black-and-white colobus females did defend food sources as a group. Considering the strong contest competition between females from different groups, socio-ecological theory would predict female philopatry in black-and-white colobus. Although, our observations can only be considered preliminary, at least some female dispersal occurred in all three species (see further discussion below).

Within a group, even if contest competition is low, scramble competition (measured as an increase in DJL and HR with group size) as a result of the mere presence of competitors is still expected to have an effect. Red colobus, with the largest groups, had longer day journey lengths than black-and-white colobus but not a larger home range. We can explain this discrepancy by incorporating the effect of food distribution and density. The relatively rare and unpredictable food sources used by black-and-white colobus require a relatively large annual home range.

Olive colobus had the largest range area per individual group member and travelled the farthest each day, despite having small food sources that occurred at high densities and the smallest groups. The ranging behavior of olive colobus is explained by their constant association with a particular Diana monkey partner group. We suggest that the selective foraging of olive colobus for high quality young leaves from abundantly available small trees, allowed them to keep up with these guenons.

In general, frugivores tend to have longer day journey lengths than folivores as a result of the more patchy and less dense distribution of fruiting trees compared to trees bearing leaves, and the readily digested high energy source that fruits present. In contrast to this pattern, black-and-white colobus had a shorter day journey length than the more folivorous red and olive colobus. To understand this result it is essential to know more about the digestive system of colobines. Colobines digest their food in much the same way as do ruminants and have an enlarged sacculated forestomach for microbial fermentation. This system allows them to get nutrients from leaves and seeds but digestion takes more time than digesting ripe fruits or insects (Kay & Davies 1994, Chivers 1994, Milton 1998). As a result of their gastro-intestinal adaptations, colobines are less able to digest ripe fruits with high sugar contents (Kay & Davies 1994). Therefore, when eating fruits, colobines extract seeds of unripe fruits, consume unripe fleshy fruits or fleshy fruits with low sugar contents. Hence, a higher percentage of fruit in the diet leads to longer daily travel distances (Clutton-Brock & Harvey 1977), unless digestively challenging seeds are consumed from those fruits.

The large number of subspecies and species (further referred to as populations) of red and black-and-white colobus offer an ideal opportunity for inter-population comparisons (Fashing 2007, Korstjens & Dunbar in press). This is not possible for the olive colobus, who is endemic to sub-Saharan West Africa and is becoming very rare. It has been studied at one other site, Sierra Leone, and the results from that study strongly resemble those from this study (Oates 1988, 1994, Oates & Korstjens in press).

African colobine populations vary widely in group size and group compositions, but black-and-white colobus groups are on average smaller than red colobus groups (Oates 1994, Fashing 2007, Korstjens & Dunbar in press). Colobines are typically expected to experience scramble but not contest competition over food within groups because of the generally even distribution of their food (van Schaik 1989). Three studies tested the importance of scramble competition within groups but found no correlation between group size and day journey length (*C. guereza* Fashing 2001a, *P. kirkii* Siex 2003, *P. tephrosceles* Struhsaker & Leland 1987). In unusually large groups, however, scramble competition may become more important (Fashing 2001a, Teichroeb *et al.* 2003). This direct test of intra-group food competition does not incorporate differences in food distribution and density, and such a correction is needed to really know the importance of scramble competition. When correcting for food density, red colobus, *P. tephrosceles*, do appear to experience increased scramble competition with increased group size (Gillespie & Chapman 2001). The generally smaller group sizes in combination with shorter day journey lengths in black-and-white colobus compared to red colobus (Fashing 2007, Korstjens & Dunbar in press) do indicate that red colobus may need to travel further than black-and-white colobus as a result of living in larger groups. In support of the idea that contest competition should be low in African colobines, most researchers report very low levels of aggression among females (Struhsaker & Leland 1979, Dunbar 1987, Fashing 2001a) the presented study on black-and-white colobus at Taï being an exception. We suggested that the high rates of aggression among black-and-white colobus females in Taï (Korstjens 2001, Korstjens *et al.* 2002) are explained by the high percentage of seeds consumed by *C. polykomos* (Korstjens & Galat-Luong in press). If this is true, we would expect the same for black colobus, *C. satanas*, at Douala-Edea, who have an even higher percentage of seeds in their diet (McKey & Waterman 1982).

Affiliative interactions are relatively rare in colobines, but have been studied in greater detail than agonistic interactions. In red colobus, the general trend is relatively little grooming among females: for *P. tephrosceles*, at Kibale and *P. pennantii* at Gbanraun grooming was most common among males, and females were more likely to groom males than other females (Struhsaker 1975, Struhsaker & Leland 1979, Werre 2000);

at Taï, *P. badius badius*, Abuko, *P. badius temminckii*, and Jozani, *P. kirkii*, males rarely groomed each other, and females more readily groomed males than other females (Starin 1991, Korstjens 2001, Korstjens *et al.* 2002, Siex 2003). In Taï, males do, however, associate together and cooperate regularly. Struhsaker and Leland suggested that the strong affiliative relationships among red colobus males were related to male philopatry. In black-and-white colobus the general trend is relatively more grooming among females than between the sexes or among males (Oates 1977, Dasilva 1989, Korstjens 2001, Fashing 2007).

Based on the generally evenly distributed food, socio-ecological theory predicts that females of different groups should rarely get aggressive in colobines. Indeed, in all African colobines studied to date, males are the more active aggressors during inter-group conflicts (reviewed in Fashing 2001b, 2007): in 17 of 18 populations males exhibited inter-group aggression, while 11 had female aggression. When information was available, male aggression was furthermore, more common than female aggression (N = 6 populations summarized by Fashing 2007). Females are only regularly involved in inter-group aggression in Abuko, *Procolobus badius temminckii* (Starin 1991), *Colobus polykomos polykomos* in Taï (Korstjens *et al.* 2005), and shamba-dwelling *P. kirkii* on Zanzibar (Siex 2003).

Socio-ecological models have always placed an important link on the connection between female involvement during inter-group aggression, as a sign of strong inter-group food competition, and female bonding. However, recent studies suggest that some of the generalizations need to be reconsidered. Either because males take on the role of food defense (reviewed in Fashing 2001b, 2007) or because strong inter-group contest competition is not associated with female philopatry and strong bonds among females (Starin 1991, Glander 1992, Strier *et al.* 1993, Pope 2000, Koenig 2002, Korstjens *et al.* 2002, Korstjens *et al.* 2005).

Although dispersal patterns have not been studied in many species, red colobus are known for their female-biased dispersal (Struhsaker 1975, Marsh 1979, Starin 1991, Decker 1994). The extra-group red colobus males in some populations do not necessarily indicate male dispersal but could reflect a surplus of males in a male-bonded society. Immigration may be very difficult for these males, and some are known to have returned to their natal group eventually (*P. tephrosceles*: Struhsaker 1975, *P. temminckii*: Starin 1994). The most intriguing exception is the *P. kirkii* population inhabiting the shambas of Jozani (Zanzibar). They differ from their neighbors in the forest in many aspects of their social organization of which male transfer is just one (Siex & Struhsaker 1999, Siex 2003). In support of socio-ecological theory, this switch to male dispersal may have resulted from the high levels of inter-group aggression among females in the shambas (Siex 2003).

Although males appear to disperse more than females in black-and-white colobus, at least some female dispersal occurs as well (Dasilva 1989, Oates 1994, Korstjens *et al.* 2002, 2005, Sicotte & MacIntosh 2004, Fashing 2007). The three females that disappeared from our study group would have benefited from emigrating through inbreeding avoidance. As predicted on the basis of their need for cooperation at the group level and the unpredictability of their food, these females seemed reluctant to leave (as measured from the increase in aggression they received before their disappearance).

In olive colobus, females and males appeared to leave quite readily and regularly even after breeding in a group (Oates 1994, Korstjens & Schippers 2003). Although the low contestability of their food, the high availability of food in the area and their ability to remain in poly-specific associations during transfer suggest that the costs of dispersal are relatively low, individuals would still need to benefit from dispersal. Korstjens and Schippers (2003) showed a preference of females for small groups, but further study is needed. Thus, in general, African colobines support the idea that a lack of contestable food allows for female dispersal.

We showed how small scale differences in the contestability of food can explain large scale differences in social organization as long as contestability is measured at various levels. The simple dichotomy between fruits and leaves is not sufficient to classify food items as contestable or not. Furthermore, it is not just differences in dietary items but also in digestive adaptations that need to be considered when testing socio-ecological theory. To determine whether food will evoke contest or scramble competition it is essential to have an independent measure of the contestability of food. Appropriate measures are handling time of food items and food patch residence time (Isbell 1991) in combination with food distribution and density (Gillespie & Chapman 2001). Furthermore, we emphasize that the differences between species should be seen as occurring on a continuum, not in classes. Lastly, future socio-ecological considerations need to incorporate the costs and benefits of dispersal that are not related to food competition (Strier 1999, Isbell & Young 2002, Isbell 2004) without ignoring the strong empirical support for the general logic behind the relationships between food traits and social organizations.

Conclusion

1. Intra-group contest competition over food was highest in black-and-white colobus. This was also the species that selected the most contestable food items as measured from handling time, food spot residence time, percentage fruit in the diet, and food density. Thus, our results support the notion that contestable food items evoke contest competition in groups.

2. The most frugivorous of the species, the black-and-white colobus, used less abundant and more variable food sources than the more folivorous red and olive colobus. As predicted based on their food distribution and density, but not on the relative size of their group, the black-and-white colobus had the largest home range. The relatively short day journey lengths for black-and-white colobus could be explained by their group size and food choice if we consider that the seeds they consumed required long processing and digesting times. The relatively long day journey lengths for the olive colobus were explained by their permanent membership in a poly-specific group that contains the frugivorous Diana monkeys.

3. Intra-group ties were not very strong in any of the species. Dispersal costs appeared highest for black-and-white colobus based on their higher levels of contest competition between groups and their use of less predictable food sources. Still, females seemed to disperse at least occasionally in red and black-and-white colobus, and regularly in olive colobus. This supports the idea that females have weak affiliative relationships and may disperse when food competition is relatively relaxed. The exception may be the occasional dispersal of black-and-white colobus females despite strong inter-group contest competition that requires cooperation at the group-level.

Acknowledgments

We thank T. Deschner, A. Schaaff, A. Moresco, J. Béné, A. Bitty, S. Kruppa, and especially C. Kremer and the Taï Monkey Project field assistants for their help in the data collection. The research was financially supported by the Max-Planck Institut für Verhaltensphysiologie, the Deutsche Forschungsgemeinschaft, the University of Utrecht, the Lucie Burgers Foundation for Comparative Behaviour Research, the Dr. J. L. Dobberke Foundation, and the Dr. Christine Buisman Foundation. A. H. Korstjens thanks E. H. M. Sterck and H. de Vries for their support and advice on many occasions. A. H. Korstjens extends special thanks to Jan van Hooff, who has been an enormous source of inspiration and support throughout her career.

References

Alexander, R. D. (1974). The evolution of social behavior. *Annual Review of Ecology and Systematics*, **5**, 325–83.

Altmann, J. (1974). Observational study of behavior: sampling methods. *Behaviour*, **49**, 229–67.

Barton, R. A., Byrne, R. W. and Whitten, A. (1996). Ecology, feeding competition and social structure in baboons. *Behavioral Ecology and Sociobiology*, **38**, 321–9.

Bergmann, K. (1998). Vergleichende Untersuchung zur Einnischung der Colobidae im Taï-Nationalpark (Elfenbeinküste) unter besonderer Berücksichtigung des Olivgrünen Stummelaffen (*Procolobus verus*). In *Biologie*, p. 103. M.Phil. Thesis, Johann Wolfgang Goethe-Universität, Frankfurt am Main, Germany.

Borries, C. (1993). Ecology of female social relationships: Hanuman langurs (*Presbytis Entellus*) and the van Schaik model. *Folia Primatologica*, **61**, 21–30.

Brugière, D., Gautier, J. P., Moungazi, A. and Gautier-Hion, A. (2002). Primate diet and biomass in relation to vegetation composition and fruiting phenology in a rain forest in Gabon. *International Journal of Primatology*, **23**, 999–1024.

Bshary, R. (1995). *Rote stummelaffen*, Colobus badius, *und Dianameerkatzen*, Cercopithecus diana, *im Taï-Nationalpark, Elfenbeinküste: Wozu assoziieren sie?* p. 193. Ph.D. Thesis, Ludwig-Maximilian-Universität München: München, Germany.

Chapman, C. A., Chapman, L. J., Rode, K. D., Hauck, E. M. and McDowell, L. R. (2003). Variation in the nutritional value of primate foods: Among trees, time periods, and areas. *International Journal of Primatology*, **24**, 317–33.

Chapman, C. A., Wrangham, R. W. and Chapman, L. J. (1995). Ecological constraints on group size: an analysis of spider monkey and chimpanzee subgroups. *Behavioral Ecology and Sociobiology*, **36**, 59–70.

Chivers, D. J. (1994). Functional anatomy of the gastrointestinal tract. In *Colobine Monkeys*, ed. A. G. Davies and J. F. Oates. New York: Cambridge University Press, pp. 205–27.

Clutton-Brock, T. H. (1989). Female transfer and inbreeding avoidance in social mammals. *Nature*, **337**, 70–2.

Clutton-Brock, T. H. and Harvey, P. H. (1977). Primate ecology and social organization. *Journal of Zoology (London)*, **183**, 1–39.

Clutton-Brock, T. H. and Harvey, P. H. (1979). Home range size, population density and phylogeny in primates. In *Primate Ecology and Human Origins: Ecological Influences on Social Organisation*, ed. I. S. Bernstein, and E. O. Smith. New York and London: Garland STPM Press, pp. 201–14.

Cords, M. (2000). Agonistic and affiliative relationships in a blue monkey group. In *Old World Monkeys*, ed. P. F. Whitehead and C. J. Jolly. Cambridge: Cambridge University Press, pp. 453–79.

Dasilva, G. L. (1989). *The ecology of the western black and white colobus* (Colobus polykomos polykomos Zimmerman 1780) *on a riverine island in southeastern Sierra Leone.* Ph.D. Thesis, University of Oxford: Oxford, England.

Dasilva, G. L. (1994). Diet of *Colobus polykomos* on Tiwai Island: selection of food in relation to its seasonal abundance and nutritional quality. *International Journal of Primatology*, **15**, 655–80.

Davies, A. G. and Bennett, E. L. (1988). Food selection by two South-East Asian colobine monkeys (*Presbytis rubicunda* and *Presbytis melalophos*) in relation to plant chemistry. *Biological Journal of the Linnean Society*, **34**, 33–56.

Decker, B. S. (1994). Effects of habitat disturbance on the behavioral ecology and demographics of the Tana river red colobus (*Colobus badius ruformitratus*). *International Journal of Primatology*, **15**, 703–37.

Deffernez, C. (1999). *Associations polyspécifiques et relations sociales dans un groupe de colobes de* van Beneden (*Colobus verus*), de cercopithèques dianes (*Cercopithecus diana*), et d'autres singes de la forêt de Taï, Côte d'Ivoire. pp. 42. M.Phil. Thesis, Université de Lausanne: Lausanne, Switzerland.

Deschner, T. (1996). Aspekte des Socialverhaltens Olivgrüner Stummelaffen *Colobus verus* (van Beneden 1838) im Taï-Nationalpark/Elfenbeinküste. In *Biology*. M.Phil Thesis, Universität Hamburg: Hamburg, Germany.

Dunbar, R. I. M. (1987). Habitat quality, population dynamics, and group composition in colobus monkeys (Colobus guereza). *International Journal of Primatology*, **8**, 299–330.

Dunbar, R. I. M. (1988). *Primate Social Systems*. London: Croom Helm.

Dunbar, R. I. M. (1991). Functional significance of social grooming in primates. *Folia Primatologica*, **57**, 121–31.

Emlen, S. T. and Oring, L. W. (1977). Ecology, sexual selection, and the evolution of mating systems. *Science*, **197**, 215–23.

Fashing, P. J. (2001a). Activity and ranging patterns of guerezas in the Kakamega Forest: intergroup variation and implications for intragroup feeding competition. *International Journal of Primatology*, **22**, 549–77.

Fashing, P. J. (2001b). Male and female strategies during intergroup encounters in guerezas (*Colobus guereza*): evidence for resource defense mediated through males and a comparison with other primates. *Behavioral Ecology and Sociobiology*, **50**, 219–30.

Fashing, P. J. (2007). African colobine monkeys. In *Primates in Perspective*, ed. C. J. Campbell, A. F. Fuentes, K. C. MacKinnon, M. Panger and S. Bearder. Oxford: Oxford University Press.

Galat, G. and Galat-Luong, A. (1985). La communauté de primates diurnes de la forêt de Taï, Côte d'Ivoire. *Revue d'Ecologie la Terre et la Vie*, **40**, 3–32.

Gillespie, T. R. and Chapman, C. A. (2001). Determinants of group size in the red colobus monkey (*Procolobus badius*): an evaluation of the generality of the ecological-constraints model. *Behavioural Ecology and Sociobiology*, **50**, 293–390.

Glander, K. E. (1992). Dispersal patterns in Costa Rican mantled howling monkeys. *International Journal of Primatology*, **13**, 415–36.

Greenwood, P. J. (1980). Mating systems, philopatry and dispersal in birds and mammals. *Animal Behaviour*, **28**, 1140–62.

Hamilton, W. J. (1964a). The genetical evolution of social behavior II. *Journal of Theoretical Biology*, **7**, 17–52.

Hamilton, W. J. (1964b). The genetical evolution of social behavior I. *Journal of Theoretical Biology*, **7**, 1–16.

Harcourt, A. H. (1989). Social influences on competitive ability: alliances and their consequences. In *Comparative Socioecology. The Behavioural Ecology of Humans and Other Mammals*, ed. V. Standen and R. A. Foley. Cambridge: Cambridge University Press, pp. 223–42.

Harvey, P. H. and Clutton-Brock, T. H. (1981). Primate home-range size and metabolic needs. *Behavioural Ecology and Sociobiology*, **8**, 151–5.

Hawkes, K. (1992). Sharing and collective action. In *Evolutionary Ecology and Human Behavior*, ed. E. A. Smith and B. Winterhalder. New York: Aldine de Gruyter, pp. 269–300.

Hochberg, Y. (1988). A sharper Bonferroni procedure for multiple tests of significance. *Biometrika*, **75**, 800–2.

Isbell, L. A. (1991). Contest and scramble competition: Patterns of female aggression and ranging behavior among primates. *Behavioral Ecology*, **2**, 143–55.

Isbell, L. A. (2004). Is there no place like home? Ecological bases of female dispersal and philopatry and their consequences for the formation of kin groups. In *Kinship and Behaviour in Primates*, ed. B. Chapais and C. M. Berman. Oxford: Oxford University Press, pp. 71–109.

Isbell, L. A. and Pruetz, J. D. (1998). Differences between vervets (*Cercopithecus aethiops*) and patas monkeys (*Erythrocebus patas*) in agonistic interactions between adult females. *International Journal of Primatology*, **19**, 837–55.

Isbell, L. A., Pruetz, J. D., Nzuma, B. M. and Young, T. P. (1999). Comparing measures of travel distances in primates: Methodological implications. *American Journal of Primatology*, **48**, 87–98.

Isbell, L. A., Pruetz, J. D. and Young, T. P. (1998). Movements of vervets (*Cercopithecus aethiops*) and patas monkeys (*Erythrocebus patas*) as estimators of food resource size, density, and distribution. *Behavioral Ecology and Sociobiology*, **42**, 123–33.

Isbell, L. A. and van Vuren, D. (1996). Differential costs of locational and social dispersal and their consequences for female group living primates. *Behaviour*, **133**, 1–36.

Isbell, L. A. and Young, T. P. (2002). Ecological models of female social relationships in primates: similarities, disparities, and some directions for future clarity. *Behaviour*, **139**, 177–202.

Janson, C. H. (1988). Food competition in brown capuchin monkeys (*Cebus apella*): Quantitive effects of group size and tree productivity. *Behaviour*, 53–76.

Janson, C. H. (1990). Ecological consequences of individual spatial choice in foraging groups of brown capuchin monkeys, *Cebus apella*. *Animal Behaviour*, **40**, 922–34.

Janson, C. H. and Goldsmith, M. L. (1995). Predicting group size in primates: foraging costs and predation risks. *Behavioral Ecology*, **6**, 326–36.

Janson, C. H. and van Schaik, C. P. (1988). Recognizing the many faces of primate food competition: methods. *Behaviour*, **105**, 165–86.

Kay, R. N. B. and Davies, A. G. (1994). Digestive physiology. In *Colobine Monkeys*, ed. A. G. Davies and J. F. Oates. New York: Cambridge University Press, pp. 229–49.

Koenig, A. (2000). Competitive regimes in forest-dwelling Hanuman langur females (*Semnopithecus entellus*). *Behavioral Ecology and Sociobiology*, **48**, 93–109.

Koenig, A. (2002). Competition for resources and its behavioral consequences among female primates. *International Journal of Primatology*, **23**, 759–83.

Korstjens, A. H. (2001). The mob, the secret sorority, and the phantoms. *An analysis of the socio-ecological strategies of the three colobines of Taï*, pp. 174. Ph.D. Thesis, Utrecht University: Utrecht, Netherlands.

Korstjens, A. H. and Dunbar, R. I. M. (in press). Time constraints limit group sizes and distribution in red and black and white colobus monkeys. *International Journal of Primatology*.

Korstjens A. H. and Galat-Luong, A. (in press). *Colobus polykomos*. In *Mammals of Africa*, ed. J. Kingdon, D. C. D. Happold and T. M. Butynski. Oxford: Oxford University Press.

Korstjens, A. H., Nijssen, E. C. and Noë, R. (2005). Inter-group relationships in western black-and-white colobus, *Colobus polykomos polykomos*. *International Journal of Primatology*, **26**, 1267–89.

Korstjens, A. H. and Noë, R. (2004). The mating system of an exceptional primate, the olive colobus (*Procolobus verus*). *American Journal of Primatology*, **62**, 261–73.

Korstjens, A. H. and Schippers, E. P. (2003). Dispersal patterns among olive colobus in Taï National Park. *International Journal of Primatology*, **24**, 515–40.

Korstjens, A. H., Sterck, E. H. M. and Noë, R. (2002). How adaptive or phylogenetically inert is primate social behavior? A test with two sympatric colobines. *Behaviour*, **139**, 203–25.

Krebs, M. (1998). Zur Rolle von Männchen in Familiengruppen des Olivgrünen Stummelaffen (*Colobus verus*) -mit Aspekten zum Sozialsystem - Taï-Nationalpark, Elfenbeinküste. In *Biologie/Chemie*, p. 83. M.Phil. Thesis, Universität Osnabrück: Osnabrück, Germany.

Kremer, C. (1999). Verteilung de Nahrungsressourcen von drei sympatrischen Colobusarten im Taï-Nationalpark, Elfenbeinküste (Westafrika). In *Biologie*, p. 62. M.Phil. Thesis, Ludwig-Maximilians-Universität: München, Germany.

Lowen, C. and Dunbar, R. I. M. (1994). Territory size and defendability in primates. *Behavioral Ecology and Sociobiology*, **35**, 347–54.

Marsh, C. W. (1979). Female transference and mate choice among Tana River red colobus. *Nature*, **281**, 568–9.

Martin, P. and Bateson, P. (1993). *Measuring Behaviour: an Introductory Guide*. Cambridge UK: Cambridge University Press.

Mathy, J. W. and Isbell, L. A. (2001). The relative importance of size of food and interfood distance in eliciting aggression in captive rhesus macaques (*Macaca mulatta*). *Folia Primatologica*, **72**, 268–77.

McKey, D. and Waterman, P. G. (1982). Ranging behavior of a group of black colobus (*Colobus satanas*) in the Douala-Edea Reserve, Cameroon. *Folia Primatologica*, **39**, 264–304.

Milinski, M. and Parker, G. A. (1991). Competition over resources. In *Behavioural Ecology: an Evolutionary Approach*, ed. J. R. Krebs and N. B. Davies. London: Blackwell, pp. 137–68.

Milton, K. (1998). Physiological ecology of howlers (*Alouatta*): energetic and digestive considerations and comparison with the colobinae. *International Journal of Primatology*, **19**, 513–48.

Mitani, J. C. and Rodman, P. S. (1979). Territoriality: The relation of ranging patterns and home range size to defendability, with an analysis of territoriality among primate species. *Behavioral Ecology and Sociobiology*, **5**, 241–51.

Mitchell, C. L., Boinski, S. and van Schaik, C. P. (1991). Competitive regimes and female bonding in two species of squirrel monkeys (*Saimiri oerstedi* and *S. sciurieus*). *Behavioral Ecology and Sociobiology*, **28**, 55–60.

Moore, J. (1988). Primate dispersal. *Tree*, **3**, 144–5.

Nijssen, E. C. (1999). Female philopatry in *Colobus polykomos polykomos*: fact or fiction? In *Ethology and Socio-ecology*, p. 109. Utrecht: Utrecht University.

Noë, R. (1992). Alliance formation among male baboons: shopping for profitable partners. In *Us Against Them: Coalitions and Alliances in Humans and Other Animals*, ed. A. H. Harcourt and F. B. M. de Waal. Oxford: Oxford University Press, pp. 282–321.

Noë, R., van Schaik, C. P. and van Hooff, J. A. R. A. M. (1991). The market effect: An explanation for pay-off asymmetries among collaborating animals. *Ethology*, **87**, 97–118.

Oates, J. F. (1977). The social life of a black-and-white colobus monkey, *Colobus guereza*. *Zietschrift für Tierpsychologie*, **45**, 1–60.

Oates, J. F. (1988). The diet of the olive colobus monkey, *Procolobus verus*, in Sierra Leone. *International Journal of Primatology*, **9**, 457–78.

Oates, J. F. (1994). The natural history of African colobines. In *Colobine Monkeys*, ed. A. G. Davies and J. F. Oates. Cambridge: Cambridge University Press, pp. 75–128.

Oates, J. F. and Davies, A. G. (1994). What are colobines? In *Colobine Monkeys*, ed. A. G. Davies and J. F. Oates. Cambridge: Cambridge University Press, pp. 1–9.

Oates, J. F. and Korstjens, A. H. (in press). The olive colobus, *Procolobus verus*. In *Mammals of Africa*, ed. J. Kingdon, D. C. D. Happold and T. M. Butynski. Oxford: Oxford University Press.

Pope, T. R. (2000). Reproductive success increases with degree of kinship in cooperative coalitions of female red howler monkeys (*Alouatta seniculus*). *Behavioral Ecology and Sociobiology*, **48**, 253–67.

Scheel, D. and Packer, C. (1991). Group hunting behavior of lions: a search for cooperation. *Animal Behaviour*, **41**, 697–709.

Schippers, E. P. (1999). Dispersal of the olive colobus (*Colobus verus*) in Taï National Park. In *Ethology and Socio-ecology*, p. 64. M.Phil. Thesis, Utrecht University: Utrecht, Netherlands.

Sicotte, P. and MacIntosh, A. J. (2004). Inter-group encounters and male incursions in *Colobus vellerosus* in Central Ghana. *Behaviour*, **141**, 533–53.

Siegel, S. and Castellan, N. J. J. (1988). *Nonparametric Statistics for the Behavioral Sciences*. New York: McGraw-Hill.

Siex, K. S. (2003). Effects of population compression on the demography, ecology, and behavior of the Zanzibar red colobus monkey (*Procolobus kirkii*). Ph.D. Thesis, Duke University: Durham NC, USA.

Siex, K. S. and Struhsaker, T. T. (1999). Ecology of the Zanzibar red colobus monkey: demographic variability and habitat stability. *International Journal of Primatology*, **20**, 163–92.

Stanford, C. B. (1991). Social dynamics of intergroup encounters in the capped langur (*Presbytis pileata*). *American Journal of Primatology*, **25**(1), 35–47.

Stanford, C. B. (1998). Predation and male bonds in primate societies. *Behaviour*, **135**, 513–33.

Starin, E. D. (1991). *Socioecology of the red colobus monkey in The Gambia with particular reference to female-male differences and transfer patterns*, p. 406. Ph.D. Thesis, City University of New York: New York, USA.

Starin, E. D. (1994). Philopatry and affiliation among red colobus. *Behaviour*, **130**, 253–70.

Starin, E. D. (2001). Patterns of inbreeding avoidance in Temminck's red colobus. *Behaviour*, **138**, 453–65.

Steenbeek, R. (1999). Tenure related changes in wild Thomas's langurs II: Between group interactions. *Behaviour*, **136**, 595–650.

Sterck, E. H. M. and Korstjens, A. H. (2000). Female dispersal and infanticide avoidance in primates. In *Infanticide by Males and its Implications*, ed. C. P. van Schaik and C. H. Janson. Cambridge: Cambridge University Press, pp. 293–321.

Sterck, E. H. M. and Steenbeek, R. (1997). Female dominance relationships and food competition in the sympatric Thomas langur and long-tailed macaque. *Behaviour*, **134**, 749–74.

Sterck, E. H. M., Watts, D. P. and van Schaik, C. P. (1997). The evolution of female social relationships in nonhuman primates. *Behavioral Ecology and Sociobiology*, **41**, 291–309.

Strier, K. B. (1999). Why is female kinbonding so rare? Comparative sociality of neotropical primates. In *Comparative Primate Socioecology*, ed. P. C. Lee. Cambridge: Cambridge University Press, pp. 300–319.

Strier, K. B., Mendes, F. D. C., Rímoli, J. and Rímoli, A. O. (1993). Demography and social structure of one group of muriquis (*Brachyteles arachnoides*). *International Journal of Primatology*, **14**, 513–26.

Struhsaker, T. T. (1975). *The Red Colobus Monkey*. Chicago: University of Chicago Press.

Struhsaker, T. T. and Leland, L. (1979). Socioecology of five sympatric monkey species in the Kibale Forest, Uganda. *Advances in the Study of Behavior*, **9**, 159–228.

Struhsaker, T. T. and Leland, L. (1987). Colobines: infanticide by adult males. In *Primate Societies*, ed. B. B. Smuts, D. L. Cheney, R. M. Seyfarth, R. W. Wrangham and T. T. Struhsaker. Chicago: The University of Chicago Press, pp. 83–97.

Teichroeb, J. A., Saj, T. L., Paterson, J. D. and Sicotte, P. (2003). Effect of group size on activity budgets of *Colobus vellerosus* in Ghana. *International Journal of Primatology*, **24**, 743–58.

Thierry, B., Alwaniuk, A. N. and Pellis, S. M. (2000). The influence of phylogeny on the social behavior of macaques (Primates: Cercopithecidae, genus *Macaca*). *Ethology*, **106**, 713–28.

Trivers, R. L. (1972). Parental investment and sexual selection. In *Sexual Selection and the Descent of Man*, ed. B. Campbell. London: Heinemann Educational Books Ltd. pp. 136–79.

van Hooff, J. A. R. A. M. and van Schaik, C. P. (1992). Cooperation in competition: the ecology of primate bonds. In *Coalition and Alliances Between Humans and Other Animals*, ed. A. H. Harcourt and F. B. M. de Waal. Oxford: Oxford University Press, pp. 357–89.

van Oirschot, B. (1999). Group fission in western red colobus (*Procolobus badius*) Taï National Park, Côte d'Ivoire. In *Institute of Evolutionary and Ecological Sciences*, p. 205. M.Phil. Thesis, Universiteit Leiden: Leiden, Netherlands.

van Oirschot, B. (2000). Group fission in Taï red colobus (*Procolobus badius*). *Folia Primatologica*, **71**, 205.

van Schaik, C. P. (1983). Why are diurnal primates living in groups? *Behaviour*, **87**, 120–44.

van Schaik, C. P. (1989). The ecology of social relationships amongst female primates. In *Comparative Socioecology. The Behavioural Ecology of Humans and Other Mammals*, ed. V. Standen and R. A. Foley. Oxford: Blackwell, pp. 195–218.

van Schaik, C. P. (1996). Social evolution in primates: the role of ecological factors and male behavior. *Proceedings British Academy*, **88**, 9–31.

van Schaik, C. P. and van Hooff, J. A. R. A. M. (1983). On the ultimate causes of primate social systems. *Behaviour*, **85**, 120–44.

van Schaik, C. P. and van Noordwijk, M. A. (1988). Scramble and contest in feeding competition among female long-tailed macaques (*Macaca fascicularis*). *Behaviour*, **105**, 77–98.

Walters, J. R. and Seyfarth, R. M. (1987). Conflict and cooperation. In *Primate Societies*, ed. B. B. Smuts, D. L. Cheney, R. M. Seyfarth, R. W. Wrangham and T. T. Struhsaker. Chicago: The University of Chicago Press, pp. 306–17.

Wasserman, M. D. and Chapman, C. A. (2003). Determinants of colobine monkey abundance: the importance of food energy, protein and fiber content. *Journal of Animal Ecology*, **72**, 650–9.

Waterman, P. G. and Kool, K. M. (1994). Colobine food selection and plant chemistry. In *Colobine Monkeys*, ed. A. G. Davies and J. F. Oates. New York: Cambridge University Press, pp. 251–84.

Werre, J. L. R. (2000). Ecology and behavior of the Niger Delta red colobus (*Procolobus badius epieni*). p. 239. Ph.D. Thesis, The City University of New York: New York, USA.

Whitten, P. L. (1983). Diet and dominance among female vervet monkeys (*Cercopithecus aethiops*). *American Journal of Primatology*, **5**, 139–59.

Wrangham, R. W. (1980). An ecological model of female bonded primate groups. *Behaviour*, **75**, 262–300.

Wrangham, R. W., Gittleman, J. L. and Chapman, C. A. (1993). Constraints on group size in primates and carnivores: population density and day-range as assays of exploitation competition. *Behavioral Ecology and Sociobiology*, **32**, 199–209.

4 The structure of social relationships among sooty mangabeys in Taï

F. Range, T. Förderer, Y. Storrer-Meystre, C. Benetton, and
C. Fruteau

Introduction

Living in groups has both advantages and disadvantages. Being in a social group may decrease vulnerability to predation or increase acquisition of certain resources, but it may also increase intra-group competition for food, mates, and sleeping sites and lead to a higher risk of disease or infanticide (reviewed in Krebs & Davies 1993). The optimal size and structure of social groups is generally thought to be a balance between the costs and benefits associated with sociality.

The evolution of sociality may be directly relevant to the evolution of cognitive skills. Recent research suggests that the primate brain evolved as an adaptation to cope with the social complexity that results from competition within a framework of kinship networks, friendships, dominance hierarchies, and triadic alliances (the "social brain hypothesis"). As group size increases, the number of triadic relations explodes and the need for triadic knowledge to choose the best behavioral strategies places high demands on individuals, which could in itself offer an explanation for the large primate brain (Seyfarth & Cheney 2001).

Sooty mangabeys live in large groups (over 100 animals), are terrestrial and forest dwelling and form-differentiated relationships among group members. All of these features make them ideal subjects with which to test the "social brain" hypothesis. However, before we consider the social intelligence of sooty mangabeys, we need to understand how dominance rank, competition, affiliation, and migration shape the pattern of interactions among females and males. In this chapter, we present the first results of our work on the social system of sooty mangabeys in their natural environment. These results form the basis for our further research on cognitive skills in this species.

Monkeys of the Taï Forest, ed. W. Scott McGraw, Klaus Zuberbühler and Ronald Noë.
Published by Cambridge University Press. © Cambridge University Press 2007.

Hypotheses and predictions

1. Several observations of sooty mangabeys in Taï, and of other *Cercocebus* species at the Tana River and in Cameroon suggest that female mangabeys are philopatric, whereas males leave their natal group and immigrate into new groups (Homewood 1976, Mitani 1989, Range & Noë 2002). As a result, when we began our study we predicted that mangabey females would be closely related to each other, and that mangabey social structure would be organized around a number of ranked matrilines as reported for other related species with the same migration patterns (e.g. baboons, vervets).

2. We examined whether sooty mangabeys form defined and uni-directional dominance relationships and if they can be ranked into a linear dominance hierarchy.

3. We assumed that if dominance relationships were established some advantages would be connected to having a high rank in the dominance hierarchy. Females' reproductive success is thought to be mainly limited by food and safety, whereas for males, competition for mates is hypothesized to be the most important factor affecting reproductive success (Trivers 1972). Thus, for the females, we predicted that rank would be correlated with acquisition of food and/or a safe position in the group, while high-ranking males should have better access to receptive females.

4. Finally, based on theoretical considerations and empirical evidence, we predicted that if dominance has an important effect on access to resources, females would form well-differentiated relationships, with preferred female partners as defined by frequent association, grooming, and the formation of alliances. Moreover, females could try to get access to important resources by forming "friendships" with adult males, for example in "exchange" for copulations.

Methods

Claudia Rutte and Ralph Bergmueller started to habituate our study group in 1996. Since then, the group has been under constant observation by at least one observer. After the first year of observation, most animals were well enough habituated that observers could follow them within a distance of five meters, allowing for the collection of detailed data on social behavior. Mangabeys are relatively easy to recognize individually, but their large group size (over 100 animals) and fragmentation into subgroups makes individual recognition difficult and time consuming. All adult animals have been known since 1998; all juveniles and sub-adults since 2001.

Data collection

The group is followed every day from dawn till dusk by at least one observer, who maintains detailed data on demographic events such as births, immigrations, disappearances, and notes the reproductive state of females (sexual swelling or lactating) on a daily basis. Swellings were rated on a four-point scale where 1 was flat and 4 maximum tumescence observed in any female.

Data on social behavior of individuals is collected by focal animal sampling in which an observer follows a selected animal for a certain amount of time and records its activity patterns and all social interactions with other individuals in the group (Altmann 1974). All focal samples are 15 minutes long with at least 60 minutes between consecutive samples of the same individual to ensure that samples are independent from each other.

During sampling we record the activity (see Table 4.1), the position of the focal animal relative to others, and the identity of the nearest adult female and male within five meters of the focal animal every minute [instantaneous sampling (Altmann 1974)]. The focal animal can be in three different kinds of positions relative to other animals. Positions are designated by the presence of other individuals in a circular area with a radius of 10 meters surrounding the focal animal. If there are other individuals on all sides, the focal is in the center position (c), if other individuals are only on one side, the focal is considered to be in a border position (b), whereas if no other group member is within 10 meters, the focal is scored to be in the periphery (p) (Figure 4.1).

If individuals are searching or feeding, we note whether or not they are inside or outside the border of a food patch. We recognize two types of food patches (1) areas of up to 10 meters in diameter on the forest floor with either mushrooms or termites (*Macrotermes spec.*), and (2) larger circular patches of seeds or fruits on the forest floor around the trunks of food trees (radius up to 10 meters).

Social interactions are recorded continuously. For definitions of the behavioral categories see Table 4.1.

When an observer is not occupied with following a certain individual, all observed agonistic interactions, grooming interactions, and copulations between identified individuals are recorded ad libitum (Altmann 1974).

Table 4.1. *Ethogram (Range & Noë 2002)*

Activity	Definition
Maintenance activities	
Feeding	Animal sits or stands at one place and puts objects in its mouth continuously, moving its jaws, emptying its cheek pouches.
Searching	Animal moves slowly forward while visually scanning the forest floor, occasionally putting objects in its mouth.
Travelling	Animal walks steadily forward without visually scanning the forest floor.
Resting	Animal is grooming, playing, sitting, or sleeping.
Social behavior	
Crouch	The belly is close to the ground. The crouch may occur during a severe physical attack, signalling complete submission.
Stare	The actor raises the eyebrows and forehead while staring directly at a target animal; the head can be rapidly lowered and raised while exhibiting the stare.
Stare and lunge	After the stare the actor darts rapidly towards the recipient, but stops before reaching the recipient at which time the actor lowers its shoulders as in preparation to jump forward.
Fighting	Any hard aggressive contact: biting, hitting, gripping, and fighting.
Taking place	The actor takes the place of the recipient after the recipient is threatened or pushed away.
Supplant	The actor approaches another individual who is occupying a resource and replaces that individual without overt aggression.
Grooming	The actor cleans the fur of the recipient with the mouth and/or hands (Altmann 1962).
Invite groom	The actor can use various behaviors to illicit grooming from another individual: the actor presents and/or exposes a part of his body to reactor while standing or sitting stiffly (Hinde & Rowell 1962).
Ventral-hug	The actor approaches a seated animal and lifts its leg onto the shoulder of the seated reactor. It moves its head towards the genital area of the seated reactor.
Hugging	The actor places the arm on the recipient's shoulder. One or both animals may rise onto two legs or remain seated and place both arms around the other's ventrum.
Touch	The actor lightly places one of its hands on the reactor.
Approach	The actor moves into the reactor's space ($r < 2m$).
Agonistic support	An intervention of a third individual in an agonistic dyad on behalf of one individual, directed against its opponent.
Coalition	The combined agonistic interaction of two animals against one opponent.

Data analysis
Dominance relationships

We defined dominance rank according to the direction of supplants for adult males and females. Several measures were used to describe characteristics of the dominance hierarchy, such as the degree

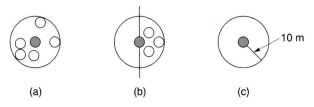

Figure 4.1. Possible positions of a focal animal. The black dot describes the position of the focal animal in relation to other group members of the group. a) center position, b) border position, and c) periphery.

of uni-directionality (van Hooff & Wensing 1987) and the degree of linearity (de Vries 1995). When no clear dominance relationship between two animals could be detected, an average rank was assigned to both of them.

Benefits of high-ranking animals
1.   Competition for food

We predicted that high-ranking females would have better access to food patches than low ranking females. Moreover, high-ranking females were expected to be less disturbed than low-ranking individuals during foraging. We quantified the degree of disturbance for each individual with the "foraging efficiency coefficient," defined as the ratio feeding time/ searching time (for definitions see Table 4.1). Time in a food patch, as well as feeding and searching time, for each female was defined as the percentages of instantaneous samples during focal animal observation the focal spent within a food patch, feeding, or searching respectively.

2.   Competition for safety

We predicted that high-ranking females would more often be in a safe position than would low-ranking females. Safety was measured as a function of the distribution and number of neighbors surrounding the focal animal. The safest position was assumed to be if the focal was in the center of several other animals and thus was "protected" from all sides. The time in the center position for each female was defined as the percentage of instantaneous samples during focal animal observation the focal was surrounded by other animals (c − position).

3.   Competition for mates

We predicted that high-ranking males would have better access to estrous females and be able to copulate more often than low-ranking males. Female sooty mangabeys exhibit sexual swellings with peak tumescence indicating ovulation, which we used as measurement for receptivity. We compared the number of copulations with adult females during peak swelling between males of different dominance rank.

Relationship patterns among group members
1.   Nearest neighbor analysis

We calculated association indices for individual dyads to analyze proximity among individuals. We predicted that if animals have preferred partners they would spend more time with these individuals than one would expect by chance alone. To calculate the probability of a chance distribution compared to the observed distribution we used permutation tests (for details see Box 4.1). We excluded samples, including grooming interactions from this analysis, which were analyzed separately to test whether differences between close associates and grooming partners existed.

2.   Affiliative relationships

We calculated hourly rates of interaction for each dyad by dividing the number of total interactions between A and B by the sum of the total

---

Box 4.1 Nearest neighbor analysis (Range & Noë 2002)

We calculate association indices using the simple ratio association index: ratio of the number of minutes two individuals are nearest neighbors, divided by the sum of the number of minutes each is observed without the other and the number of minutes they are neighbors (Cairns & Schwager 1987). Association indices can vary from 1 (nearest neighbor present all the time) to 0 (never nearest neighbor). To test whether adult females have preferred companions or if each individual associates with others by chance alone, we use a permutation test in the SOCPROG software. The tests are based on the Monte Carlo procedure and have been modified by Whitehead (Whitehead & Dufault 1999) (software is available at http:/www.dal.ca/ ~-hwhitehe/social.htm). The test compares the observed association indices with the results of a random set of data generated by 20,000 permutations of the original data set. To generate permutations, the total numbers of observations for each individual and the total number of association partners per individual are drawn from the matrix of actual observations. The 20,000 permutations are used as the null-hypothesis against which observed values are tested. We use the mean, the standard deviation, and a p-value to compare the two matrices with each other.

observation time (h) that A and B were observed. The duration of grooming bouts was recorded to the nearest minute during focal sampling. The total amount of time spent grooming was estimated for each dyad as the proportion of all sample intervals during which grooming occurred. Minutes per hour were calculated for each female dyad.

To test whether the distribution of grooming among female sooty mangabeys differed from the expected distribution, we used the permutation test in the SOCPROG software (see Box 4.1).

Statistics

Statistical analyses were performed with the SPSS (Version 7.5.1) statistical program for Windows 2000, SOCPROG (Version 1.3), and with MATMAN (Version 1.0, Noldus Technologies). Spearman's test of correlation between ranks was used to test for a correlation between foraging efficiency and the rank order, between position and rank order as well as number of copulations and rank order. These tests were one-tailed, as the predictions were directional. The results were considered significant when $p < 0.05$. (Alpha was set at 0.05 unless we corrected for multiple tests with the same data set.)

Results
Birth and mating season

During the three years of sooty mangabey observation in the Taï National Park, births were recorded from October through March, with a peak in the months December through February (Figure 4.2). The peak coincides with the dry season in Taï, which starts in the beginning of December and lasts until the end of February. Of the 52 infants that were born during three birth seasons, two infants (3.8 per cent) died within the first month, three others (5.8 per cent) survived for two months only. One infant was neglected by its nulliparous mother. Although the mother still nursed her infant, she ceased carrying it and it was instead carried by sub-adult females. After about two weeks the infant vanished. Another infant died after what seemed to be a disease because no injuries could be detected. Another infant was observed with a large wound on its back the day before it vanished. The distribution of births per female and the seasonality of births, suggest an interbirth interval of approximately 2 years.

In accordance with birth seasonality, we have found that mangabeys in Taï exhibit a mating season beginning in June and lasting till October. The mating season is defined by the occurrence of sexual swellings, which was highly correlated with copulations and mate guarding of adult males.

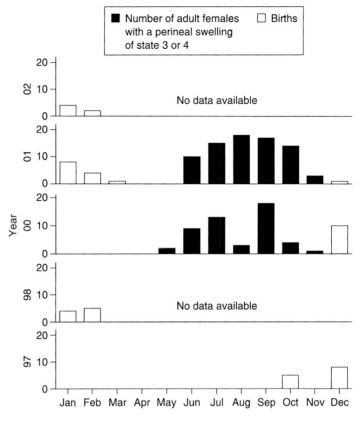

Figure 4.2. Birth and mating seasons of sooty mangabeys. Data were collected from 1997 until 2002 in the Taï National Park. Sexual swellings of adult females can vary between 1 (flat) and 4 (maximal tumescence).

Dispersal patterns and group membership

Dispersal is usually difficult to observe, since we often know only that an animal has vanished, and not where it has gone or whether it is still alive. Moreover, unhabituated animals from other groups are often afraid of human observers and will not transfer into groups where humans are present. In Taï, several groups of mangabeys are used to human observers so migration events can be observed more easily. Several observations suggest that males are the dispersing sex in sooty mangabeys. First, several known sub-adult females (10) have stayed in their natal group and have started to reproduce. Second, nine sub-adult males were observed to transfer into the group, while six males vanished; one of them was seen in another group. We have never observed new females joining the study

Actor (⟶ ascending dominance)

	Mul	Pis	Ven	Oul	Fal	Fer	Sam	Mar	Sim	Ste	Alf	Total
Mul		2	3	2	2		2	3		1		15
Pis			16	19	21	7	17	10	7	20	18	135
Ven	1			5	8	30	19	23	10	10	12	118
Oul		1			3		12	3	7	8	16	50
Fal						15	16	17	5	10	26	89
Fer		4		1			33	20	8	23	10	99
Sam						2		18	11	10	27	68
Mar							1		15	11	13	40
Sim							1	3		14	20	38
Ste							1				28	29
Alf							1	1				2
Total		3	24	26	35	57	100	98	63	107	170	683

Figure 4.3. Matrix based on 683 interactions among adult males in 2000 and 2001 involving submissive behaviors (avoid and yield) recorded during both focal animal and ad libitum sampling. Individuals are noted by three-letter codes and the order is chosen by minimizing the circular triads.

group, although several old females disappeared. Third, lone males are frequently encountered in the home range of the study group, but never lone females.

Male mangabeys exhibit at least three different patterns of group membership. Some males stayed in the group for long periods of time (>6 months). Other males joined a group for one to four months then vanished for good or at least for several weeks to months. Sometimes these males were subsequently observed in neighboring groups. Finally, other males remained in the group only for a few hours or days and stayed mainly in the periphery. Whereas the third behavior was mainly observed during the mating season, the second was observed all year round. The fluidity of male group membership further supports the hypothesis that male mangabeys transfer between groups.

Dominance hierarchy

Adult males and adult females can be arranged in a linear dominance hierarchy (Figures 4.3 and 4.4). Based on 683 submissive interactions between adult males, we could determine the dominance relationships in 40 of 55 possible adult male dyads (72.73 per cent). For adult females present in 2000 and 2001, we observed 1199 submissive interactions and could define 250 of 300 possible dominance relationships (83.33 per cent). In the female hierarchy one circular triadic relationship (Va-Ti-Bi) was observed (Figure 4.4). Three female dyads (Co-Vi; Ka-Gi; Fu-Di) were

Actor (⟶ ascending dominance)

	Co	Vi	Ro	So	Ka	Gi	Fa	St	Cl	Si	Ol	Po	Va	Ti	Bi	Ma	Lu	Em	Ri	Fu	Di	Hi	Lo	Fe	Sa	Total
Co		3	3	5	7	6	1	2	1	9	5	3	4	9	2	3	8	13	4	4	5	4	3	2		106
Vi	9		18	1	6	6	4	6	12	8	6	1	4	4	5	7	1	2		1	1	1	1		3	107
Ro		1		6	5	3	1	6	2	3	5	5	6	3	2	1	3	2	5	4			1			66
So					3	5	1	4	3	4	5	4	1	7	4		4	3	1	1	2	1	2	3		58
Ka		1				1	6	4	6	11	6	6	3	5			3	3	4	1	1	3	2	3		72
Gi		1	1				2	6	3	1	2	16	4	1	5	5	1	2	1	6	2		5			64
Fa			1					1	1	5	3	4	1	3	1	1	3	5	6	1	4	2	4			46
St									8	7	3	4	9	2	5		4	1	5	1	5		2			56
Cl	1									1	3	2	3	1	3	1		2	1	1	1	1				21
Si		1	1					1			4	1	3	2	5	1	4	3	1	4	1	1	5		1	39
Ol		1	1		1							9	3	4	2	3	7	1	5	3	5	2	1			49
Po													3	6	5	7	5	3	4	1	7	1	6			48
Va		1								1				5	2	3	3	1	3	5	1	5		2		33
Ti															3	1	3	3	5	4	5	4	4	1		33
Bi													2	1		13	8	4	4	6	3	4	11	4	4	64
Ma																	3	1	5	2	1	4		1		17
Lu																		5	5	5	7	6	5	2	4	39
Em						1													21	12	8	8	9	5	9	73
Ri																				19	9	17	12	4	12	73
Fu																					1	18	15	4	1	39
Di																		1				9	9	4	1	24
Hi																		1					15	6	14	36
Lo																		1						7	14	22
Fe																									14	14
Sa																										
(1.rang)																										
Total	1	14	26	18	25	24	23	35	48	45	59	31	54	45	45	53	59	76	82	62	19	113	48	14		1199

Figure 4.4. Matrix based on 1,199 interactions among adult females present in 2000 and 2001 involving submissive behaviors (avoid and yield) recorded during both focal animal and ad libitum sampling. Individuals are noted by two-letter codes and the order is chosen by minimizing the circular triads. Fe is a subadult female in 2000 but already often involved in agonistic interactions and thus included in this table.

assigned equal rank in this rank order, because no interactions or an equal number of submissive interactions were observed between them.

The degree of uni-directionality of dominance relationships for adult males and females is high (DC = 0.95 and DC = 0.97 respectively). The degree of linearity is slightly higher in males (h' = 0.83) than females (h' = 0.77). The probability that the observed linearity results from a random process is $p < 0.001$ for males and females. Relative ranks of at least 14 females were stable from the first study conducted in 1997 till 2001. For the other females, data from the first study were not sufficient to construct a well-defined dominance hierarchy that would allow a comparison between the relative ranks of these females for the entire study periods. For males no comparative data over a longer period of time are currently available.

Benefits of rank
1.   Competition for food

Female sooty mangabeys attribute the majority of their time to activities related to foraging – on average over 74 per cent of the observation time.

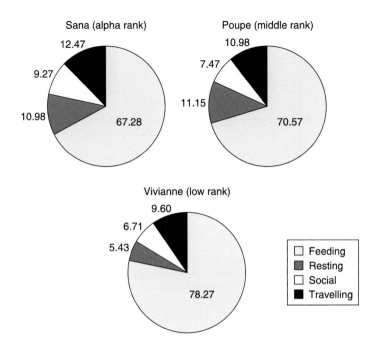

Figure 4.5. Activity budgets of a high, middle, and a low-ranking adult female.

The other time they spend mostly resting and grooming other individuals (Figure 4.5). High-ranking females spend significantly more time in food patches than low-ranking females (Spearman rank correlation: $r_s = -0.6$, $N = 24$, $p < 0.001$). Moreover, the foraging efficiency coefficient (ratio feeding/searching) is correlated with rank (Spearman rank correlation: $r_s = -0.634$, $N = 24$, $p < 0.01$). The highest-ranking females have the highest scores for foraging efficiency (Range & Noë 2002).

2.   Competition for safety

High-ranking females spend significantly more time surrounded by other animals than low ranking females (Spearman rank correlation: $r_s = 0.773$, $p < 0.001$) (Figure 4.6). Low ranking individuals are often in a border position or in the periphery.

3.   Competition for mates

High-ranking males copulate significantly more often with estrous females (swelling 4) than low ranking males (Spearman rank correlation: $r_s = 0.833$, $p < 0.01$). The male Mul was excluded from this analysis since no data were available for him during the mating season.

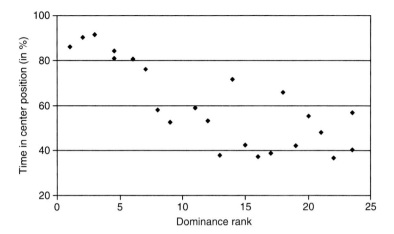

Figure 4.6. Relationship between position and rank. We defined the time in the center position for each female as the percentage of instantaneous samples during focal animal observation the focal was surrounded by other animals. Spearman's rank coefficient: $r_s = $ Spearman's rho $= 0.773$, $p < 0.001$.

Relationships between group members
1.   Nearest neighbor analysis

Figures 4.7 and 4.8 summarize the most frequent associations among adult male and among adult female sooty mangabeys in 2000–2001. Adult males associated mainly with two other partners except the α-male, who associated with four other males. The highest association rate was observed between the α- and β-male (Figure 4.7). Lower-ranking males that showed no high association indices with high-ranking individuals (Ven, Pis, and Mul) were all natal, slightly younger individuals. Fer, who did not associate much with other males either, was a male of type 2 (see above), who joined the group several times for up to two months but in between these stays vanished for weeks at a time.

Adult females exhibit a different association pattern. The seven highest-ranking females associated more often with each other than with any other lower-ranking females in the group (Figure 4.8) (Range & Noë 2002). However, no clear pattern can be observed between middle and low-ranking individuals. Figure 4.9 shows association indices between adult males and females, which indicate that high-ranking males associated mainly with high-ranking females. The α-female had the highest association index with the α-male. However, while the α-male associated as well with other high-ranking females, the α-female had no other

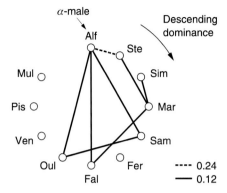

Figure 4.7. Association among adult males. The simple ratio association index was used to calculate associations between two males A and B using frequencies of the total observation time of A and B that A and B were nearest neighbors. Association indices could vary from 1 (nearest neighbor all the time) to 0 (never nearest neighbor). Only association indices, which are higher than 0.10, are represented. Ranks decline in clockwise direction (Alf has the highest rank).

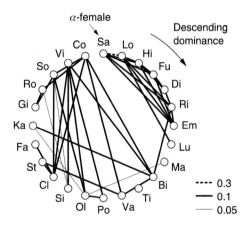

Figure 4.8. Association among adult females. The simple ratio association index was used to calculate associations between two females A and B using frequencies of the total observation time of A and B that A and B were nearest neighbors. Association indices could vary from 1 (nearest neighbor all the time) to 0 (never nearest neighbor). Only association indices, which are higher than 0.05, are represented. Ranks decline in clockwise direction (Sa has the highest rank).

male associates. Several high-ranking females, especially Di and Fu, were frequently observed with different high-ranking males. In contrast to the high-ranking females, hardly any association was observed between low-ranking females and adult males.

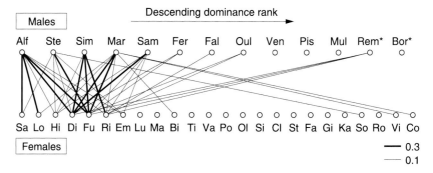

Figure 4.9. Association between adult males and females. Males and females are ordered according to their dominance rank. (*Bor and Rem left the group before their dominance rank could be determined.) The simple ratio association index was used to calculate associations between two individuals, male A and female B using frequencies of the total observation time of A and B that A and B were nearest neighbors. Association indices could vary from 1 (nearest neighbor all the time) to 0 (never nearest neighbor). Only association indices, which are higher than 0.10, are represented.

To test whether these association patterns (**m**ale-**m**ale, **f**emale-**f**emale, and **m**ale-**f**emale) differ significantly from what would be expected if each individual associated with other individuals at random, we generated several random sets of data and conducted permutation tests (see Box 4.1). We found that there were significant differences in the mean association indices for dyads between the observed matrices and the generated matrices (**mm**: 0.07 versus 0.075, $p < 0.000$; **ff**: 0.079 versus 0.081, $p < 0.01$; **mf**: 0.021 versus 0.022, $p < 0.001$), indicating that individual dyads associated less than expected. Moreover, the standard deviation for dyads was significantly lower for the random than for the observed matrices (**mm**: 0.056 versus 0.058, $p < 0.05$; **ff**: 0.035 versus 0.056, $p < 0.001$; **mf**: 0.029 versus 0.036, $p < 0.001$), suggesting that the observed data contain some very high and some very low values. We would expect this result if animals chose to associate with certain individuals, and avoid others.

2.   Affiliative relationships among females

Grooming among adult females was not evenly distributed. Most females tended to restrict their grooming to preferred partners. In 2000, 11 females devoted 50 per cent or more of their grooming bouts towards a single female partner. Figure 4.10 summarizes the distribution of grooming among adult females in 2000 (from Range & Noë 2002). We tested if the observed distribution differs significantly from a chance distribution by generating

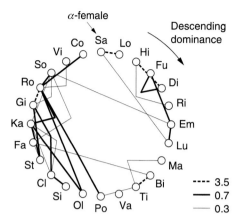

Figure 4.10. Grooming network among adult female sooty mangabeys. Presented here is the grooming duration calculated as rates per hour. Rates lower than 0.3 are not shown. Females are shown in decreasing rank order reading clockwise from the top (Range & Noë 2002).

a random data set as described above and comparing it with the observed grooming matrix (from Range & Noë 2002). We found a significant difference between the mean as well as the variance for dyads between the observed and the generated matrix (mean = 0.024, STD = 0.083 versus mean = 0.23, STD = 0.043; p < 0.01) indicating that grooming was not equally distributed among group members. Female sooty mangabeys groomed less than expected (lower mean in observed matrix), but the higher standard deviation in observed matrix compared to the generated matrix suggest that they had preferred female grooming partners.

3.   Coalitions among adult females

Forty-six coalitions between adult females were recorded using ad libitum and focal animal data (Figure 4.11). Most coalitions were observed between the eight highest-ranking females, especially between the two highest-ranking females, Sa and Lo. Agonistic support was mainly given from Sa to Lo (six times), while Lo supported Sa only once in a conflict. Interestingly, we observed several coalitions between these two highest-ranking females as early as 1997/98, when Lo was still a sub-adult female suggesting that these two animals are mother and daughter. Another dyad that was observed to support each other in conflicts frequently was Fu and Di. Fu supported Di twice, while Di helped her three times in agonistic encounters. However, it was not clear who is the higher-ranking female in this dyad (see Figure 4.4). Among most other dyads, coalitions were

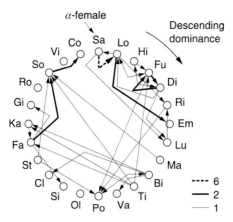

Figure 4.11. Agonistic support given and received among adult female sooty mangabeys. Presented is the total number of agonistic supports for each female recorded during focal and ad libitum sampling in 2000 and 2001.

observed only once or twice. Most agonistic support (37) was given when the receiver was higher-ranking than the supporter.

Discussion

Many species of Old World Monkeys such as baboons, vervets, and macaques show a similar pattern of social organization. In these species, males transfer from their natal groups into new groups, high intra-sexual competition for mates results in linear hierarchies and males often form friendships with adult females. On the other hand, females are philopatric and form well differentiated relationships often based on relatedness. The support of relatives in coalitions usually results in matrilineal-based, stable, and linear dominance hierarchies. The results of our study on the social behavior of sooty mangabeys in their natural environment suggest that they resemble these other old world primate species that live in multi-male multi-female groups.

Seasonality, group membership, and transfer pattern

Our observations of male group transfer support the hypothesis that males are the dispersing sex and that female are philopatric, which is likely to have implications on female relatedness. In theory, females in these groups should be more closely related to each other than males facilitating the formation of a matrilineal-based social system among female sooty mangabeys. However, so far no DNA-analysis has been conducted to test this hypothesis.

Sooty mangabeys in Taï exhibit a mating and birth season, which is likely to have profound consequences on the social behavior of adult males. If estrus in adult females is synchronized it will be impossible for a single high-ranking male to monopolize mating allowing the development of alternative strategies for lower-ranking males. The different types of group memberships that males engage in imply these alternative strategies. Some males stay with a group for long periods of time, form close relationships with females and males, while others join the group for weeks at a time, never fully integrating (no association with other males). And then during the mating season, we observe a third strategy: males only come for hours or days at a time. Interestingly, these "visitors" have been observed to copulate with estrus females that were mate guarded by constant group males. Currently, we are trying to understand the characteristics of these different strategies and the benefits for the males.

Dominance and benefits of rank

We found that all adult females could be ranked in a linear dominance hierarchy that has remained stable for several years. But are there advantages correlated with a high dominance rank? The diet of mangabeys in Taï mainly consists of fruits and seeds (68 per cent) and invertebrates (26 per cent) (Bergmüller 1998). Especially in the summer, the seeds of *Sacoglottis gabonensis* represent the dominant food resource, occurring in large circular patches around trees. These food patches are very large, but still cannot provide space for all group members (~120 animals). Our study showed that high-ranking females are more often in food patches than low-ranking females. Foraging efficiency was also positively correlated with rank (Range & Noë 2002); a further indication that high-ranking females have better access to food resources, which could have consequences for their reproductive success.

Safety from predators has been argued to be an important factor influencing survival in adult females and especially the position within a group has been shown to be an important predictor of predation risk (Ron *et al.* 1996). Predation pressure in Taï is relatively high (Chapters 10 & 11) and both eagles and leopards are known to prey upon mangabeys. Although it is difficult to observe predation events and highly risky to allocate the vanishing of a female to predation, it is theoretically safer in a position surrounded by other animals than being at the border or in the periphery of the group. High-ranking females in our study group were more often surrounded by group members than low-ranking females presumably decreasing their risk of predation.

Adult males form linear dominance hierarchies as well and access to estrous females is rank-dependent. Interestingly, even the male that joined the group only for a few months at a time and was not considered a long-term resident male, had access to estrous females. However, until paternity tests have been conducted it will not be possible to really evaluate the influence of rank on reproductive success in sooty mangabeys. Moreover, more studies have to be conducted to elucidate the role of visiting males and other short-term group members to conclude about costs and benefits of different male strategies.

Affiliative relationships

Female sooty mangabeys had well differentiated social relationships in regard to association, grooming as well as coalition partners. Moreover, we showed in another study that females not only groomed their association partners, but especially females with positions close in rank (Range & Noë 2002) which suggests a matrilineal based social system. However, we cannot test the hypothesis that female bonds are based on relatedness with behavioral data alone, but need to conduct a genetic analysis (currently under way). Theoretically, the observed interaction patterns could be predicted even without underlying nepotistic mechanisms if females were attracted to high-ranking females (Seyfarth 1977), or to females of similar rank (de Waal & Luttrell 1986).

In Taï, association pattern of adult males differed according to their age and the duration of their group memberships. Young males had no close associates among high-ranking males, neither had the male (Fer) who stayed with the group only a limited amount of time. Grooming was never observed between adult males (personal observations) and so far, no detailed data are available on the formation of coalitions among adult males. However, coalitions have been observed frequently especially during the mating season and could explain the association patterns observed among males. Young males leave the group, thus investing in the formation of relationships in their natal group might not be a good strategy, which could explain the absences of associations with high-ranking males. Long-term group members, in contrast, associated with only a few other males indicating differentiated relationships. Further research will elucidate if these associations are correlated with agonistic support in conflicts against other group males or intruders.

High-ranking females associated closely especially with high-ranking males in the group. Grooming and several coalitions have been observed within the same dyads (personal observations). Moreover, adult males often tolerate high-ranking females in food patches, while low-ranking females are chased away (personal observation). "Friendships" with adult males could have several advantages for the adult females, but adult males could also benefit from these relationships if it increases their probability of siring the females' offspring. We are currently studying this hypothesis in Taï.

Current and future research
Research on sooty mangabeys in their natural environment provides us with the background knowledge to investigate questions concerning cognitive skills of sooty mangabeys. To date, several experimental studies show that primates engage in a number of complex interactions that demonstrate an understanding of third-party relationships implying some kind of triadic knowledge. For example, (1) in conflicts with higher-ranking individuals, animals solicit help from group members higher ranking than the opponent (Silk 1992, 1999a), (2) redirect aggression preferentially toward the kin of their former opponent (Cheney & Seyfarth 1986, 1989, Aureli & Schaik 1991a), (3) react differently compared to controls if dominant individuals behave submissively towards lower-ranking individuals (Cheney *et al.* 1995). Evidence from field studies on vervets and baboons and captive studies on macaques suggest that primates understand categories of third-party relationships and that they can make inferences about the behavior of other pairs of individuals (e.g. Cheney & Seyfarth 1980, 1986, 1999, Dasser 1988a, 1988b, Cheney *et al.* 1995, Silk 1999b). Considering the size of our study group ($\sim$120 animals) compared to baboons ($\sim$80 animals) and vervets ($\sim$25 animals), the question of what sooty mangabeys know about each other's dominance and kin relationships becomes intriguing. After all, they would have to know over 180,000 triadic relationships, twice as much as a baboon. Moreover, sooty mangabeys live in a forest habitat, where an animal is hardly ever able to see many interactions between others. Considering these circumstances, it seems an almost impossible task to learn to differentiate relationships between all other group members. Thus, do sooty mangabeys really know all these relations or do they use other, simpler strategies to achieve their goals? Currently, we are investigating this question with observational data such as the pattern of solicitation of agonistic support in conflicts and the pattern of supplant of grooming partners between females. Moreover, we conduct playback experiments to investigate experimentally if sooty mangabeys know third-party relationships.

Conclusions

1.   Observations of male group transfer in the absence of female dispersal in Taï imply that female mangabeys are philopatric, whereas males leave their natal group and transfer into new groups.

2.   Adult male and female sooty mangabeys form linear dominance hierarchies, based on the direction of approach-retreat interactions. Relative ranks of several females remained stable over the entire study period.

3.   Several benefits of high rank have been shown for females as well as for males. High-ranking females have better access to food patches as well as a higher foraging efficiency. Moreover, they occupy more often positions surrounded by other group members, implying lower predation risk. Both access to food as well as to safe positions are thought to be closely linked with reproductive success in primate females. Males' access to estrous females was rank-dependent as well as hinting towards higher reproductive success of high-ranking males compared to low-ranking males.

4.   Finally, females and males formed well-differentiated relationships with preferred partners. Females especially had a limited number of partners with whom they frequently associated, groomed, and in conflicts against others supported. Moreover, high-ranking females formed close associations with high-ranking males, however, no data are available yet on the function of the relationships between the sexes.

References

Altmann, S. A. (1962). A field study of the sociobiology of the rhesus monkey, *Macaca mulatta. Annals of the New York Academy of Science*, **102**, 338–435.

Altmann, J. (1974). Observational study of behaviour: Sampling methods. *Behaviour*, **49**, 227–65.

Aureli, F. and Schaik, C. P. V. (1991a). Post-conflict behaviour in long-tailed macaques (*Macaca fascicularis*). I. The social events. *Ethology*, **89**, 89–100.

Bergmüller, R. (1998). Feeding ecology of the sooty mangabey *(Cercocebus torquatus atys). A key for its social organization?* Dipl. Friedrich-Alexander-Universitat Erlangen-Nurnberg, Germany.

Cairns, S. J. and Schwager, S. J. (1987). A comparison of association indexes. *Animal Behaviour*, **35**, 1454–69.

Cheney, D. L. and Seyfarth, R. M. (1980). Vocal recognition in free-ranging vervet monkeys. *Animal Behaviour*, **28**, 362–7.

Cheney, D. L. and Seyfarth, R. M. (1986). The recognition of social alliances among vervet monkeys. *Animal behaviour*, **34**, 1722–31.

Cheney, D. L. and Seyfarth, R. M. (1989). Redirected aggression and reconciliation among vervet monkeys, *Cercopithecus aethiops*. *Behaviour*, **110**, 258–75.

Cheney, D. L. and Seyfarth, R. M. (1999). Recognition of other individuals' social relationships by female baboons. *Animal Behaviour*, **58**, 67–75.

Cheney, D. L., Seyfarth, R. M. and Silk, J. B. (1995). The responses of female baboons (*Papio cynocephalus ursinus*) to anomalous social interactions – evidence for causal reasoning. *Journal of Comparative Psychology*, **109**, 134–41.

Dasser, V. (1988a). Mapping social concepts in monkeys. In *Machiavellian Intelligence: Social Expertise and the Evolution of Intellect in Monkeys, Apes and Humans*, ed. R. Byrne and A. Whiten, Oxford: Clarendon Press, pp. 85–93.

Dasser, V. (1988b). A social concept in Java monkeys. *Animal Behavior*, **36**, 225–30.

Hinde, R. A. and Rowell, T. E. (1962). Communication by postures and facial expressions in the rhesus monkey (*Macaca mulatta*). *Procedures of the Zoological Society of London*, **138**, 1–21.

Homewood, K. M. (1976). Ecology and behavior of the Tana Mangabey (*Cercocebus galeritus galeritus*), Ph.D. Thesis, University of London, UK.

Krebs, J. R. and Davies, N. B. (1993). *An Introduction to Behavioural Ecology*. Blackwell: Oxford.

Mitani, M. (1989). *Cercocebus torquatus*: Adaptive feeding and ranging behaviors related to seasonal fluctuations of food resources in the tropical rain forest of south-western Cameroon. *Primates*, **30**(3), 307–23.

Range, F. and Noë, R. (2002). Familiarity and dominance relations in female sooty mangabeys in the Taï National Park. *American Journal of Primatology*, **56**, 137–53.

Ron, T., Henzi, S. P. and Motro, U. (1996). Do female chacma baboons compete for safe spatial position in a southern woodland habitat. *Behaviour*, **133**, 475–90.

Seyfarth, R. M. (1977). A model of social grooming among adult female monkeys. *Journal of Theoretical Biology*, **65**, 671–98.

Seyfarth, R. M. and Cheney, D. L. (2001). Cognitive strategies and the representation of social relations by monkeys. In *Evolutionary Psychology and Motivation*, ed. J. A. French, A. C. Kamil and D. W. Leger, Lincoln: University of Nebraska Press, pp. 145–78.

Silk, J. B. (1992). Patterns of intervention in agonistic contests among bonnet macaques. In *Coalitions and Alliances in Humans and Other Animals*, ed. A. H. Harcourt and F. B. M. deWaal, Oxford: Oxford University Press, pp. 215–32.

Silk, J. (1999a). Male bonnet macaques use information about third-party rank relationships to recruit allies. *Animal Behaviour*, **58**, 45–51.

Silk, J. B. (1999b). Male bonnet macaques use information about third-party rank relationships to recruit allies. *Animal Behaviour*, **58**, 45–51.

van Hooff, J. A. R. A. M. and Wensing, J. A. B. (1987). Dominance and its behavioural measures in a captive wolf pack. In *Man and Wolf*, ed. H. Frank, Dordrecht: Junk Publishers, pp. 219−52.

Trivers, R. L. (1972). Parental investment and sexual selection. In *Sexual Selection and the Descent of Man*, ed. B. Campbell, Chicago: Aldine, pp. 136−79.

Vries de, H. (1995). An improved test of linearity in dominance hierarchies containing unknown relationships. *Animal Behaviour*, **50**, 1375−89.

Waal de, F. B. M. and Luttrell, L. M. (1986). The similarity principle underlying social bonding among female rhesus monkeys. *Folia Primatology*, **46**, 215−34.

Whitehead, H. and Dufault, S. (1999). Techniques for analyzing vertebrate social structure using identified individuals: Review and recommendations. *Advances in the Study of Behavior*, **28**, 33−74.

II *Anti-predation strategies*

5 Interaction between leopard and monkeys

K. Zuberbühler and D. Jenny

Introduction

Although predation is clearly a crucial factor in the evolution of primates its actual effects as a selective force are not well understood. Predation is thought to have affected various traits such as body size, group size and composition, vigilance, ecological niche, as well as vocal and reproductive behavior (van Schaik 1983, Cheney & Wrangham 1987, Cords 1990, Hill & Dunbar 1998, Stanford 1998, Uster & Zuberbühler 2001). However, there are reasons to remain cautious about many of the proposed relationships. In particular, little is known about the hunting pressure exerted by the various primate predators and the selective pressure they impose on a primate community. Nevertheless, predation is often treated as a homogeneous evolutionary force even though predators differ considerably in their hunting behavior. For instance in the Taï forest, monkeys are hunted by chimpanzees *Pan troglodytes*, crowned eagles *Stephanoaetus coronatus*, and leopards *Panthera pardus*. Predatory chimpanzees locate monkey groups by acoustic cues and hunt for individuals in the high canopy (Boesch & Boesch-Achermann 2000). Not surprisingly, the presence of chimpanzees reliably elicits cryptic behavior in nearby monkeys (Zuberbühler *et al.* 1999). Crowned eagles, in contrast, hunt by sweeping through the canopy to surprise their prey (Gautier-Hion & Tutin 1988, Shultz 2001) and their discovery typically elicits loud and conspicuous alarm calling and sometimes even mobbing behavior (Zuberbühler 2000b). Because they differ fundamentally in their hunting strategies, the selective force of chimpanzees, leopards, and eagles — as predators — is not homogeneous. Predation, in other words, is a heterogeneous selective force.

In this chapter, we review studies on the hunting behavior of the first Taï predator: the leopard. Leopards occur in a wide variety of habitats ranging from open savannah to closed rainforests (Kitchener 1991). To date, most information on leopard ecology and behavior has been

Monkeys of the Taï Forest, ed. W. Scott McGraw, Klaus Zuberbühler and Ronald Noë.
Published by Cambridge University Press. © Cambridge University Press 2007.

Figure 5.1. The adult male Cosmos passing a photo-trap (Photo: D. Jenny).

collected from individuals living in the African savannah (Hamilton 1981, Bailey 1993), and little is known about forest leopards (Hart *et al.* 1996). As the largest carnivore predator, leopards are a key component in the forest ecosystem and are likely to play an important role in the evolution of primates and other animal groups.

The hunting behavior of the Taï leopards

To investigate the behavior and ecology of forest leopards, four adult individuals were captured with a cable snare and sedated with a mixture of Domitor/Ketamine administered with a syringe fired from a carbon-dioxide gun (Jenny 1996, Dind *et al.* 1996). The animals were then fitted with a radio-collar, anti-sedated, and liberated back into the forest (Jenny 1996, Figure 5.1, Table 5.1).

In this chapter, we mainly review data collected from two of the study animals (Cosmos, Adele) that were monitored from three different platforms installed in the high forest canopy. This permitted monitoring activity patterns as a function of time of day, month, and amount of rain (Jenny & Zuberbühler 2005). Readings were taken every 15 minutes both during the day and at night. Activity was scored as either "moving" or "resting," depending on whether the impulses of the received signal were fluctuating or stable. Both radio-tracked individuals were significantly

Table 5.1. *Morphometric data on the study animals (Jenny 1996)*

Individual	Capture date	Sex	Age (years)	Weight (kg)
Cosmos	5 Feb 93	Male	3–5	56
Adele	16 Aug 93	Female	3–5	34
Cora	16 Jun 94	Female	2–3	32
Arthur[a]	11 Oct 94	Male	3–4	49

[a]Dind *et al.* (1996), Dind, F. (1995) (unpublished M.Sc. thesis, University of Lausanne)

more active during the day than during the night (Adele: $mean_{day} = 46.9$ per cent, $N = 53$, $mean_{night} = 26.3$ per cent, $N = 43$, $z = 6.34$, $p < 0.001$; Cosmos: $mean_{day} = 49.3$ per cent, $N = 53$, $mean_{night} = 30.3$ per cent, $N = 43$, $z = 4.384$, $p < 0.001$; Mann-Whitney U-tests, two-tailed; Zuberbühler & Jenny 2002; see Figure 5.2).

At night, two distinct activity patterns could be distinguished. Either the individuals remained completely inactive throughout the night or they moved continuously, often traveling great distances. Daytime activity was more evenly distributed and inactive periods during the day never lasted more than five hours. This pattern was comparable to that observed in Asian forest leopards (Karanth & Sunquist 1995, 2000), but contrasted strongly with savannah leopards that were reported to be predominantly nocturnal (e.g. Bailey 1993), suggesting that habitat type determine circadian activity patterns.

Overall moving activity was significantly correlated with season. We found that the lowest monthly moving rates were observed during the main rainy season in October, whereas the highest moving rates were recorded during the dry season in January (see Figure 5.3). Per cent moving activity per month was significantly negatively correlated with amount of rainfall (Spearman-Rank correlation, $N = 11$, $r_s = -0.718$, $z = -2.271$, $p < 0.03$). During heavy rains, it may be more difficult for prey to detect an approaching or hiding leopard which likely leads to greater hunting success and decreased traveling time during the rainy season (Jenny 1996).

Although direct observations of leopards was not possible in the forest, two individuals (Cosmos and Adele) were also followed on a regular basis at a close distance ranging from 30 to 150 meters. Focal animal follows of Adele and Cosmos were conducted between February 1993 and August 1994 for 15 and 11 months, respectively (Jenny 1996). The other two collared animals, Cora and Arthur, were subjects of a follow-up study (Dind *et al.* 1996) and are not included here (Table 5.1).

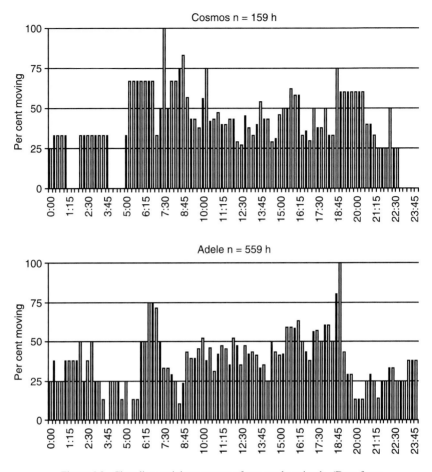

Figure 5.2. Circadian activity patterns of two study animals. (Data from Jenny & Zuberbühler 2005.)

Once a focal animal was located from platforms by triangulation, one observer moved quickly to the area and began following the animal. These indirect follows revealed additional information about leopard hunting behavior.

Forest leopards are thought to be ambush predators that hide and attack their prey by surprise. Our study supports this view. On one occasion, we were able to observe a successful attack by Adele on a *C. atys* after a prolonged period of hiding. After making this kill, she remained active in the same area for a few consecutive days. In general we noted that Adele appeared to approach monkey groups selectively. Once close to a group,

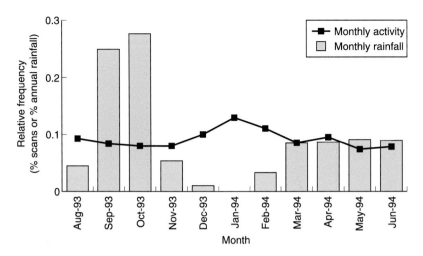

Figure 5.3. The relationship between average monthly activity and the amount of rainfall in the adult female Adele. Monthly rates of moving were significantly negatively correlated with amount of rainfall. (Data from Jenny & Zuberbühler 2005.)

she hid in dense understory on or near the ground, presumably waiting for monkeys to descend. In two out of nine direct sightings, Adele was observed sitting in the lower canopy on the lower branches of a tree, but never in the high canopy. All eight species of monkeys occasionally come to the ground to forage or play (McGraw 1998). To determine whether Adele selectively approached monkey groups, Jenny and Zuberbühler (2005) analyzed whether at least one monkey group could be located within 50 m from Adele's various hiding locations. This occurred approximately four times as often as when the observer David Jenny (DJ) was sitting alone at randomly selected points throughout the study area. In 60 out of 97 hiding bouts (7.4 per cent; N = 7821 min) a monkey group came within 50 meters of the hiding leopard. In contrast, when sitting at one of 10 different observation points throughout the study area, monkeys came within 50 meters only four times (1.9 per cent; N = 5940 min), a statistically significant difference (z = − 3.092; p < 0.01; binomial test; two-tailed, Jenny & Zuberbühler 2005). This finding suggested that Adele selectively chose her hiding spots close to monkey groups. In contrast, Adele clearly avoided chimpanzee parties. After the occurrence of drumming or screaming from nearby chimpanzee parties, she inevitably altered her course in the opposite direction if already moving. We never observed her approaching chimpanzee parties (Zuberbühler & Jenny 2002).

Table 5.2. *Home range sizes*

Individual	Study period 1[a] (February 1993–August 1994)		Study period 2[b] (January 1995–June 1995)	
	N	Estimated home range size (km^2)	N	Estimated home range size (km^2)
Cosmos	159	85.6	93	41.5
Adele	342	28.5	161	25.3
Cora	53	22.2	79	22.4
Arthur	–	–	136	35.4
Total/mean	554	45.4	469	31.2

N = number of independent triangulations; [a]Jenny (1996); [b]Dind *et al.* (1996)

Leopard encounters and predation risk

Predation pressure can be measured at two different levels, predation rate and predation risk (e.g. Dunbar 1988, Janson 1998). Predation risk represents the animals' own perception of the likelihood of being attacked by a predator, regardless of whether the attack is successful (Hill & Dunbar 1998). This risk can be operationalized by the likelihood of a group encountering a predator (Hill & Dunbar 1998). Encounter rates between leopards and the different primate species will depend on factors such as density, home range, and daily travel distance of both predator and prey. Additional factors, such as the predators' prey preference and searching abilities as well as the preys' ability to predict and avoid the predator will also play a role, but they are more difficult to quantify. It is beyond the scope of this chapter to provide an accurate assessment of the predation pressure exerted by the Taï leopards. Nevertheless, a rough estimate is possible due to information available concerning the home range size of four radio-collared animals (Jenny 1996, Dind *et al.* 1996). Home range size was determined by two observers sitting on platforms in the high canopy simultaneously locating the focal individuals by triangulation (location accuracy ± 0.01 km^2). Distance was determined by the strength of the signal using a reference table (see Table 5.2).

The analysis provides important information on ranging behavior. First, male home ranges overlapped strongly with female home ranges while the ranges of same-sex individuals showed little overlap (Dind *et al.* 1996). Second, the home range size of the adult male Cosmos decreased dramatically between 1994 and 1995, most likely due to the appearance of a new male, Arthur, who began to occupy much of Cosmos' home range. Third, the actual density of leopards is likely to be higher because information collected from the photo traps suggests that at least three more individuals frequented the roughly 100 km^2 study area. The addition of these individuals yields a density estimate of 7–11 individuals per 100 km^2 (Jenny 1996). As a consequence, a particular primate group will encounter not only the 1–2 resident leopards, but also occasional trespassers or newly settled individuals. Assuming a home range size of 22 km^2 for female and 86 km^2 for male leopards (Jenny 1996), the average monkey group is likely to have one of the two resident leopards within its own home range once every 15–30 days.

Leopard prey spectrum and predation rates

In contrast to predation risk, predation rate refers to the successful predation events a predator can actually achieve. To determine the predation rates of leopards on the Taï monkeys we review two studies that have analyzed leopard feces in the Taï forest (Hoppe-Dominik 1984, Zuberbühler & Jenny 2002). In the second study, a total of 200 fecal scat samples were collected systematically along trails and throughout the study area. Samples were collected regularly between June 1992 and June 1994. We assumed that each fecal sample corresponded to one predation event. A day's search rarely led to the recovery of more than one fecal sample. All samples were inspected for the presence of hairs, bones, teeth, nails, and other remains. Hairs were identified using a reference collection and reference photographs (Hoppe-Dominik 1984, Bodendorfer 1994).

Roughly 140 mammal species are known to be present in the park and at least 12 of them are endemic. Table 5.3 illustrates the wide variety of prey species found in leopard feces, most of them mammals weighing less than 10 kg. The 200 feces analyzed contain remains of at least 23 different prey species. The large proportion of monkeys and duikers is particularly noteworthy, which reflects the species diversity of the rainforest habitat. Our results are comparable to those of an earlier study (Hoppe-Dominik 1984) in which a large number of samples were collected from the eastern side of the park where disturbed secondary forest prevailed and poaching pressure was much more intense. Still, the prey profiles are not equivalent. For example, *Colobus polykomos* and *Procolobus badius* were under-represented in the 1984 study, perhaps as the result of lower population densities of these two species in the eastern side of the park due to high poaching pressure.

Table 5.3. *Prey spectrum of Taï leopards*

Scientific name	Common name	Zuberbühler & Jenny (2002)	Hoppe-Dominik (1984)
Procolobus badius	Red colobus	21	8
Colobus polykomos	Black-white colobus	16	5
Procolobus verus	Olive colobus	1	0
Cercopithecus diana	Diana monkey	5	17
Cercopithecus petaurista	White-nosed monkey	1	5
Cercopithecus campbelli	Campbell's monkey	3	4
Cercopithecus nictitans	Putty-nosed monkey	0	0
Cercocebus atys	Sooty mangabey	6	9
Cercopithecidae	Unknown monkeys	10	3
Pan troglodytes	Chimpanzee	1	0
Perodicticus potto	Potto	0	1
Primates total		**64**	**61**
Cephalophus spp total	Duikers	82	82
Manis spp.	Pangolins	43	10
Sciuridae (undet.)	Squirrels	8	9
Panthera pardus	Leopards	6	6
Other Mammals	Other mammals	18	62
Mammalia (undet.)	Unknown mammals	6	26
Non-primates total		**163**	**195**
Aves total		**2**	**2**

Are leopards opportunistic or selective hunters?

Leopards are generally described as opportunistic predators, implying that they hunt prey species in proportion to abundance. Much of the evidence supporting this contention comes from open savannah habitats, however a study conducted in the Congolese Ituri Forest suggested that leopards are selective hunters with a particular bias towards l'hoest's guenons, *Cercopithecus lhoesti* (Hart *et al.* 1996). Moreover, studies using fecal data routinely average over a large number of individuals to calculate the prey spectra that masks the effects of individual differences (e.g. Zuberbühler & Jenny 2002). If the difference between savannah and forest leopards is real, then this could be the outcome of decreased competition from other predators in the forest habitat (Ray & Sunquist 2001). Similarly, because of the spatial overlap in home ranges, individuals are likely to develop individual prey preferences to reduce competition with more dominant competitors with whom they share a home range.

Opportunistic and selective predators are likely to differ in the evolutionary pressure they exert on a prey population. Selective hunters tend to increase species diversity in an ecosystem, particularly if their preference is directed towards a competitively dominant prey (Begon *et al.* 1996, p. 809). Furthermore, selective predators are predicted to increase the behavioral flexibility in prey. If predators develop individual preferences for certain species, then members of different species are forced to compete with each other to avoid preference formation. Individual or kin based anti-predator strategies thus might not be sufficient to avoid preference formation, because individuals should be interested in avoiding predation on any conspecific group members, regardless of the degree of relatedness. However, individuals responding to a generalist predator that does not develop preferences should be mainly concerned about their own survival and that of close relatives. Living in large groups provides a good strategy against opportunistic predators because individuals can benefit from a dilution effect where costs are shared among unrelated conspecifics. In sum, selective predators are expected to favor the evolution of flexible and cooperative defense behaviors, whereas generalist predators are expected to favor the evolution of simple selfish or kin selected defense strategies.

Considering the potential implications of selective hunting for the evolution of prey defense strategies, we attempted to determine whether Taï leopards are best classified as selective or generalist hunters. In a recent study (Jenny & Zuberbühler 2005) we compared the overall prey spectrum of Taï leopards (see Table 5.3) with the prey spectrum using three sets of data. First, local variations in prey spectra were used as an indicator of individual differences in prey selectivity. Second, an infrared-triggered photo-trap was installed along one trail that was frequently used by leopards, allowing us to assign a large number of feces to particular individuals. This was possible when individuals were photographed by the photo-trap, identified, and when its spoor could be followed to a fresh fecal sample. Third, while following a radio-collared individual it was possible to assign a further set of fecal samples to identified individuals.

Pangolin remains were frequently found in leopard feces throughout the 100 km^2 study area (see Table 5.3) but frequencies varied both regionally and temporarily, suggesting individual differences in the hunting behavior of the various leopards (Jenny & Zuberbühler 2005). In the southeast part of the study area, 40 per cent of feces contained pangolin remains. In the northwest part the number was lower: between June 1992 and March 1993 33 per cent of all feces contained pangolin remains. This rate dropped to less than 5 per cent between March 1993 and June 1994. No changes in the percentage of feces containing pangolins were observed in the southeast part during the same time periods. Interestingly, the sudden drop in feces containing pangolin remains in the northwest coincided with the death of a resident leopard, perhaps a specialized pangolin hunter, in early March.

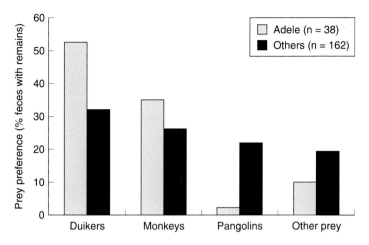

Figure 5.4. Prey selectivity of the focal animal Adele compared to other leopards in the Taï forest. (Data from Jenny & Zuberbühler 2005.)

In contrast, Adele, whose home range was also in the northwest part, did not usually hunt pangolins. With the aid of a photo-trap it was possible to assign a large number of feces to Adele, only one (3.6 per cent) of which contained pangolin scales. This is in comparison to 26.7 per cent for the other leopards contributing to the whole sample (see Table 5.3). Adele preferred to hunt duikers and monkeys significantly more often than other leopards (see Figure 5.4).

In sum, the different rates of pangolin predation in geographically distinct areas, the sudden change of these rates following the death of a resident leopard and the difference between Adele's prey spectrum and that of the rest of the leopard population strongly suggest that forest leopards developed highly idiosyncratic prey preferences among the large spectrum of possible prey species. In an earlier study, Boesch (1991) suggested that one individual specialized in preying on his chimpanzee study group, causing unusually high mortality rates during some time. In the two studies reviewed in this chapter chimpanzee remains were only found exceptionally, suggesting that preference formation for chimpanzees is an exception.

Primate anti-predator strategies

Recent molecular studies estimate the origin of modern leopards occurred approximately 500,000 years ago (Uphyrkina et al. 2001), suggesting that leopards have been a significant factor in the recent evolutionary history of non-human primates. To assess the potential impact of leopards on primates, we compared rates of leopard predation on the Taï monkeys with a number of behavioral, demographic, and morphological traits that are commonly viewed as anti-predation adaptations: body size, group size, group composition, female reproductive rate, and use of forest strata. The general prediction was that if a trait had evolved as an adaptation to leopard predation, there would be a negative relationship between the expression of the trait and the individual's vulnerability to leopard predation.

It has been suggested that large body size is an adaptation to predation (e.g. Isbell 1994). If true, then the larger Taï primates should be under-represented in the leopards' prey spectrum compared to the smaller ones. Similarly, it has been suggested that individuals living in large groups are less susceptible to predation than individuals living in a small group, due to dilution effect and increased vigilance (e.g. van Schaik 1983). Taï primates that live in larger groups should thus be less susceptible to predation and therefore underrepresented in the leopard's prey spectrum. It has also been argued that the formation of groups containing several adult males is an adaptation to predation pressure, particularly in species where males engage in cooperative defense against predators (Stanford 1998). According to this hypothesis, Taï primates living in multi-male groups should be better protected against predation and therefore underrepre-sented in the leopard's prey spectrum relative to single-male groups. Another hypothesis states that natural selection can lead females to accept higher levels of predation if their potential reproductive rate is high enough to compensate for the losses incurred from predation (Hill & Dunbar 1998). In that case, rather than evolving predator-specific defense mechanisms, natural selection favors females who shorten their inter-birth intervals to increase their lifetime reproductive success. Species with short inter-birth intervals should thus be overrepresented in the leopards' prey spectrum. Finally, Taï primates show species-specific preferences for particular forest strata (McGraw 1998), presumably as a result of interspecies competition. Hence species living in the lower forest strata should be more exposed to ground predators (Dunbar 1988, Plavcan & van Schaik 1992) and should therefore be overrepresented in the leopards' prey spectrum. We should note that since the different primate species in the Taï forest vary dramatically in their population density this variable might be important in explaining variation in leopard hunting success. To assess how the various primate traits affected leopard hunting success, we compiled a data set for the Taï primate species, using several sources of information (see Table 5.4).

Table 5.4. *Data on population density, group size, body weight, strata use, number of males per group, birth rate, and usage of the lower forest strata for the Taï primates (Data from Zuberbühler & Jenny 2002)*

Species	Density	Body size	Group size	N males	Reproduction	Habitat
Cercopithecus diana	48.2	3.9	20.2	1	0.62	6.1
C. campbelli	24.4	2.7	10.8	1	0.63	36.8
C. petaurista	29.3	2.9	17.5	1	0.52	9.9
C. nictitans	2.1	4.2	10.5	1	0.50	0.7
Procolobus badius	123.8	8.2	52.9	10.1	0.42	0.4
Colobus polykomos	35.5	8.3	15.4	1.42	0.59	1.3
Procolobus verus	17.3	4.2	6.7	1.43	0.61	13.2
Cercocebus atys	11.9	6.2	69.7	9.0	0.40	88.9
Pan troglodytes	2.6	47.5	61.1	6.7	0.23	85.0

Density: estimated number of individuals per square kilometer
Body size: adult female body weight in kg (from Oates *et al.* 1990)
Group size: average number of individuals per group
N males: average number of adult males per group
Reproductive rate: average number of infants per adult female per year
Habitat: per cent time observed in lower forest strata (Data from McGraw 1998, 2000, Eckardt 2002.)

Data were natural log transformed $[y = LN (x + 1)]$ to ensure normality before performing linear regression analyses (Hill and Dunbar 1998). Univariate analyses of the six variables using data of the eight monkey species showed that predation rate was significantly related to population density ($r^2 = 0.583$, $F_{1,6} = 8.383$, $p = 0.028$) and body size ($r^2 = 0.572$, $F_{1,6} = 8.011$, $p = 0.030$, Figure 5.5). However, contrary to predictions, body size and predation rates were positively related because the larger monkey species were preyed upon more often than smaller ones. The relationships between predation rate and group size and the number of adult males per group were also positive (group size: $r^2 = 0.390$, $F_{1,6} = 3.836$, $p = 0.098$; number of males: $r^2 = 0.353$, $F_{1,6} = 3.277$, $p = 0.120$), although they did not reach statistical significance. Predation rates were unrelated to the reproductive rate of adult females ($r^2 = 0.054$, $F_{1,6} = 0.340$, $p = 0.581$) and to a species' use of the lower forest strata ($r^2 = 0.030$, $F_{1,6} = 0.188$, $p = 0.680$). A stepwise multiple regression analysis using all six variables indicated that population density and body size combined accounted for a significant proportion of the overall variance of the leopard predation rate ($F_{2,5} = 18.347$, $p = 0.006$). Figure 5.5 summarizes the main findings.

These data show that leopard predation was most reliably associated with density, suggesting that leopards hunt primates according

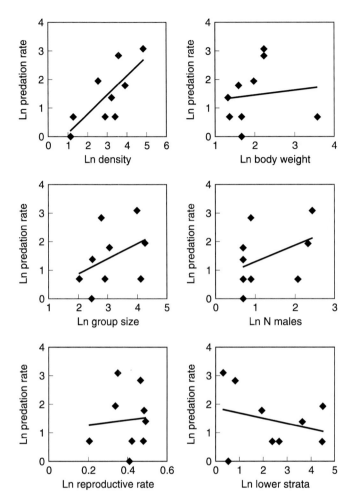

Figure 5.5. The relationship between leopard predation rates and various primate traits commonly interpreted as anti-predator adaptations. (Data from Zuberbühler & Jenny 2002.) Predation rate has been estimated by using the natural logarithm of the number of feces that contained remains of a particular species in a sample of 200 leopard feces collected over a period of two years from a 100 km² study area. *Density*: estimated number of individuals per square kilometer; *Body size*: adult female body weight in kg (from Oates *et al.* 1990); *Group size*: average number of individuals per group; *N males*: average number of adult males per group; *Reproductive rate*: average number of infants per adult female per year; *Habitat*: per cent time observed in lower forest strata. (Data from McGraw 1998, 2000, Eckardt 2002.)

to abundance. Contrary to predictions, leopard predation rates increased significantly with body size and was positively related to group size and the number of males per group, suggesting that predation by leopards did not drive the evolution of these traits in the predicted way.

In the next section we discuss these findings in light of some recent experimental data and suggest that the principal effect of leopard predation may have been on primates' cognitive evolution.

Interactions between leopards and monkeys

To investigate the primates' responses to the presence of leopards, one of us (DJ) collected data on the monkeys' responses to detection of a leopard's presence while following the radio-tagged leopard. Monkeys reacted strongly when detecting a leopard by giving a myriad of alarm calls and by approaching the predator in the lower canopy. Anti-predator behavior of this kind appeared to have striking effects on the leopard's hunting behavior: individuals typically gave up their hiding positions and moved on to find another group. To investigate this empirically, we performed two kinds of analyses using focal data collected from the adult female Adele (Zuberbühler *et al.* 1999). First, we tested whether detection had an effect on Adele's hiding behavior by comparing the duration of hiding before and after detection by the monkeys. Between August 1993 and June 1994, Adele was followed on 27 days (310 h) at a distance of 30 to 150 m. We scored the following variables based on changes in the strength and constancy of the received signal: encounter, detection, departure, leopard resting, leopard movement. An "encounter" between the leopard and a group of monkeys started when the leopard came to rest within about 50 m of a monkey group. The observer (DJ) then identified the monkey species present according to their vocalizations and remained concealed at a distance of 50–100 m from the leopard and monkeys. As long as a constant signal was received, the leopard was scored to be "resting," presumably hiding from the monkeys. A changing signal indicated "movement," which could be anything from body movements to changing the hiding spot. Movement that increased the distance from the monkey group beyond the 50 m radius was scored as "departure," which ended an encounter. The observer monitored the vocal behavior of the monkey group. As soon as the monkeys started to give loud and conspicuous vocalizations at high rates, it was assumed that the group had noticed the leopard and "detection" was scored.

We witnessed 24 different encounters between the leopard and a mono- or polyspecific monkey group (Zuberbühler *et al.* 1999). In 18 cases we were able to determine the encounter's exact duration (median: 61 min,

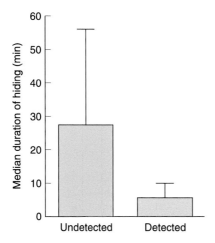

Figure 5.6. Median duration of hiding behavior in minutes of the focal animal before and after detection by a group of monkeys. (Data from Zuberbühler *et al.* 1999.)

range: 7−285 min). In all cases the monkeys detected the leopard at some point and subsequently vocalized at high rates. This affected the hunting behavior of the leopard. Detection by a group of monkeys appeared to terminate hunting behavior by the leopard because the time spent hiding underneath a monkey group was significantly shorter after detection than before (see Figure 5.6; Wilcoxon-test, one-tailed: $z = 2.112$, $n = 18$, $p < 0.02$). As mentioned before, one prolonged period of hiding led to a successful attack by Adele on a *C. atys*.

A qualitative analysis of the data suggested that the relationship between monkey alarm calls and the leopard's departure was causal because the leopard's stay after detection was short regardless of the time already spent hiding (see Figure 5.7). Adele not only gave up the hiding spot after detection but also was more likely to move on and leave the group.

Encounters with leopards generally led to extraordinarily high alarm call rates in all primate species. At first, this appears paradoxical because instead of remaining cryptic, individuals deliberately make their presence known to a highly dangerous predator. Conspicuous behavior in the presence of predators has been described in a number of species (e.g. skylarks, *Alauda arvensis*, Cresswell 1994). In these cases, it is typically argued that conspicuous behavior has evolved because predators rely on unaware prey for successful hunting; it is to the advantage of prey to signal detection and the futility of further hunting attempts

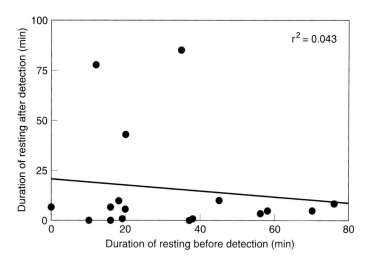

Figure 5.7. Relationship between the leopard's resting behavior in minutes close to a monkey group before and after detection (N = 18 encounters).

(the perception advertisement or detection-signalling hypothesis). Although the mechanism has been well described for some time, primate alarm calls have not been conceptualized this way. Instead, kin selection arguments have usually been put forward to explain why primates vocalize in the presence of predators (e.g. Cheney & Seyfarth 1981).

To investigate the alarm call behavior of Taï monkeys more systematically, predator presence was simulated by broadcasting typical vocalizations of the two major ground predators of Taï monkeys, leopards, and chimpanzees, from a concealed speaker (Zuberbühler *et al.* 1999). Various monkey groups were tested throughout the study area, but never more than once on each stimulus type. Once a group was located, usually by auditory cues, the speaker was hidden about 50 m away and a trial was conducted provided no monkey had detected the observer or part of the equipment and no predator alarm calls had occurred for at least 30 minutes. The focal group's vocal response was recorded on audiotape and it was determined whether the group had approached. All monkey species tested gave significantly higher rates of alarm calls to playbacks of leopard growls than to playbacks of chimpanzee pant hoots (see Figure 5.8). Groups occasionally approached the speaker after hearing playback stimuli, but only during playback of leopard growls and never after playback of chimpanzee pant hoots. The latter stimulus typically caused flight away from the speaker.

Among other things, our analyses show that monkey alarm calls affected Adele's hunting behavior because she tended to give up her hiding spot to

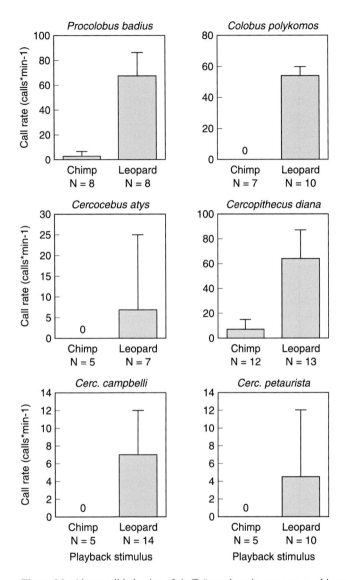

Figure 5.8. Alarm call behavior of six Taï monkeys in response to chimpanzee pant hoots and leopard growls. (Data from Zuberbühler *et al.* 1999.)

move on and leave the area (see Figure 5.6). Although no systematic data were collected, qualitative observations on the other three radio-tracked individuals revealed the same pattern, suggesting that Adele's behavior represented a general pattern of forest leopard hunting behavior.

In contrast, Taï chimpanzees are clearly not deterred by monkey vocalizations and they may even use them to locate a group (Boesch & Boesch 1989). Differences in predation pressure exerted by chimpanzees and leopards are unlikely to account for the differences in calling behavior since both predators prey upon the six species investigated in this study. In sum, data support the hypothesis that monkey alarm calls to leopards have a predator deterrence function because leopards, in contrast to chimpanzees, elicited conspicuously high alarm call rates, which drove the leopards away.

As discussed earlier, the fact that forest leopards can develop individual preferences may also explain why all monkey species engage in this cooperative form of acoustic anti-predator behavior. Successful predation on a conspecific is likely to increase preference formation, which will increase future predation pressure on the individual and its kin. Since leopards defend geographically stable home ranges (Jenny 1996, Dind *et al.* 1996), a particular monkey group will encounter the same few resident individuals repeatedly, perhaps even over several years, suggesting that preference formation can have fatal consequences for the affected individuals. Clearly, this hypothesis will require further and more rigorous testing. In particular, it needs to be demonstrated that individual leopards vary in their preference for different monkey species, and that this preference is the result of past hunting success.

The results presented in Figure 5.5 question the impact of leopards as a selection factor for the evolution of body size, group size, and other variables commonly thought to be anti-predator adaptations. Instead, we suggest that the main impact of leopard predation has been to enhance the primates' behavior flexibility to deal with this predator (Stoddard 1999). Support for this notion comes from at least four lines of evidence. First, several guenon species have evolved acoustically distinct alarm calls to warn each other about the presence of specific predators, including leopards (Seyfarth *et al.* 1980, Zuberbühler 2000a, 2001, Zuberbühler *et al.* 1997) and it is likely that similar findings will emerge from other species. Second, the suspected presence of a leopard appears to trigger complex cognitive processes (Zuberbühler 2000b, 2001, Zuberbühler *et al.* 1999). For example, Diana monkeys distinguish between chimpanzee screams given in a social setting and chimpanzee screams given to a leopard (Zuberbühler 2000d) suggesting that these calls are meaningful and inform the monkeys about the presence of a leopard. Diana monkey groups living near the periphery of a chimpanzee territory are less likely to understand variations in chimpanzee screams than groups living in the core area of a chimpanzee group, suggesting that meaning is individually acquired

(Zuberbühler 2000d). Third, meaning is not always rigidly attached to specific acoustic structures. Instead, it can be generated according to pragmatic information obtained from the environment (Zuberbühler 2000c). This is exemplified by the Diana monkeys' response to the alarm calls of crested Guinea fowl (*Guttera pulcheri*). Guinea fowl forage in large groups and when chased, produce conspicuously loud alarm calls that can be heard over long distances. Guinea fowl are not hunted by chimpanzees but may be taken by leopards and human poachers. Diana monkeys respond to recordings of Guinea fowl alarm calls as if a leopard were present. Playback experiments have shown that Diana monkeys are able to recognize that Guinea fowl alarm calls can be caused by both leopards and humans, and can also determine the most likely cause of the birds' alarm calls (Zuberbühler 2000c). Finally, recent research suggests that monkeys are able to alter the meaning of alarm calls by simple combinatorial rules in their vocal repertoire (Zuberbühler 2002). It is clear that the hypothesis that leopard predation has favored the cognitive evolution of primates, rather than body size or group composition, will require more rigorous testing using various empirical approaches. Nevertheless, we think that predation has generally been underestimated as a factor in primate cognitive evolution, a position that enjoys increasing support (e.g. Grimes 2002).

Conclusions

Although predation is believed to be an important driving force of natural selection its effects on primate evolution are still not well understood, mainly because little is known about the hunting behavior of the primates' various predators. A number of studies demonstrate that forest leopards primarily hunt for monkeys on the ground and during the day. The behavior of forest leopards differs in many aspects from that of individuals living in the savannah. Most strikingly, individuals are most active during the day and individuals show idiosyncratic prey preferences suggesting that at the individual level, leopards are better classified as selective hunters. Fecal analyses confirm that primates account for a large proportion of the leopards' diet and reveal in detail the predation pressure exerted on the eight different monkey species. Relating the species-specific predation rates to various morphological, behavioral, and demographic traits, usually considered adaptations to predation, reveal that leopard predation is most reliably associated with primate densities and body size although opposite the direction predicted. Body size, group size, and the number of males per group were all positively correlated to predation rates suggesting that predation by leopards does not drive the evolution of these traits in the manner predicted. At the same time, a variety of evidence shows that monkeys have evolved a number of behavioral strategies that appear to be based on complex cognitive capacities. In particular, all monkey species use their alarm call behavior to interfere with the leopards' hunting techniques. Signallers appear to gain fitness benefits in directly communicating to the predator by advertising perception and unprofitability. In these cases, leopards give up their hiding spot and leave the group significantly faster than expected by chance, suggesting that the high vocalization rates to leopards are part of an anti-predator strategy in primates that may have evolved to deter predators who depend on surprise. In light of the studies discussed, we conclude that leopard predation has acted as a major selection factor, enhancing individuals' ability to predict predator presence and behavior and to interfere with its hunting technique. Predation by leopards, it appears, has acted as a major natural selection factor in the evolution of primate intelligence.

References

Bailey, T. N. (1993). *The African Leopard*. New York, Columbia University Press.

Begon, M., Harper, J. L. *et al.* (1996). *Ecology*. Oxford, Blackwell.

Bodendorfer, T. (1994). Zur Ernährungsbiologie des Leoparden *Panthera pardus* und des Löwen *Panthera leo* im Comoé- und Marahoué Nationalpark, Elfenbeinküste – Eine Untersuchung anhand von Kotproben. M.Sc. Thesis, University of Vienna, Australia.

Boesch, C. (1991). The effects of leopard predation on grouping patterns in forest chimpanzees. *Behaviour*, **117**(3–4), 220–42.

Boesch, C. and Boesch, H. (1989). Hunting behavior of wild chimpanzees in the Taï National Park. *American Journal of Physical Anthropology*, **78**, 547–73.

Boesch, C. and Boesch-Achermann, H. (2000). *The Chimpanzees of the Taï Forest. Behavioural Ecology and Evolution*. Oxford, Oxford University Press.

Cheney, D. L. and Seyfarth, R. M. (1981). Selective forces affecting the predator alarm calls of vervet monkeys. *Behaviour*, **76**, 25–61.

Cheney, D. L. and Wrangham, R. W. (1987). Predation. In *Primate Societies*, ed. B. Smuts, D. L. Cheney, R. M. Seyfarth, R. W. Wrangham and T. T. Struhsaker, Chicago: University of Chicago Press, pp. 227–39.

Cords, M. (1990). Vigilance and mixed-species association of some East African forest monkeys. *Behavioral Ecology and Sociobiology*, **26**(4), 297–300.

Cresswell, W. (1994). Song as a pursuit-deterrent signal, and its occurrence relative to other anti-predation behaviors of skylark (*Alauda arvensis*) on attack by merlins (*Falco columbarius*). *Behavioral Ecology and Sociobiology*, **34**(3), 217–23.

Dind, F., Jenny, D. and Boesch, C. (1996). Ecologie et prédation du leopard en forêt de Taï (Côte d'Ivoire). Rapport d'activité 1992–4, ed. J. F. Graf, M. Zinsstag and J. Zinsstag. Abidjan, Centre Suisse de Recherches Scientifiques en Côte d'Ivoire, 22–6.

Dunbar, R. I. M. (1988). *Primate Social Systems*. New York: Comstock Publishing Associates.

Eckardt, W. (2002). Nischenvergleich der Grossen Weissnasenmeerkatze (*Cercopithecus nictitans*) und der Dianameerkatze (*Cercopithecus diana*) im Taï Nationalpark (Elfenbeinküste) under besonderer Berücksichtigung interspezifischer Konkurrenz. M.Sc. Thesis, University of Leipzig, Germany.

Gautier-Hion, A. and Tutin, C. E. G. (1988). Simultaneous attack by adult males of a polyspecific troop of monkeys against a crowned hawk eagle. *Folia Primatologica*, **51**, 149–51.

Grimes, K. (2002). Hunted. *Scientific American*, **2338**, 34–7.

Hamilton, P. H. (1981). The leopard *Panthera pardus* and cheetah *Acinonyx jubatus* in Kenya; ecology, status, conservation and management. The African Wildlife Leadership Foundation, and the Government of Kenya.

Hart, J. A., Katembo, M. and Punga, K. (1996). Diet, prey selection and ecological relations of leopard and golden cat in the Ituri forest, Zaire. *African Journal of Ecology*, **34**, 364–79.

Hill, R. A. and Dunbar, R. I. M. (1998). An evaluation of the roles of predation rate and predation risk as selective pressures on primate grouping patterns. *Behaviour*, **135**, 411–30.

Hoppe-Dominik, B. (1984). Etude du spectre des proies de la panthère, *Panthera pardus*, dans le Parc National de Taï en Côte d'Ivoire. *Mammalia*, **48**, 477–87.

Isbell, L. (1994). Predation on primates: ecological patterns and evolutionary consequences. *Evolutionary Anthropology*, **3**, 61–71.

Janson, C. H. (1998). Testing the predation hypothesis for vertebrate sociality: prospects and pitfalls. *Behaviour*, **135**(4), 389–410.

Jenny, D. (1996). Spatial organization of leopards (*Panthera pardus*) in Taï National Park, Ivory Coast: Is rain forest habitat a tropical haven? *Journal of Zoology*, **240**, 427–40.

Jenny, D. and Zuberbühler, K. (2005). Hunting behavior in West African forest leopards. *African Journal of Ecology*, **43**, 197–200.

Karanth, K. U. and Sunquist, M. E. (1995). Prey selection by tiger, leopard, and dhole in tropical forests. *Journal of Animal Ecology*, **64**, 439–50.

Karanth, K. U. and Sunquist, M. E. (2000). Behavioural correlates of predation by tiger (*Panthera tigris*), leopard (*Panthera pardus*) and dhole (*Cuon alpinus*) in Nagarahole, India. *Journal of Zoology*, **250**, 255–65.

Kitchener, A. (1991). *The Natural History of the World's Cats*. Ithaca: Cornell University.

McGraw, W. S. (1998). Comparative locomotion and habitat use of six monkeys in the Taï Forest, Ivory Coast. *American Journal of Physical Anthropology*, **105**, 493–510.

McGraw, W. S. (2000). Positional behavior of *Cercopithecus petaurista*. *International Journal of Primatology*, **21**, 157–82.

Oates, J. F., Whitesides, G. H., Davies, A. G. *et al.* (1990). Determinants of variation in tropical forest primate biomass – new evidence from West Africa. *Ecology*, **71**, 328–43.

Plavcan, J. M. and van Schaik, C. P. (1992). Intrasexual competition and canine dimorphism in anthropoid primates. *American Journal of Physical Anthropology*, **87**(4), 461–77.

Ray, J. C. and Sunquist, M. E. (2001). Trophic relations in a community of African rainforest carnivores. *Oecologia*, **127**(3), 395–408.

Seyfarth, R. M., Cheney, D. L. and Marier, P. (1980). Monkey responses to three different alarm calls: evidence of predator classification and semantic communication. *Science*, **210**, 801–3.

Shultz, S. (2001). Notes on interactions between monkeys and African crowned eagles in Taï National Park, Ivory Coast. *Folia Primatologica*, **72**, 248–50.

Stanford, C. B. (1998). Predation and male bonds in primate societies. *Behaviour*, **135**, 513–33.

Stoddard, P. K. (1999). Predation enhances complexity in the evolution of electric fish signals. *Nature*, **400**, 254–6.

Uphyrkina, O., Johnson, W. E., Quigley, H. *et al.* (2001). Phylogenetics, genome diversity and origin of modern leopard, *Panthera pardus*. *Molecular Ecology*, **10**, 2617–33.

Uster, D. and Zuberbühler, K. (2001). The functional significance of Diana monkey "clear" calls. *Behaviour*, **138**(6), 741–56.

van Schaik, C. P. (1983). Why are diurnal primates living in groups? *Behaviour*, **87**(1–2), 120–44.

Zuberbühler, K. (2000a). Causal cognition in a non-human primate: field playback experiments with Diana monkeys. *Cognition*, **76**, 195–207.

Zuberbühler, K. (2000b). Causal knowledge of predators' behaviour in wild Diana monkeys. *Animal Behaviour*, **59**(1), 209–20.

Zuberbühler, K. (2000c). Interspecific semantic communication in two forest monkeys. *Proceedings of the Royal Society of London, B*, **267**(1444), 713–18.

Zuberbühler, K. (2000d). Referential labelling in Diana monkeys. *Animal Behaviour*, **59**(5), 917–27.

Zuberbühler, K. (2001). Predator-specific alarm calls in Campbell's guenons. *Behavioral Ecology and Sociobiology*, **50**, 414–22.

Zuberbühler, K. (2002). A syntactic rule in forest monkey communication. *Animal Behaviour*, **63**(2), 293–9.

Zuberbühler, K. and Jenny, D. (2002). Leopard predation and primate evolution. *Journal of Human Evolution*, **43**, 873–86.

Zuberbühler, K., Jenny, D. and Bshary, R. (1999). The predator deterrence function of primate alarm calls. *Ethology*, **105**(6), 477–490.

Zuberbühler, K., Noë R. and Seyfarth, R. M. (1997). Diana monkey long-distance calls: messages for conspecifics and predators. *Animal Behaviour*, **53**, 589–604.

6 Interactions between red colobus monkeys and chimpanzees

R. Bshary

The question how red colobus monkeys try to avoid predation by chimpanzees was central from the start of the Taï monkey project. Boesch and Boesch (1989) had documented the hunting behavior of Taï chimpanzees and their prey spectrum. Their major results were that Taï chimpanzees: (1) hunt in a very cooperative coordinated way in which individuals play different roles, i.e. a "chaser" tries to chase monkeys out of their hiding trees into neighboring trees where "blockers" try to block the escape routes; (2) often decide to hunt before they have singled out a group of monkeys for attack and actively search for groups; (3) stalk monkey groups silently and only start screaming after being detected, probably to increase confusion; (4) mainly hunt red colobus and black and white colobus monkeys, while guenons and non-primate species are rarely caught; (5) hunt all age/sex classes of red colobus. Studying the anti-predation behavior of Taï red colobus thus offered a wonderful opportunity to study a predator-prey system where both predator and prey are primates that live in large permanent individualized social groups supposedly selecting for an enlargement of the neocortex (Dunbar 1992; note that data on red colobus neocortex size are lacking). We therefore hoped that the complexity of Taï chimpanzee hunting behavior reflected adaptations (genetic or learned) to complex red colobus counter adaptations (genetic or learned). This perspective was exciting because evidence that predator-prey systems can indeed evolve like an "arms race" (Dawkins & Krebs 1979) in which one adaptation by one side is matched by a counter adaptation and so on was scarce (Abrams 1986, Endler 1991). In particular, it was unclear whether this apparent scarcity reflected reality or whether it was due to the problem of showing its existence in systems where strategies are mostly one-dimensional (i.e. running as fast as possible). Futuyma (1986) had argued that predator-prey arms races might be rare because typically, several predator species with varying hunting

Monkeys of the Taï Forest, ed. W. Scott McGraw, Klaus Zuberbühler and Ronald Noë.
Published by Cambridge University Press. © Cambridge University Press 2007.

Box 6.1 Clarifying the terminology used in this chapter

The project's approach to red colobus-chimpanzee interactions was dealing with functional questions rather than asking how behavioral decisions are made. Studying mechanisms according to scientific laboratory standards under natural conditions is virtually impossible. We thus do not know the extent that the arms race between red colobus and chimpanzees reflects genetically determined or learned behavior. The underlying mechanisms, however, would be important for the correct terminology that should be used. I use the evolutionary terminology and write about strategies and coevolution, which implies that genetic components guide the behavior. I do not imply, however, that this is actually the case. The behavior of both chimpanzees and red colobus is likely to be shaped by learning, in which case one should use the term "tactics" rather than strategies.

---

strategies interact with several prey species with varying escape strategies. Under these circumstances, an adaptation to one opponent or strategy may decrease efficiency against another one. Taï red colobus have four main predators that partly differ in their hunting strategies. Chimpanzees and human poachers are pursuit hunters that may continue to hunt even after early detection. Leopards rely on surprise and attack from the forest floor, while eagles rely on surprise as well but attack through the canopy. Thus, if red colobus show specific strategies to each type of predator, Futuyma hypothesis can be disregarded.

Fortunately, we were not the only ones who started to study red colobus-chimpanzee interactions in more detail. Almost parallel to us but slightly ahead, Stanford and colleagues (Boesch 1994, Stanford *et al.* 1994a, 1994b, Stanford 1995) investigated similar questions at Gombe, Tanzania. Their results were strikingly different from our results for Taï (see below), allowing us to search for factors that may explain these differences (Bshary & Noë 1997b). At present, additional data on the interactions between chimpanzees and red colobus at Kibale/Uganda and Mahale/Tanzania have been studied in detail (Uehara 1997, Mitani & Watts 1999, 2001, Chapman & Chapman 2000). Here I incorporate this new information as a test of how well our original conclusions still stand up to the evidence. I first describe our major experimental results on the anti-predation behavior of Taï red colobus with respect to chimpanzees.

We tested the following predictions.

1. As chimpanzees actively search for monkey groups, red colobus should counter this strategy by becoming silent when chimpanzees are heard in the vicinity and move away from them, thereby reducing the probability of being detected.

2. The coordinated hunts of chimpanzees seem to be an adaptation to a prey that is superior in moving through the highest parts of the canopy where thin branches do not support the weight of chimpanzees. Therefore, we expected that red colobus move up into the highest parts during close encounters or hunts. In addition, each individual should hide so that the chimpanzee chasers might select another tree for their hunt.

3. The interspecific associations with Diana monkeys might be part of the red colobus anti-predation strategy with respect to chimpanzees. As Diana monkeys rarely fall victim to chimpanzees, the major advantage of being with them might be to receive an early warning.

Box 6.2 Methods

We conducted playback experiments that simulated the presence of chimpanzees and other predators, combined with presentations of control stimuli. The experiment that tested the behavior of red colobus when chimpanzees are almost underneath the group was conducted with our first red colobus study group. For all other playback experiments, nine different groups were chosen, and each group was tested only once with each stimulus presented. Most groups were unhabituated to the presence of humans and had to be followed out of sight. The playback design matched recommendations of the NATO life sciences symposium as closely as possible (McGregor 1992) to avoid variance caused by the experimental design (which would have worked in favor of the null-hypotheses). To find out how well different monkey species detect predators that stalk along the forest floor, I approached 48 different monkey groups of variable species composition and noted which species detected me first, the distance of the alarming individual at the moment of detection and the distances of the nearest individuals of all species present.

Avoidance behavior of red colobus monkeys during early stages of an encounter

Chimpanzees are usually noisy primates. Their pant hoots and drumming on tree buttresses can be heard over several hundred meters. Therefore, monkey groups are often informed about their presence. It is important to note that chimpanzees are rarely within the home range of each single monkey group. This is because the home range of chimpanzees comprises about $27 \, km^2$ (Boesch & Boesch 1989), while there are about 2.4 red colobus groups per km^2 (Noë & Bshary 1997). Hearing chimpanzees thus tells the monkeys that the chance of being hunted by chimpanzees is much higher than on average. At such an early stage of an encounter where the monkeys are probably not even detected by the chimpanzees, the red colobus can reduce the chance of being detected by becoming silent. In addition, they can move away from the chimpanzees, thereby reducing the probability that the chimpanzees will move close by chance. To test this idea, we conducted playback experiments in which we used pant hoots of several chimpanzee males to simulate the presence of a potential hunting party within the home range of nine different red colobus monkey groups. Prior to the playbacks, we used the vocalizations of the monkeys and branch movements to locate the group's center of activity, i.e. where the highest concentration of individuals was. The speaker was placed approximately $100 \, m$ from the nearest individual(s) and the playback followed. Eight out of nine groups responded with movements, and the resulting vector of these movements was almost exactly in the opposite direction to the speaker (177 degree). Usually, group movements in red colobus are accompanied by an increase in vocalization rates. In their response to the playbacks, however, red colobus were almost completely silent during the 5 min interval following the playbacks, except for some initial calls. Similar data were obtained during natural encounters of our first two study groups and chimpanzees (Bshary & Noë 1997a). Thus, red colobus indeed reduce the probability of being detected either by vocalizations or by a chance approach when chimpanzees are nearby by falling silent and by moving away from the chimpanzees. This behavior of red colobus is specific to pursuit predators that approach over the forest floor, i.e. chimpanzees and humans. In contrast, playbacks of leopard growls lead to an increase in calling frequencies and groups do not move away from this stimulus (Bshary & Noë 1997a). Similar responses are elicited by eagle shrieks (personal observations).

Hiding behavior of red colobus when chimpanzees are (almost) underneath the group

In response to playbacks of a group of male chimpanzee pant hoots, individuals of our first study group moved up into the high canopy or emergent trees, fell silent and the exposure to the forest floor, when corrected for height, was lower than before playbacks. It became much more difficult for the observer to actually spot individuals after the chimpanzee playbacks (personal observation), indicating that the exposure of the monkeys to the forest floor was even more reduced than the data suggest. Individuals of all age/sex classes behaved in a similar way. This was expected as Taï chimpanzees hunt all age/sex classes (Boesch & Boesch 1989), so red colobus males that try to defend other group members would be at high risk of predation themselves. We obtained similar data during natural close encounters between chimpanzees and our first two study groups (Bshary & Noë 1997a). Again, this behavior of red colobus is specific to pursuit predators that approach over the forest floor, i.e. chimpanzees and humans. In contrast, placing a cloth with the color pattern of a leopard underneath the group or placing an artificial crowned eagle (*Stephanoaetus coronatus*) in a tree does not elicit movements away from the stimuli. On the contrary; on four occasions, groups of 4−5 red colobus males attacked the eagle model and continued to bite it after it fell to the forest floor (Bshary & Noë 1997a). Wrapping the leopard cloth around the body was an ideal way to film unhabituated red colobus monkeys because the monkeys were looking and shouting at the cloth (personal observation).

Interspecific associations with Diana monkeys

Observational evidence strongly suggested that red colobus seek the presence of Diana monkeys to reduce predation pressure from chimpanzees (Bshary & Noë 1997b, Noë & Bshary 1997). Most importantly, association rates peaked during September through November, the chimpanzee-hunting season (Boesch & Boesch 1989), due to the initiative of red colobus. Even more pronounced was the observation that association rates were particularly low between June and August, the time of the year when members of chimpanzee communities are dispersed. If chimpanzees are dispersed, that means that there is no group of males together that could start a cooperative hunt. Predation risk from chimpanzees should therefore be particularly low during this season. As the associations with Diana monkeys most likely reflect a balance between the benefits of reduced predation risk and the costs of deviation from an optimal foraging pattern, one would predict low association rates when predation risk is low. The only predator in Taï, which clearly has a seasonal pattern in hunting activity, is the chimpanzee. It could therefore be considered unlikely that the low association rates in June through August were caused by low predation risk from other predators. However, an alternative explanation was that the same environmental constraints that promote dispersed feeding of chimpanzees also constrain high association rates of red colobus and Diana monkeys.

The hypothesis that the low association rates in June to August were due to low predation pressure by chimpanzees was tested with playback experiments. Hearing pant hoots of several chimpanzee males would indicate to the monkeys that the momentary risk of being (successfully) attacked by chimpanzees is very high relative to the average risk in this season. Note that the playbacks merely simulated the presence of a nearby group of male chimpanzees (100–200 m from the group) rather than an actual hunting attempt. However, given that our playbacks successfully fooled the monkeys into believing that chimpanzee males are nearby, the monkeys should have some knowledge that if the chimpanzees decided to hunt at some point, it is likely that they would target the nearest red colobus group, i.e. them. If the associations with Diana monkeys indeed serve as a flexible strategy to reduce mortality due to chimpanzee predation, we had two predictions concerning the association pattern. (1) We predicted that associated groups should stay together for longer throughout the day after hearing a playback of chimpanzee calls in the morning than after other playback stimuli. (2) We predicted that red colobus-Diana monkey groups that were about to split up should reunite immediately in response to playbacks of chimpanzee calls but not in response to other playback stimuli.

The results were clearly in line with the hypothesis that the associations with Diana monkeys serve as a flexible strategy to reduce predation from chimpanzees (Noë & Bshary 1997). Six out of nine groups that were associated in the morning stayed together with their Diana monkey group until the end of observation at 17:00 following a short playback of male chimpanzee calls in the morning. The three groups that eventually split up did so only late in the afternoon. Red colobus-Diana monkey associations thus lasted significantly longer after playbacks of chimpanzee calls than after control playbacks of a generator engine or empty tape. Playbacks of leopard growls yielded intermediate results in that associations did not last significantly shorter than after chimp calls and not significantly longer than after control playbacks. The results for our second prediction were even more clear-cut. All nine red colobus groups that had a Diana monkey group nearby on the verge of moving away were intermingled with them again shortly after the playbacks of male chimpanzee calls. In most cases, the red colobus had made the more significant move towards the Diana monkeys though Diana monkeys contributed on two occasions and were fully responsible for the intermingling on one occasion. Intermingling after playbacks of chimpanzee calls was more likely than intermingling after both control stimuli (generator and empty tape) and intermingling after playbacks of leopard growls (Noë & Bshary 1997). In particular the response after leopard growls was strikingly different from the response following playbacks of chimpanzee calls as five out of nine groups split up further. In conclusion, the association pattern of red colobus and Diana monkeys was flexibly adjusted to perceived momentary predation pressure from chimpanzees. This adjustment is not a general strategy against perceived presence of predators, as leopard growls did not elicit an increase in association duration or rate compared to control stimuli. It is important to note that the absence of an influence of leopard growls on association patterns does not show that associations do not serve to reduce predation pressure from leopards. A major difference between chimpanzees and leopards is that chimpanzees are noisy unless they start searching for prey, while leopards are usually silent. Therefore, hearing or not hearing chimpanzees provides monkeys with some information about momentary predation risk with respect to these predators. Not hearing a leopard, however, does not offer any information about the whereabouts of these predators. Thus, even if associations serve to reduce predation pressure from leopards, monkeys cannot adjust their association rates to momentary predation risk with respect to leopards but have to base their decision to stay together or to split up on an average risk value. The only way to test this idea is to search for a site where leopards are the only predators of monkeys and see whether or not association rates are still above chance levels or not.

What is the advantage of being with Diana monkeys?

Until now, we have only shown that red colobus adjust the duration and frequency of associations with Diana monkeys according to momentary predation risk from chimpanzees; we have yet to show why this should be advantageous. Diana monkeys rarely fall victim to chimpanzees (Boesch & Boesch 1989), making dilution advantages for red colobus very unlikely. On the contrary, a major advantage for Diana monkeys with respect to predation by chimpanzees could be to be associated with a prey species that is preferred by the predators. For example, Grant's gazelles are often associated with Thompson's gazelles for apparently this reason: cheetahs mainly hunt the latter when attacking a mixed species association (FitzGibbon 1990). Thompson's gazelles benefit from the presence of Grant's gazelles because the latter are more vigilant and therefore more likely to spot a stalking cheetah in the first place (FitzGibbon 1990). We assumed that, as with the gazelles' system, Diana monkeys might provide red colobus with an early warning when chimpanzees stalk an associated group and try to surprise them. Diana monkeys use the lower strata more frequently than red colobus (Bshary & Noë 1997b), they move around more during foraging, searching for insects (McGraw 1996) and are more often at outer branches than red colobus (McGraw 1996). All these factors should facilitate detection of predators approaching over the forest floor, as chimpanzees, humans, and leopards do. To test this hypothesis, I approached 48 different monkey groups that consisted of variable species composition. I had a leopard cloth wrapped around my body because all monkey species (with the possible exception of olive colobus) react with clear alarm vocalizations to this stimulus while their response following the detection of a human is to fall silent and flee, making it difficult to assess which individual actually detected me first. I noted which species were present, to which species the individual that alarmed first belonged, and my distance to the nearest individual of each species. In particular, I was interested whether Diana monkeys would alarm before red colobus monkeys in situations where red colobus were closer to me.

As predicted, Diana monkeys were significantly better than red colobus in spotting me first. On 16 out of 19 occasions, a Diana monkey called first. On 10 of these occasions, a red colobus had been as close or even closer to me. On those three occasions where a red colobus alarmed first, individual Diana monkeys were still quite far away from me. Diana monkeys detected me from significantly larger distances than red colobus. This difference persisted when I controlled for the stratum where the alarming individual was. Diana monkeys alarmed first more often than any other monkey species with a similar home range (olive colobus, black and white colobus, lesser spot nosed monkey, Campbell's monkey). Perhaps due to small sample sizes, these differences were not significant, however. Still, the results provide a clue as to why Diana monkeys are the main partner species of all other monkey species with similar home range sizes in the Taï forest. Note that sooty mangabeys are even better watchmen for ground predators (McGraw & Bshary 2002) however, they are not reliable partner groups as their home ranges are much larger than those of the other monkey species. Therefore, mangabeys visit each group of other species irregularly at low frequencies (5–10 per cent of daylight hours;

McGraw & Bshary 2002). In conclusion, in the absence of mangabeys, red colobus benefit from being with Diana monkeys because the latter often provide an early warning for ground predators. Note that Diana monkeys do not give alarm calls for the red colobus but for their own group members. However, there is increasing evidence that different monkey species recognize each other's alarm calls (Zuberbühler 2000). The warning of red colobus by Diana monkeys is thus a typical case of a by-product mutualism (Brown 1983) in which help is given without any costs involved. If chimpanzees are heard in the vicinity, red colobus increase association rates with Diana monkeys because of the risk that if the chimpanzees decide to hunt, they are likely to stalk a red colobus group that is nearby. Being with the Diana monkeys increases the probability of receiving a timely warning, allowing the red colobus to move into the high canopy and hide.

Measuring fitness consequences of associations with Diana monkeys with respect to predation from chimpanzees

The results from the playback experiments and from the approach experiment make a convincing case that red colobus associate with Diana monkeys to reduce predation pressure from chimpanzees. However, the data do not prove that red colobus are indeed at lower predation risk when associated than when on their own. Showing such fitness advantages for red colobus is not just important to complete the evidence but also because there is an alternative hypothesis. Boesch (1994) had argued that chimpanzees use calls of Diana monkeys to detect interspecific monkey groups, which they approach to hunt red colobus monkeys. He therefore proposed that associations with Diana monkeys are *disadvantageous* for red colobus with respect to predation by chimpanzees. According to Boesch's hypothesis, one would predict that red colobus should avoid Diana monkeys when chimpanzees are nearby, the opposite of what we observed.

We re-examined Boesch's original data (Bshary & Noë 1997a). While following chimpanzees, he had noted what monkey species he had heard (first if there was a mixed species group) whether or not the chimpanzees had gone to that group and which species they had hunted (Boesch 1994). It was true that *if* chimpanzees approached a group where he had heard Diana monkeys first, the chimpanzees most often hunted red colobus that were associated. However, chimpanzees rarely approached groups when he had heard Diana monkeys first (note that no information is given on how often groups of other species were present) i.e. in only 13 out of 236 occasions. In contrast, chimpanzees approached groups after Boesch had heard red colobus calls first on 41 out of 143 occasions. In conclusion, these data clearly show that chimpanzees avoid approaching groups when they hear Diana monkeys (Bshary & Noë 1997a). The results become even more impressive if one considers that Diana monkeys are rarely not in association with either red colobus or black and white colobus. Our project data revealed that our first Diana monkey study group spent almost 90 per cent of daylight hours intermingled with groups of either colobus species. Diana monkeys are thus a very good indicator of the presence of large colobus groups and, according to Boesch (1994), chimpanzees even prefer to hunt black and white colobus to red colobus. Black and white colobus are much more difficult to detect, however, due to their cryptic life style (Bshary & Noë 1997a). Thus, there is even greater reason for chimpanzees to follow Diana monkey calls to detect their favorite prey, however they rarely did so.

In their book on Taï chimpanzees, Boesch and Boesch (2000) provided additional data collected during a different time period than those reported in the Boesch (1994) paper. The new data include data on whether red colobus groups were in association (or on their own) and appear to contradict the previous ones. Red colobus in association were approached and hunted more frequently than were red colobus on their own (in association: approached 64 times, hunted 33 times out of 106 encounters; on their own: approached 18 times, hunted 14 times out of 143 encounters). While these new data include more information than the previous data and suggest that associations with Diana monkeys might indeed be costly for red colobus with respect to predation from chimpanzees, there are several potentially confounding variables. First, the new data set is conservative with respect to our hypothesis that associations are advantageous because on some occasions, chimpanzees may have refrained from approaching red colobus groups that were scored being on their own but actually were associated with Diana monkeys. One can only positively say that Diana monkeys are present; their absence is more difficult to confirm. This is particularly true if one follows chimpanzees since Diana monkeys become silent in their presence (Boesch 1994, Zuberbühler et al. 1997). Second, no association criterion is mentioned. Thus, there might have been several data points where Diana monkeys were in audible distance to a red colobus group, without being intermingled. According to data on the vigilance behavior of red colobus, only intermingled is the appro-priate criterion for an association (Bshary 1995). Third, the data were collected a) during the chimpanzee-hunting season when red colobus spend about 65 per cent of daylight hours in association and, b) in May to July — that part of the year where chimpanzees rarely hunt and red colobus are much less associated (Noë & Bshary 1997). It is thus conceivable that the many data points where encounters between chimpanzees and monospecific red colobus groups that did not lead to approach and hunt may have been collected in the non-hunting season. Note that data in Boesch (1994) were collected not only during the hunting season, but also during the following dry season. However, the addition of data collected during the dry season mainly increases the variance as more encounters do not lead to approach and hunt but do not confound predictions as association rates of red colobus are very similar during the data collection period (Noë & Bshary 1997). Ideally, the data should be analyzed separately for the various seasons or at least corrected for variation in both chimpanzee hunting frequencies and red colobus association rates.

As it stands, a way to analyze the new data without having to worry about potential confounding variables is to investigate whether there is a difference in how frequently a chimpanzee approach results in an actual hunt when red colobus are associated compared with when they are on their own. The distinction between approach and hunt can be made because Boesch (1994) scored a hunt only if at least one chimpanzee was as high in the trees as the red colobus. When red colobus were in association, 64 approaches resulted in 33 hunts. In comparison, when red colobus were on their own, 18 approaches resulted in 14 hunts (Boesch & Boesch 2000). The probability that an approach resulted in a hunt was thus significantly higher when red colobus were on their own than when associated (G-Test, $G = 4.1$, $N = 82$, $df = 1$, $p = 0.044$). This result supports our original hypothesis and our observational/experimental data suggesting that being with Diana monkeys yields fitness advantages for red colobus with respect to predation from chimpanzees due to improved early detection of chimpanzees.

Comparison of red colobus-chimpanzee interactions at Taï, Gombe, Kibale, & Mahale

The arms race between red colobus and chimpanzees at Taï appears quite different from the arms race between red colobus and chimpanzees at Gombe (Table 6.1, see Bshary & Noë 1997a). Both predator hunting strategies and prey escape strategies vary considerably between the two sites. We argued that two factors account for these differences (Bshary & Noë 1997a). First, the relative body weights of predator and prey are much more in favor of chimpanzees at Taï (6:1) than at Gombe (about 4:1) (Struhsaker 1975, Morbeck 1989, Oates *et al.* 1990). These differences in relative weight between sites may explain why Taï chimpanzees dare to

Table 6.1. *Qualitative description of features of red colobus-chimpanzee arms races at four different study sites*

	Taï	Kibale	Gombe	Mahale
Chimp body size	large	small	small	small
Colobus body size	small	large	large	large
Canopy structure	high closed	high closed	low broken	low broken
Chimps: search	yes	yes	no	no
silent approach	yes	yes	no	no
collaborate	yes	no(?)	no	no
share meat	yes	yes	no	no
kill colobus males	yes	no	no	no
avoid associations	yes	?	no	no
Colobus: hide	yes	yes	no	no
male defense	no	yes	yes	yes
associations	yes	yes	no	no

hunt red colobus males while Gombe red colobus males dare to defend other group members against chimpanzees. A second major difference between Taï and Gombe is the structure of the canopy. At Taï, the canopy is high and relatively closed while it is low and open at Gombe (Wrangham 1975, Whitmore 1990). The forest structure at Taï allows red colobus to behave cryptically and to use escape routes during hunts that are inaccessible for the heavier chimpanzees. In return, the coordination between chimpanzee males during hunts overcomes the problem of how to follow the mobile red colobus through the canopy. In addition, Taï chimpanzees can silently stalk their prey, trying to surprise it. As a final counter adaptation, red colobus associate with Diana monkeys to improve early detection of stalking chimpanzees. At Gombe, the low broken canopy allows single chimpanzees to hunt successfully if they manage to get past the red colobus males. Silent escapes or surprise attacks are not possible, thereby considerably simplifying the arms race between red colobus and chimpanzees at that site.

The arms races between red colobus and chimpanzees at Kibale forest and the Mahale mountains provide a good test for the contrast we made based on the Taï and Gombe data. The forest structure of Kibale is very similar to the one in Taï with a high closed canopy (Mitani & Watts 1999). In contrast, the Kibale red colobus are relatively heavy compared to the Kibale chimpanzees, a feature that closely resembles the situation at Gombe. Mahale resembles Gombe very closely in that the canopy structure is broken and red colobus are relatively heavy compared to the chimpanzees. Thus, the following predictions can be made about the predator-prey arms race at Kibale.

1. Chimpanzees search for red colobus groups and approach them silently for a surprise hunt.
2. Chimpanzees coordinate their hunts and play different roles during a hunt. As a consequence, they share meat with co-hunters.
3. Chimpanzees hunt mainly immature and juvenile red colobus.
4. Red colobus males defend the group against chimpanzees.
5. Red colobus frequently associate with other species to improve early warning, allowing the males to form a defense line and other group members to move up to the high closed canopy to have many possible escape routes.
6. As a consequence of improved early warning in mixed species associations, chimpanzees preferentially hunt monospecific red colobus groups. The predator-prey arms race at Mahale should closely resemble the one at Gombe.

Comparisons of red colobus-chimpanzee interactions at the four study sites by Mitani and Watts (1999) support our original scenario quite well (see Table 6.1). Hunting behavior of chimpanzees and red colobus defense behavior are very similar at Gombe and Mahale. At Kibale, most existing data fit our predictions: red colobus frequently associate with other monkey species, mainly redtail monkeys *Cercopithecus ascanius*, and association rates correlate positively with chimpanzee densities (Chapman & Chapman 2000). It is still not clear how these associations improve early warning or whether chimpanzees preferentially hunt monospecific red colobus groups. Two factors about chimpanzee behavior do not fit our predictions. First, chimpanzees do not seem to obviously coordinate their hunts, but they nevertheless often share meat (Mitani & Watts 1999, 2001). This interpretation is based on observations by Mitani and Watts (1999) who state that assessing the degree of cooperation/ coordination was often difficult. These authors do mention that chimpanzees "sometimes collaborated" by encircling red colobus groups, blocking potential escape routes, or "driving" prey down hill slopes from taller to shorter trees (Mitani & Watts 1999, p. 446). Thus, Kibale chimpanzees could be seen as more cooperative than the authors conclude. An additional factor that may influence the degree of collaboration during a hunt is demography. The chimpanzee community studied in Kibale is extremely large and included 26 males in 1998 (Mitani & Watts 1999). Hunting party sizes included on average 13.3 males and are thus much larger than at the other study sites. Mitani and Watts (1999) argue that because of the large hunting party size, chimpanzees are extremely successful without elaborate coordination. Meat sharing may function for male-male bonding that is used in social interactions (Mitani & Watts 2001). In conclusion, our original scenario (Bshary & Noë 1997a) should be broadened to include a demographic factor, namely chimpanzee group size.

Conclusions

1. The anti-predation behavior of Taï red colobus corresponds in several features to the predation behavior of Taï chimpanzees. Thus, the two species appear to be involved in an arms race with several adaptations and counter-adaptations. The role of learning in this arms race is still unexplored.

2. A major component of the red colobus anti-predation strategy is to associate with Diana monkeys. Diana monkeys do not provide dilution advantages but rather an early warning against silently approaching chimpanzees.

3. As a consequence of the watchmen abilities of Diana monkeys, chimpanzees appear to avoid associated red colobus groups and/ or are less likely to start a hunt after approach.

4. Recent data from Kibale and Mahale generally fit the idea that forest structure and relative body sizes between chimpanzees and red colobus may account for most of the observed differences in the arms races between study sites. In addition, demographic effects have to be considered.

References

Abrams, P. A. (1986). Is predator-prey coevolution an arms race? *Trends in Ecology and Evolution*, **1**, 108–10.

Boesch, C. (1994). Chimpanzees-red colobus monkeys: a predator-prey system. *Animal Behaviour*, **47**, 1135–48.

Boesch, C, and Boesch, H. (1989). Hunting behaviour of wild chimpanzees in the Taï National Park. *American Journal of Physical Anthropology*, **78**, 547–73.

Boesch, C. and Boesch, H. (2000). *The Chimpanzees of the Taï Forest: Behavioural Ecology and Evolution*. Oxford: Oxford University Press.

Brown, J. L. (1983). Cooperation-a biologist's dilemma. In *Advances in the Study of Behaviour*, ed. J. S. Rosenblatt, New York: Academic Press, pp. 1–37.

Bshary, R. (1995). Rote Stummelaffen (*Colobus badius*) und Dianameerkatzen (*Cercopithecus diana*) im Taï National park, Elfenbeinküste: warum assoziieren sie? Ph.D. thesis, University of Munich, Germany.

Bshary, R. and Noë, R. (1997a). Anti-predation behaviour of red colobus monkeys in the presence of chimpanzees. *Behavioral Ecology and Sociobiology*, **41**, 321–33.

Bshary, R. and Noë, R. (1997b). Red colobus and Diana monkeys provide mutual protection against predators. *Animal Behaviour*, **54**, 1461–74.

Chapman, C. A. and Chapman, L. J. (2000). Interdemic variation in mixed species association patterns: common diurnal primates of Kibale National Park, Uganda. *Behavioral Ecology and Sociobiology*, **47**, 129–39.

Dawkins, R. and Krebs, J. B. (1979). Arms races between and within species. *Proceedings of the Royal Society of London, B*, **205**, 489–511.

Dunbar, R. I. M. (1992). Neocortex size as a constraint on group size in primates. *Journal of Human Evolution*, **22**, 469–93.

Endler, J. A. (1991). Interactions between predator and prey. In *Behavioural Ecology*, An Evolutionary Approach, ed. J.R. Krebs and N.B. Davies, 3rd ed. Oxford: Blackwell Scientific Publications, pp. 169–96.

FitzGibbon, C. D. (1990). Mixed-species grouping in Thomson's and Grant's gazelles: the antipredator benefits. *Animal Behaviour*, **39**, 1116–26.

Futuyma, D. J. (1986). *Evolutionary Biology*. Sunderland, Massachusetts: Sinauer Associates.

McGraw, W. S. (1996). Cercopithecid locomotion, support use and support availablity in the Taï Forest, Ivory Coast. *American Journal of Physical Anthropology*, **100**, 507–22.

McGraw, W. S. and Bshary, R. (2002). Association of terrestrial mangabeys (*Cercocebus atys*) with arboreal monkeys: experimental evidence for the effects of reduced ground predator pressure on habitat use. *International Journal of Primatology*, **23**, 311–25.

McGregor, P. K. ed. (1992). Playback and studies of animal communication. *NATO ASI Series A, Life Sciences* Vol. **228**. New York: Plenum Press.

Mitani, J. C. and Watts, D. P. (1999). Demographic influences on the hunting behavior of chimpanzees. *American Journal of Physical Anthropology*, **109**, 439–51.

Mitani, J. C. and Watts, D. P. (2001). Why do chimpanzees hunt and share meat? *Animal Behaviour*, **61**, 915–24.

Morbeck, M. E. (1989). Body size and proportions in chimpanzees, with special reference to *Pan troglodytes schweinfurthii* from Gombe National Park, Tanzania. *Primates*, **30**, 369–82.

Noë, R. and Bshary, R. (1997). The formation of red colobus-Diana monkey associations under predation pressure from chimpanzees. *Proceedings of the Royal Society of London, B*, **264**, 253–9.

Oates, J. F., Whitesides, G. H., Davies, A. G. *et al.* (1990). Determinants of variation in tropical forest primate biomass: new evidence from West Africa. *Ecology*, **71**, 328–43.

Skorupa, J. P. (1989). Crowned eagles *Stephanoaetus coronatus* in rainforest: observations on breeding chronology and diet at a nest in Uganda. *Ibis*, **131**, 294–8.

Stanford, C. B. (1995). The influence of chimpanzee predation on group size and anti-predator behaviour in red colobus monkeys. *Animal Behaviour*, **49**, 577–87.

Stanford, C. B., Wallis, J., Mpongo, E. and Goodall, J. (1994a). Hunting decisions in wild chimpanzees. *Behaviour*, **131**, 1–18.

Stanford, C. B., Wallis, J., Matama, H. and Goodall, J. (1994b). Patterns of predation by chimpanzees on red colobus monkeys in Gombe National Park, 1982–1991. *American Journal of Physical Anthropology*, **94**, 213–28.

Struhsaker, T. T. (1975). *The Red Colobus Monkey.* Chicago: The University of Chicago Press.

Uehara, S. (1997). Predation on mammals by the chimpanzee (*Pan troglodytes*). *Primates*, **38**, 193–214.

Whitmore, T. C. (1990). *An Introduction to Tropical Forests.* Oxford: Clarendon Press.

Wrangham, R. W. (1975). The behavioural ecology of chimpanzees in Gombe National Park, Tanzania. Ph.D. thesis, Cambridge University, Cambridge, UK.

Zuberbühler, K. (2000). Interspecies semantic communication in two forest primates. *Proceedings of the Royal Society of London, B*, **267**, 713–18.

Zuberbühler, K., Noë, R. and Seyfarth, R. M. (1997). Diana monkey long-distance calls: messages for conspecifics and predators. *Animal Behaviour*, **53**, 589–604.

7 Interactions between African crowned eagles and their prey community

S. Shultz and S. Thomsett

Introduction

One explanation for the evolution of sociality among vertebrates as an adaptive response to predation pressure. Individuals in groups are able to invest less in vigilance behaviors, are more likely to detect a predator, less likely to be the victim of an attack, and can more effectively mount a defense against a predator. Although there is evidence that individual vigilance levels decline with increasing group sizes in birds and primates (Lima 1987, Cowlishaw 1994), documenting benefits of sociality in terms of predation rates has been difficult in either field or laboratory studies. Group size has been shown to reduce predator capture success in only a few controlled laboratory experiments (e.g. Krause & Godin 1994) or opportunistic field observations (e.g. Lindstrom 1989). Thus despite numerous models predicting how sociality and group size impacts predation risk, there remains little evidence from natural systems to support the theoretical predictions. Through comparing predator diet composition with prey behavioral characteristics we can assess how anti-predator behaviors influence predation rates by different predators.

The different primate species in Taï National Park exhibit a variety of social systems, making the community in Taï an ideal situation to test theories about the relationship between group size, composition, and predation risk. Crowned eagles (*Stephanoaetus coronatus*), leopards (*Panthera pardus*), chimpanzees (*Pan troglodytes*), and humans (*Homo sapiens*) are the four main predators of primates in Taï. The hunting strategy of each of these predators varies and each predator should show biases towards different prey depending on the habitat use and anti-predator strategies employed by the prey species. The rationale driving the study of the crowned eagles was to understand the role of predation

Monkeys of the Taï Forest, ed. W. Scott McGraw, Klaus Zuberbühler and Ronald Noë.
Published by Cambridge University Press. © Cambridge University Press 2007.

by raptors in shaping the behavioral adaptations and polyspecific associations that are observed in the monkey species in Taï. The key advantage of Taï as a site to study crowned eagles is that not only have the ecology and behavior of primate prey been studied, but also that of their other main predators. Thus the primate community in Taï National Park represents the first primary forest site where there is a comprehensive picture of predation pressure from all main predators.

This chapter has several goals: (1) to describe crowned eagle natural history in Taï and relate diet preferences to prey behavior and ecology, (2) compare diet selection of the main predators, (3) outline the behavioral interactions between crowned eagles and the monkeys in Taï, and (4) explore the behavioral responses of monkeys to predation risk.

Crowned eagle natural history

African crowned eagles are found from Sub-Saharan equatorial forests south to South African subtropical and temperate habitats (Brown *et al.* 1982). Their genus, *Stephanoaetus*, is monotypic and closely related to the pan-tropical hawk eagles of the *Spizaetus* genus. Crowned eagles have been extensively studied in savanna and woodlands in East and South Africa (Brown 1971, Daneel 1979, Vernon 1984, Boshoff *et al.* 1994). However, in their primary lowland forest habitat, the few previous studies in primary forest have involved only bone collection from under nest sites (Skorupa 1989, Struhsaker & Leakey 1990, Mitani *et al.* 2001) or opportunistic observations of hunting incidents (Leland & Struhsaker 1993, Maisels *et al.* 1993). This study was the first time adjacent eagle nests have been concurrently monitored, providing a population level picture of the impact of eagle predation on the monkey community.

African crowned eagles are resident year-round and actively defend territories around their nest sites. In South Africa, some populations of African crowned eagles regularly breed annually (Steyn 1982), but throughout most of their range they breed in biennial cycles. Immature crowned hawk-eagles exhibit very prolonged dependency periods; young can remain dependent for up to 300 days after fledging (Brown *et al.* 1982). With such a long dependency period, an entire reproductive cycle from nest repair to independence lasts 20−22 months (Brown *et al.* 1982), making annual breeding impossible. In populations where eagles do breed annually, the dependency period appears to be shorter. Crowned eagles are non-migratory and new breeding cycles commence after previous young achieve independence or after brood failure; single nest sites have been known to be active continuously for several generations over a period of more than 50 years (Brown *et al.* 1982). During incubation and while there are dependent young at the nest, adults visit the nest sites regularly to bring prey items to attendant adults or dependent young.

Crowned eagles in Taï

The crowned eagle study in Taï began in October 1998 and ran through August 2000. During this time nests were located by several different means: primate researchers and technicians were asked to report nest specific vocalizations encountered while following primate groups, forest blocks without known nest sites were systematically searched, and aerial transects were conducted over parts of the park without adequate trail systems. In total, 12 nests were found within an approximate 80 km^2 block surrounding the primary research station. An additional 12 nests were located using aerial transects in other areas of the park. Detailed information on behavior, breeding biology, and diet composition was only collected from the nests near the research station.

Crowned eagle population density is extremely variable in different habitats; in Kenyan woodlands Brown *et al.* (1982) estimated a density of one pair every 100 km^2. At the other extreme is the estimated density in the relatively mature forest of Taï. Eagle density here is estimated to be one pair approximately every 6.5 km^2, and the average nearest-neighbor nest distance around the research station is 1.81 ± 0.43 s.d. km (Shultz 2002). Nests can be several meters across and are generally found in an exposed fork or branch of an emergent tree. In Taï, the average height of trees with nests was 52 ± 14 m and had an average DBH of 234 ± 66 cm; the nests were 36 ± 9 m high and were located on either the lowest side branch or in the central fork (Malan & Shultz 2002). The nests were also found exclusively in emergent trees, with the main fork above the surrounding canopy. This allows both easy access and protection from non-flying nest predators.

In Taï breeding is highly seasonal, nest repair begins in July/August and eggs are laid during the beginning of the dry season in late November or December (Shultz 2002). Chicks fledge in March before the rainy season recommences (see Figure 7.1). In all studied populations, crowned eagles have a clutch size that varies between one and two, but if the younger sibling hatches it is always killed by the older (Brown *et al.* 1977). Clutch size in Taï is also likely to vary between one and two, but all documented breeding attempts had a final brood size of one. Provisioning is frequent for young nestlings (1.53/day in the first four months), but nest visits reduce in frequency as the chicks age and are very infrequent when the chicks are nearly a year old (c. 0.13/day) (Shultz 2002).

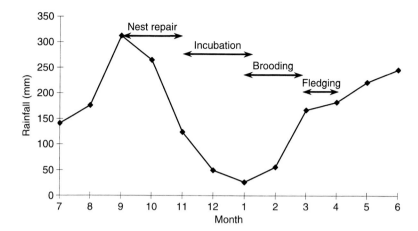

Figure 7.1. The timing of the crowned eagle breeding cycle in relation to rainfall in Taï National Park.

Ranging and activity

Crowned eagles are territorial and regularly display above the canopy near their nest sites. Despite these obvious territorial displays, little is known about patterns of home range use in any habitat. They are expected to use fairly small home ranges, because nest sites are evenly spaced and neighboring pairs are aggressive towards one another (Brown *et al.* 1982). Provisioning adults regularly return with food to the nest located near the center of their home range and thus are central place foragers (CPF) (Orians & Pearson 1978). This bias in space use means they should be found more often near the center of their home range than on the periphery and this will have implications on the encounter rate with different prey groups across their home range. In Taï, most primate home ranges are much smaller than the eagles' home range, and those groups that are found near the center of an eagle's territory should have higher encounter rates than primate groups on the periphery. We used radio telemetry to determine patterns of space use and the area exploited by each breeding eagle pair.

Methods

Two adult females and two juvenile eagles were captured and fitted with either tail-mounted or backpack radio transmitters with activity sensors. Eagles were captured using nooses attached to a piece of goat flesh. The piece of meat was then raised on a pulley system to a branch in a tree adjacent to the nest tree. Once captured, the eagles were lowered to the ground, hooded, and jessed for the duration of handling.

Ranging data were collected using two methods, depending on the signal strength of each radio signal. The transmitter for one female was fairly weak with a transmitting radius of no more than 250 m and it was not possible to maintain constant contact with the individual as she moved around her home range. In order to avoid biasing the locations we set up point locations across a predetermined grid that covered 16 km^2 and recorded presence or absence of individuals at each point. For the two juveniles, and the second female, we located individuals by searching across systematic transects. Once an individual's signal was heard, its exact location was determined using triangulation, and individuals were followed and locations recorded every ten minutes. If the individual remained inactive and the signal had not changed direction, only one point was registered until the individual moved and then two points were taken at 50 m intervals every five minutes. For evaluating space use, we took locations at hourly intervals to insure independence of the different point locations.

We also monitored eagle nest visits and eagle calls around each nest site to get a picture of the activity pattern of the eagles. Six nests were observed at least one day per month between December 1998 and December 1999. An observer sat under each eagle nest from approximately 7:00 to 17:00 hours on 26 days, totalling 312 hours of observation. Arrivals, departures, eagle vocalizations, and continuous behavioral observations were recorded during the observation period.

Results

The adult females were found more often near their nest sites than on the periphery of their range (see Figure 7.2). Females were found no further than 1625 m away from their nest on 61 hourly point locations (Shultz & Noë 2002). The juveniles were located at 183 hour samples across 31 days. Adult visits to the nests were more frequent in the mornings and in the late afternoon, whereas calls were most frequently recorded around mid-day (Shultz & Noë 2002). The juvenile eagles stayed within 200 meters on average (range 0–1200 m) of their nest site for the first ten months after fledging. After this period they were found increasingly far from the nests until their signals were rarely heard. During this period, regular searches for their signals were made across the research area. Although the data are insufficient to document where the juveniles were traveling, it is likely that they dispersed into areas of the park other than the research area. The lifespan of the transmitters was estimated to be approximately three to four years and at least one of the transmitters should have continued to transmit during the period when the locations became less frequent.

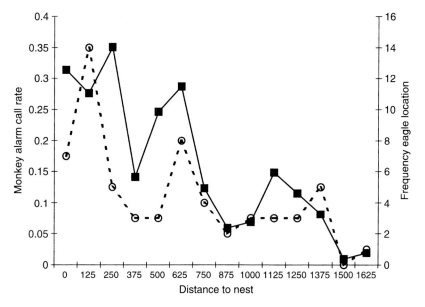

Figure 7.2. The relationship between eagle locations (circles), monkey alarm call rates (squares), and distance to the closest crowned eagle nest. Eagle locations and alarm call rates are significantly correlated ($r^2 = 0.47$, $p < 0.001$). Taken from Shultz & Noë (2002).

Eagle hunting behavior

Predation is an evolutionary contest, predators make behavioral decisions to maximize capture success, while prey employ counter-strategies to minimize their predation risk. The success of each strategy is dependent on the response given by the opponent. Predator hunting behavior can be broadly classified into two coarse strategies: sit-and-wait or searching. Sit-and-wait predators rely on surprise to catch their prey, hiding themselves until prey are within striking distance, whereas searching predators move through their environment until they encounter potential prey items. Many predators do not adhere exclusively to one hunting strategy, rather they employ a mixture of sit-and-await and searching behaviors. Of the main predators of primates in Taï, it has been suggested that crowned eagles and leopards ambush their prey and thus both rely on surprise by either hiding in dense vegetation or approaching a monkey group undetected and waiting for the optimal opportunity to capture an unwary individual. Once their presence has been detected, or an attempted attack has been unsuccessful, they move away from a group to find other less wary or alert individuals (Zuberbühler et al. 1999).

At other sites, crowned eagles have been observed to behave as sit-and-wait predators, dropping down onto an unsuspecting monkey group, as well as attacking on the wing from an unobserved vantage. In Taï, we have seen several instances where eagles position themselves in front of an approaching group and wait to attack until individuals have moved into trees directly below their perch (Shultz 2001). During two encounters, we have followed eagles as they track a monkey group, flying in large arcs around the group and then perching in the path of the approaching group. One occasion the eagle we were following changed its location three times as the monkey group changed direction. When it first flew over the monkey group, a male Diana gave a low intensity alarm call. The eagle then flew off in an oblique angle from the group for about 100 meters, then turned and perched in the path of the monkey group. Each time the group's trajectory changed, the eagle again flew away from the center of the group, then turned back and waited in the new anticipated path of the group. The monkeys appeared to be unaware of the eagle's movement, as they did not alarm call after the first encounter. The entire encounter lasted for more than an hour and a half.

Anecdotal evidence from Kibale forest (e.g. Leland & Struhsaker 1993) and Taï suggest crowned eagles may use cooperative hunting techniques to increase capture success, a trait seen in very few raptors. Research assistants in Taï have seen pairs of eagles attack monkey groups. However, as it is very difficult to observe two attacking eagles, it is unclear whether the two individuals are an adult pair, or a parent and offspring. It is generally believed that an attacking pair of eagles are both mature, but it is also possible that one individual is an older juvenile following a hunting parent.

Such behavior implies a fair amount of awareness in the eagles, as they were not only able to judge the direction of travel by the monkey group, but were able to respond and predict direction changes made by the monkey group. As eagles rely on surprise to attack successfully, it is necessary for them to be able to approach a monkey group undisturbed. An interesting implication of these observations is the apparent lack of effectiveness of vigilance by the monkeys. In the cases where we were able to follow individual eagles, the monkeys were unable to detect the eagle unless it was either attacking or flying directly over the group. Most models of sociality stress the adaptive importance of increased effectiveness of vigilance or decreased investment by individuals in vigilance behavior. Treves (2000) reviews the lack of a consistent group size affect in the primate literature. Two reasons may explain this inconsistency; one is that primate groups are generally larger than models predict for there to be an

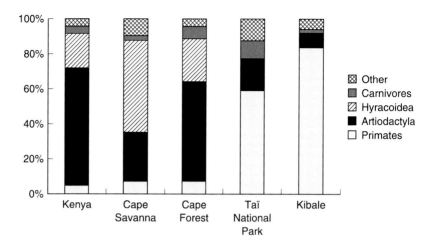

Figure 7.3. Variation in crowned eagle diets in different habitats across
sub-Saharan Africa. Data are from Brown *et al.* (1982), Kenya; Struhsaker &
Leakey (1990), Uganda; Boshoff *et al.* (1994), South Africa. Other category
includes reptiles, birds, rodents, pangolins, and unidentified individuals.
Figure from Shultz (2002).

effect of group size on individual vigilance levels and the second may be
that vigilance is not the most important benefit of sociality. If individuals
are not able to detect predators reliably in dense habitats, the benefits of
sociality may lie more in communal defense and dilution rather than in
increased vigilance.

Predator diet composition

Most predators eat more than one type of prey, but do not necessarily take
each type in direct proportion to their abundance in the environment.
Apostatic selection is commonly observed, where predators take a
proportionally higher amount of common prey than rare prey (Murdoch
1969). However, in addition to selecting prey on abundance alone,
predators prefer prey based on ecological characteristics such as body
size, habitat use, activity pattern, and group size. Predators that
preferentially prey on only a few of the available prey spectra are known
as specialist predators, whereas predators that take prey roughly according
to abundance are generalist predators.

Comparative data about crowned eagle diets in different habitats
show them to be relatively opportunistic rather than specialized predators
(see Figure 7.3). Hyrax and antelope dominate eagle diets in savannah
habitats, as densities of medium sized primates are generally low; where

Plate 1 Western red colobus (*Procolobus badius*) live in large, noisy groups and frequently associates with other monkey species. Their anti-predation strategy is one of safety-in-numbers (Photo: Florian Möllers).

Plate 2 Red colobus rely on mature, primary forest and are very sensitive to habitat disturbance (Photo: Scott McGraw).

Plate 3 The King colobus (*Colobus polykomos*) is the largest arboreal monkey at Taï.
King colobus avoid associating with other monkey species, adopting an
anti-predator strategy of crypsis (Photo: Scott McGraw).

Plate 4 King colobus are low energy strategists that spend much of their time
resting and digesting. Liana leaves and seeds form a large portion of their diet
(Photo: Scott McGraw).

Plate 5 King colobus feeding on the seeds and pods from *Pentaclethera macrophylla*. The large, brown, boomerang-shaped fruits can be seen suspended from various supports (Photo: Scott McGraw).

Plate 6 The olive colobus (*Procolobus verus*) is the smallest colobine. This monkey's near permanent association with Diana monkeys (*Cercopithecus diana*) is an evolved anti-predation strategy (Photo: Florian Möllers).

Plate 7 Olive colobus frequent vine tangles and other areas of dense vegetation in the understory (Photo: Florian Möllers).

Plate 8 Olive colobus females develop large peri-anal swellings during estrus (Photo: Scott McGraw).

Plate 9 The olive colobus diet is dominated by young leaves, a feeding strategy that requires extended periods of rest for digestion (Photo: Florian Möllers).

Plate 10 The Diana monkey (*Cercopithecus diana*) is the most common guenon in the Taï forest. This adult female is feeding on a fruit from *Saccoglottis gabonensis* (Photo: Scott McGraw).

Plate 11 Diana monkeys groups have higher levels of agonistic interactions than do groups of Campbell's monkeys or lesser spot-nosed monkeys (Photo: Florian Möllers).

Plate 12 Diana monkeys exploit all layers of the forest. This individual is walking across a large liana below the forest's main canopy (Photo: Scott McGraw).

Plate 13 The acrobatic Diana monkey is equally at home in the main canopy and on supports near the forest floor. This individual is foraging for fruit and insects on a tree fall (Photo: Scott McGraw).

Plate 14 Diana monkeys, like all cercopithecines, have cheek pouches for storing food (Photo: Scott McGraw).

Plate 15 The alert nature and constant movement of Diana monkeys throughout all layers of the forest translate into high vigilance levels (Photo: Scott McGraw).

Plate 16 The behavior of diana monkeys is an excellent indicator of poaching pressure. In areas of heavy hunting by humans, Diana monkeys adjust their behavior and become cryptic (Photo: Florian Möllers).

Plate 17 A young Diana monkey suckles from its mother (Photo: Friederike Range).

Plate 18 Groups of Campbell's monkeys (*Cercopithecus campbelli*) contain a single adult male (Photo: Scott McGraw).

Plate 19 The drab coat color of Campbell's monkeys blends in well with the dark understory (Photo: Florian Möllers).

Plate 20 The bright blue scrotal sac of adult male Campbell's monkeys is easily visible amid the understory shadows (Photo: Scott McGraw).

Plate 21 Two juvenile Campbell's monkeys feed on the residue of termite tunnels adhering to tree trunk (Photo: Scott McGraw).

Plate 22 Close-up of young Campbell's monkey feeding on termite tunnel residue (Photo: Scott McGraw).

Plate 23 Campbell's monkey (right) presenting to olive colobus for grooming on a fallen trunk (Photo: Scott McGraw).

Plate 24 Adult male lesser spot-nosed monkey (*Cercopithecus petaurista*) scanning the understory. Groups usually contain a single adult male, however following a male take-over, one lesser spot-nosed group had two males for five months (Photo: Scott McGraw).

Plate 25 Lesser spot-nosed monkeys prefer to travel on branches and boughs while branches and twigs are used for foraging (Photo: Scott McGraw).

Plate 26 Female sooty mangabeys (*Cercocebus atys*) remain in their natal group while males transfer to new groups (Photo: Florian Möllers).

Plate 27 Sooty mangabeys are carriers of a lentivirus (SIVsm) that is the probable origin of HIV-2 in humans (Photo: Florian Möllers).

Plate 28 *Cercocebus* mangabeys are close relatives of mandrills and drills. These large, terrestrial, forest-dwelling monkeys share a suite of features that reflects a common reliance on aggressive manual foraging and hard object feeding. Here, a subadult sooty mangabey uses its strong jaws and teeth to crack open a hard fruit (Photo: Florian Möllers).

Plate 29 Grooming forms an integral part of sooty mangabey daily life
(Photo: Florian Möllers).

Plate 30 Juvenile female sooty mangabeys occupy similar ranks as their mothers
and interact preferably with their mothers and close associates of their mothers
(Photo: Florian Möllers).

Plate 31 Sooty mangabeys and zebra duikers (*Cephalophus zebra*) feed on fruits and nuts that fall to the forest floor (Photo: Friederike Range).

Plate 32 The future of the Taï monkeys is in doubt due to widespread poaching within the park (Photo: Florian Möllers).

primates occur at high density, such as in primary forest, they make up the majority of the eagles' diet. Crowned eagles have been known to take prey species up to 40 kilograms, but larger bodied primates such as baboons and chimpanzees are rarely preyed upon. The difference in diet composition between Taï and Kibale in Figure 7.3 is especially interesting as both sites are considered to be of similar habitat types. Kibale has slightly lower rainfall than Taï, but both are considered to be tropical moist forests and there is no a priori reason to expect the prey communities to vary greatly between the two sites. However, the density of terrestrial species (e.g. duikers) is much lower in Kibale, resulting not only in a different diet for the eagles, but also a lower density of leopards (Struhsaker 1997).

The main predators of primates, leopard, eagles, and chimpanzees differ widely in their natural history. Their population densities are very different, with leopards at a density of $0.07-0.11/km^2$ (Jenny 1996), eagles at a density of $0.30-0.47/km^2$ (Shultz 2002), chimpanzees at a density of $1.84/km^2$ (Herbinger *et al.* 2001). Leopards and chimpanzees are mostly terrestrial, climbing into trees to hunt and forage but moving through their territories across the forest floor. Chimpanzees are omnivorous, but they tend to concentrate their hunting within a few months of the year, whereas eagles and leopards are almost exclusively carnivorous with little seasonal variation in hunting pressure. The composition of different predators' diets was compared to prey ecology and behavior to determine how prey behavior influences resulting predation rates.

Methods

Eagle diet composition was determined by removing prey remains from underneath 12 nests near the research station from October 1998 until July 2001. Individual primate prey were identified to species, sex, and age by S. McGraw, using his skeletal reference collection from Taï National Park. Non-primate remains were identified using collections at the Muséum National d'Historie Naturelle in Paris. The minimum number of prey individuals represented at each nest site was then determined. Of the approximately 1200 prey remains collected at the nest sites, a minimum of 333 prey individuals were represented. Amongst primates, there were a large number of specimens that could not be identified as either Campbell's or spot-nosed monkeys, so we combined these two species in the analyses as they have similar body sizes and group sizes. Diet composition for the other main predators of primates were compiled from previous studies (chimpanzees, Boesch & Boesch-Ackermann 2000, leopards, Zuberbühler & Jenny 2002).

We then calculated a standardized selection index for the composition of prey species in the diets of the different predators versus their proportion in the community using the following equations:

forage ratio

$$\hat{w}_i = \frac{o_i}{p_i}$$

standardized forage ratio

$$B_i = \frac{\hat{w}_i}{\sum\limits_{i=1}^{n} \hat{w}_i}$$

Where $\hat{w}_i =$ forage ratio for species i
$o_i =$ Proportion of species i in diet
$p_i =$ Proportion of species i available in the environment

We also examined whether eagle prey selection differed from what would be expected at random using the equation suggested by Manly *et al.* (1993):

$$\chi^2 = 2 \sum_{i=1}^{n} \left[u_i \ln \left(\frac{u_i}{U p_i} \right) \right]$$

where u represents the number of each prey item in the diet, p represents the proportion of each prey item in the environment, and U is the total number of prey individuals recorded.

The standardized forage ratio measures bias towards or against prey on a scale of zero to one. A ratio equal to 1/(# of prey species) indicates no bias, while numbers above this indicate preference and numbers below this indicate avoidance. For example, in this analysis with ten prey taxa, the neutral selection ratio would be 0.10. To understand these differences in preference, we related ecological and behavioral characteristics of the prey to the eagles' preferences. The categories included were group size, group density (groups/km^2), body size, diet composition (folivorous > 50 per cent leaves in diet) and habitat use (arboreal versus terrestrial). Prey preferences were calculated over all possible prey species and for primates alone. We also used information on caloric requirements and population density to estimate an overall predation rate for the different predators.

Results

Except where values are presented, the following results are summarized from Shultz *et al.* (2004). Based on the collection of prey remains, crowned eagles do not select their prey randomly, but show preferences for several of the prey types (G-test, $\chi^2 = 76.44$, d.f. $= 6$, p < 0.001); mangabeys and pottos are relatively preferred, olive colobus, and the two guenons are

Table 7.1. *Standardized prey preferences for the three predator species*

Prey species	Standardized predator preferences		
	Eagle	Leopard	Chimpanzee
Diana Monkey	0.05	0.02	0.09
Lesser spot-nosed guenon	0.12	0.02	0.08
Campbell's monkey			
Red colobus	0.03	0.04	0.44
Black and white colobus	0.03	0.11	0.24
Olive colobus	0.08	0.02	0.10
Sooty mangabey	0.17	0.07	0.01
Potto	0.18	0.04	0.16
Maxwell's duiker	0.12	0.23	0.00
Other duikers	0.03	0.27	0.00
Cusimanse	0.20	–	0.00
Chimpanzee	–	0.17	–

Table 7.2. *Species differences in likelihood of joining mixed-species associations, alarm call rates, detecting and approaching eagles and diet preference measures for crowned eagles. Standardized preference coefficients in this table are calculated only for primates prey species*

	Joining likelihood (Z-test)	p	Alarm call rate (calls/hour)	% Calls following other group	% of encounters eagle approached (n)	Standardized preference coefficient
Diana monkey	–	–	0.396	22%	46% (24)	0.07
Lesser spot-nosed guenon	−3.300	0.001	0.012	40%	0% (5)	0.183
Campbell's monkey	0.000	1.000	0.103	33%	0% (14)	
Red colobus	0.000	1.000	0.060	14%	15% (20)	0.046
Black and white colobus	−2.500	0.012	0.066	43%	80% (5)	0.048
Olive colobus	−1.000	0.317	–	–	0% (7)	0.127
Sooty mangabey	−1.732	0.083	0.047	25%	20% (5)	0.254
Potto	–	–	–	–	–	0.274

relatively closer to neutral, while Diana and the remaining colobus are relatively avoided (see Table 7.1). Looking only at primates, mangabeys were selected well out of proportion to their abundance (see Table 7.2, Figure 7.3) as compared to the arboreal monkeys. For all primates, neither group size nor body size were significantly related to prey preference. However, when the analysis only includes arboreal species, group size ($r^2 = -0.836$, $F_{1,4} = 20.45$, $p = 0.01$), but not body size ($r^2 = -0.469$, $p = 0.088$) was significantly related to eagle prey preferences.

When non-primate prey were also included in the analysis, eagles did not show a bias to primate prey over non-primate prey ($F_{1,8} = 0.24$, $p = 0.65$). They also did not prefer arboreal to terrestrial prey ($F_{1,8} = 1.27$, $p = 0.29$). There was a bias towards non-folivorous prey species, but this trend was not quite significant ($F_{1,8} = 4.59$, $p = 0.06$). There was no significant relationship between group size and prey preference for terrestrial species, but as there is an effect of group size amongst arboreal prey, there is a significant interaction term between habitat use and group size for prey preference (Shultz *et al.* 2004).

Comparison with other predators

The diets of all three predators differed significantly from random (Shultz *et al.* 2004). Each predator responded differently to prey habitat use: leopards strongly preferred terrestrial prey, chimpanzees preferred arboreal, whereas eagles did not show a preference for either. Standardized prey preference was negatively associated with increasing body size for eagles, but not for leopards or chimpanzees (Shultz *et al.* 2004). For chimpanzees, the strongest predictor of prey preference was prey diet, as they preferred folivorous prey (one-way ANOVA, $F_{1,9} = 13.79$, $p = 0.005$), but diet was not associated with preference for either eagles ($F_{1,9} = 4.59$, $p = 0.06$) or leopards ($F_{1,9} = 2.56$, $p = 0.144$).

When all prey species were combined, there were no significant effects of average prey group size on selectivity (see Figure 7.4a). However, when prey preferences and prey group size were separated by prey habitat use, clear patterns of prey preference emerged. For eagles, there was strong negative relationship between preference and group size for arboreal prey, but no clear group size effect among terrestrial prey taxa (see Figure 7.4b, Table 7.2). In contrast, for leopards there was a negative relationship between prey group size and prey preference for terrestrial but not arboreal prey (see Figure 7.4c). As group size was important only for either arboreal or for terrestrial prey for eagle and leopard prey preferences, there was a significant habitat by group-size interaction for both predators (see Table 7.2). In contrast, there was no clear relationship between prey group size and prey preference for chimpanzees among either terrestrial or arboreal prey (see Figure 7.4d, Table 7.2). Prey group density and group size were significantly correlated, potentially obscuring causal relationships between the two factors and prey preferences.

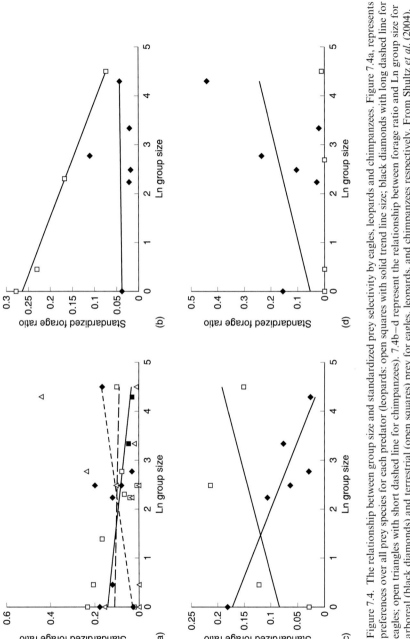

Figure 7.4. The relationship between group size and standardized prey selectivity by eagles, leopards and chimpanzees. Figure 7.4a, represents preferences over all prey species for each predator (leopards: open squares with solid trend line size; black diamonds with long dashed line for eagles; open triangles with short dashed line for chimpanzees). 7.4b–d represent the relationship between forage ratio and Ln group size for eagles, leopards, and chimpanzees respectively. From Shultz *et al.* (2004).

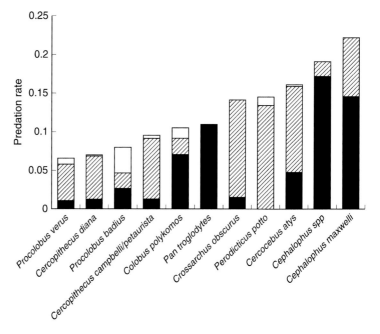

Figure 7.5. Estimated overall annual predation rates for different prey species by each of the three main predators of primates (filled = leopard; hatched = eagle, white = chimpanzee). Preferences were calculated using the proportion of each species in the predator's diet divided by the proportion of each species in the community. These were then converted into predation rates using predator density (individuals/km^2) and individual metabolic requirements (calculated per annum). From Shultz *et al.* (2004).

Predation rates

Total estimated predation rates for the different prey species ranged between 6.5 per cent and 22 per cent of the total population per year (see Figure 7.5). The relative impact of the different predators varied between the prey populations, with some species preyed upon relatively evenly by all three predators, whereas others were only found in the diet of one predator (see Figure 7.5). The overall arcsine-transformed proportion of prey population removed did not differ between leopards and eagles but the proportion of prey removed by chimps was lower than that of both eagles and leopards (Shultz *et al.* 2004).

Overall estimated predation rates (proportion of population removed per year) were higher for terrestrial than for arboreal prey species (see Figure 7.6). Over all prey species, group density was significantly correlated with total predation rate, but this relationship did not remain when prey were separated according to their habitat use. Neither group

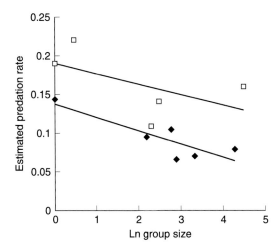

Figure 7.6. The relation between overall predation rate and group size.

size nor group density had a significant effect on total estimated predation rate for arboreal and terrestrial prey. Body size was not correlated to predation rate either over all prey species nor when habitat use was used to separate prey species. However, when either group size and habitat, or group density and habitat, were incorporated into a general linear model, both variables had a significant effect on predation rate (see Figure 7.6). Overall predation rates were also significantly correlated with the maximum intrinsic growth rate for the different prey ($r^2 = 0.79$, $p = 0.02$).

Monkey anti-predator behavior

Mortality is not the only cost associated with predation; behaviors that mitigate predation risk are costly for individuals. Predation risk, as opposed to predation rate, can be defined as the likelihood that a prey individual will encounter a predator. Primates reliably give alarm calls to many of their predators (Seyfarth *et al.* 1980), and thus it is possible to use alarm calls as a measure of encounter rates and thus predation risk. Assuming this premise to be valid, it is possible to look at spatial and temporal patterns of alarm calling as a proxy for variation in predation risk. The logical extension is that if alarm calls are an accurate measure of predation risk, individuals should show consistent behavioral responses following alarm calls. We employed these ideas in Taï to look at spatial and temporal variation in risk across the home range of a crowned eagle pair and to look at behavioral responses associated with elevated alarm calling rates.

The monkeys in Taï need to be able to cope with predation pressure from each of their different predators. They reliably give alarm calls to the two ambush predators, eagles and leopards, but not to chimpanzees or man, the two predators that do not rely on surprise for a successful attack. Zuberbühler *et al.* (1999) have demonstrated that leopards are more likely to move away from a monkey group after an alarm call has been given. Chimpanzees and man, on the other hand, are more proactive in their hunting behavior, locating monkey groups and approaching them regardless of whether they have been detected (Boesch & Boesch-Achermann 2000). In response to these predators, the monkeys typically employ a cryptic strategy and try to move away from the predator without being singled out as a potential prey item (Zuberbühler *et al.* 1997). Thus, alarm calling, as a signal to a predator that it has been detected is an effective strategy against an ambush predator, but may be a very costly strategy against predators such as chimpanzees or man that would use the calls to more easily locate prey individuals. Previous work has documented semantic differences between alarm calls given to different predator stimuli (Zuberbühler *et al.* 1997, Zuberbühler 2000). Monkeys are expected to change their behavior accordingly depending on the attacking species. Most importantly they need to be able to distinguish between an attack from an aerial and a terrestrial predator. However, the differences between the alarm calls, although discernible after acoustic analysis, are difficult to detect by human observers (Zuberbühler 2000).

Mixed species associations are a common response of animals as a strategy to mitigate predation risk. Aggregations of conspecifics result in high levels of competition between individuals because all members of the group exploit roughly the same resource base. However, individuals can increase the size of their group with proportionally smaller increases in intragroup competition by associating with individuals or groups of other species. In such associations, the degree to which individuals of different species compete will depend on the amount that their diet or resources overlap. Associations between species with very different diets will result in low levels of competition between individuals, but may result in conflict about habitat use and patch utilization. Mixed-species groups are often documented in primate studies. Just as in single-species groups, individuals in mixed-species groups benefit by increases in vigilance levels, overall dilution effects, and communal defense. However, not all species are equally good partners. If some partner species are reliably better at both detection and defense, other species can effectively parasitize their efforts. Such a situation may result in a producer-scrounger scenario, whereby the costs and benefits are disproportionate for the different partners.

In Taï National Park, the monkeys regularly form mixed-species aggregations, with some species (e.g. olive colobus) nearly always in association with other species groups. Diana monkeys have been shown to be the most reliable sentinels for detecting eagles and together with black and white colobus are the most likely to approach or attack a model (Bshary & Noë 1997, Table 7.2). Where actual eagle attacks have been observed, these counter-attacks seem to be effective in dissuading the eagles from pursuing the attack (Cordiero 1992, Maisels *et al.* 1993). Having multiple males to defend the group may be a selective pressure for associations to form as males in multi-species groups have been observed mobbing eagles (Gautier-Hion & Tutin 1988). The other species may therefore associate with the Diana to exploit their vigilance.

The final objective of the crowned eagle/predation study was to examine patterns of alarm calling and anti-predator behavior to determine how predator behavior may impact patterns of predation risk for their prey and how primates respond to elevated predation risk.

Methods

We first documented spatial and temporal patterns of alarm call frequency and compared these patterns to patterns of space use by eagles (see Shultz & Noë 2002). To look at spatial variation in alarm calling, we set up a predetermined grid of recording locations across the home range of an eagle pair. During each 45 minute sample, an observer recorded the approximate distance and direction of all alarm calls heard at each location. Additionally all audible primate vocalizations were recorded every 20 minutes to provide an index of primate activity. The approximate location of each alarm call was determined by plotting a point in the direction and distance from the recording post. These point locations were then converted into calling rates at different distance intervals to the nest (in 125 m classes).

In an effort to determine the stimuli causing the alarm calls, we documented vocal responses after alarm calls in one of the Diana groups (Shultz *et al.* 2003). Group scans were completed every half hour. During this period the group spread (defined as the longest chord across the group), the presence of any associated groups of other species and the location of the group within the study grid were noted. Additionally during each group scan, the behavior, height and exposure of all visible individuals were recorded. At 20-minute intervals, and immediately following any male long-distance call, a five-minute sample of all audible vocalizations was recorded. Ten-minute focal samples were collected. At one-minute intervals, behavior, exposure, height, and nearest-neighbor distance and age-class were recorded. Movement rates were calculated by the distance moved across the study grid between successive group scans (see Shultz *et al.* 2003 for more detailed description of the methodology). We predicted that there should be observable changes in movement rates, exposure level, height in the canopy, individuals engaging in less risky behaviors and increases in association rates following male alarm calls.

Results

We documented 611 alarm calls from six of the species during visits to recording sites as well as an additional 138 alarm calls for one Diana monkey group. For all recorded species, with an observer at a listening post and not with a given focal group, 2.2 per cent of calls were given in response to a tree fall, 5.3 per cent in response to an eagle seen or heard, and 24 per cent in response to another species call. For male alarm calls recorded from the focal Diana group, 13 per cent were given in response to tree falls, 17 per cent in response to a call from another Diana group, 6 per cent in response to an eagle seen or heard, 16 per cent in response to another species call, and 13 per cent in response to an initial call from a female. In neither data set were any of the calls given in response to leopards or leopard vocalizations that were seen by the observer.

Spatial variation in alarm calling

Crowned eagles do appear to act as central place foragers in that they are found more often near the center of their range than on the periphery. Additionally, alarm calling rates of the monkeys are correlated with both distance from an eagle nest as well as the frequency of eagle locations (Shultz & Noë 2002, Figure 7.2). Thus, monkey groups near the center of an eagle pair's home range should experience higher predation risk levels than those on the periphery as they encounter eagles more often.

Behavioral responses to elevated call rates

S. Shultz and C. Faurie recorded behavioral and vocal responses to 134 male alarm calls in the habituated DIA 3 group from April through June 2000. In the 30 minutes following alarm calls, the group travelled significantly farther than in the 30 minutes before the alarm. The cohesion (measured as group spread) increased, nearest neighbor distance decreased and the number of associated species was higher following alarm calls (Shultz *et al.* 2003). These results are similar to those of Noë and Bshary (1997) where both association rates and group travel rates have also been shown to increase in response to elevated risk of predation by chimpanzees. Intra-group vocalization rates increased, but interestingly there was no change in either the average height or exposure level of the group members. There was also a change in behavior, with more individuals recorded in traveling or scanning behaviors and fewer individuals engaged being social or resting. Travel rate of the group was significantly associated with both alarm call and association rates.

Mixed species associations and eagles

The number of species in association with Diana groups was higher following a Diana alarm call (Shultz *et al.* 2003). Species may join after alarm calls either because predation risk is higher, or because loud calls allow other groups to locate a potential Diana partner. The monkey species showed different responses to Diana alarm calls; black and white colobus and lesser spot-nosed guenons were significantly more likely to join a Diana group than to leave it after an Diana alarm, whereas this effect was not significant for the other species (see Table 7.2). The species also varied in their likelihood to alarm following a call from another species (see Table 7.2). Those species that rarely called first, may use other species as an early warning system. Both lesser spot-nosed monkey and the black and white colobus were significantly more likely to join a Diana group following an alarm call and more likely to give alarm calls in response to calls of other groups.

Discussion

Eagles are selective with regards to the body size, group size, and the degree of terrestriality of their primate prey species. However, given the population abundance of primates in Taï especially relative to the population abundance of their predators, the predation rate on the primates is relatively low. This is because of the effectiveness of primate anti-predator behaviors, such as vigilance, mobbing and confusion. However, these strategies are costly to employ as social individuals incur higher levels of competition, whether in mono-specific or polyspecific groups. Thus predation risk, or the pressure to employ anti-predator strategies, is probably a more deterministic cost of predation than mortality. We can measure the relative investment in these strategies through fitness correlates, such as ranging distance, behavioral changes, or changes in group cohesion and between species association patterns. We were able to show, using spatial patterns of alarm call rates and vocal responses to these calls, that most of the male alarm calls are given in response to eagle encounters. In a Diana monkey group, these calls are associated with consistent changes in behavior, the group moved farther following calls, group cohesion increased as did association rates with other species.

Thus, we can assume that elevated predation pressure results in monkeys adopting costly anti-predator behavior. Many of these behavioral responses are facultative, meaning individuals have short-term responses to changing levels of predation risk.

One of the difficulties in documenting the effect of predation pressure on monkey behavior is that the predators are less habituated than the focal monkey groups. Monkey groups followed continuously by observers have lower predation rates than non-focal groups. Isbell and Young (1993) documented the "Nairobi" effect whereby predation rates increased during periods when a focal group was not being observed. It is not possible, therefore, to accurately record all potential predation events or encounters with predators. Semi-terrestrial predators, such as leopards and chimps are likely to be more affected by the presence of observers, however, even eagles high in the canopy will flush when they see humans on the forest floor. If we want to document the effect of predation on behavior, it is necessary to consider how the presence of observers can affect both predator and prey behavior.

It was interesting that not all of our predictions were supported with the Diana group. Moving into lower areas in the canopy, into higher density vegetation or reducing exposure level were not observed to be consistent responses to elevated risk levels. There may be several explanations for this. We measured behavioral changes no less than five minutes following an alarm call as we were documenting vocal responses to alarm calls for the five minutes following an alarm. Behavioral responses such as changing exposure level, canopy height may be very short term responses that are not detectable even five minutes after the alarm. Alternatively, there may be more semantic information in either the alarm call or the response of other group members that give an indication of the level of risk. Individuals may only seek refuges under threat of immediate attack.

Perhaps the most exciting result from the compilation of predator studies in Taï is that we finally have an idea of how both prey ecology and social system tie into predation risk. Before these studies the risk level for terrestrial forest species was underestimated, as the logical assumption is that arboreal predators will be a negligible risk factor for terrestrial species compared to terrestrial predators. However, understanding how arboreal predators hunt and select prey explains why terrestrial species are such a large component of their diet. These results have a fairly profound impact on our understanding of predation risk. In both the Paleo- and Neo-tropics there are few terrestrial primates in primary forest environments. These species are typically large bodied and/or found in large groups. When we consider hominid evolution, it may be necessary to take into account that predation risk is elevated not only by increased exposure to terrestrial predators but also aerial ones, once the safety of the trees has been left (McGraw *et al.* 2006).

Primates are only one component of the mammalian prey community. Leopards and eagles share a number of other prey species, including ungulates, small carnivores and even some birds and reptiles. However, when we look at all prey classes, rather than limiting ourselves to one taxonomic group, we see consistent relationships between prey behavior and ecology, and realized predation rates. Most notably, we see that terrestriality is inherently risky, however, regardless of habitat use group size is an effective strategy to reduce predation risk. However, if we looked at predation from only a single predator, we would have missed these overall patterns. Thus, future studies on the relationship between anti-predator behavior and predation rates should either include all major predators within a community or consider that a two-species interaction may not reflect the overall predator-prey relationships in the community.

Acknowledgments

This project would not have been possible without the assistance of many kind and supportive people. Roger Kami, Sio Theodore, Charlotte Faurie and Jeffrey Jennings were invaluable as research assistants and colleagues. Financial support from WCS, Leakey Foundation, the Raptor Research Foundation and the Peregrine Fund made the project possible. The support of the Centre Suisse, PACPNT, and the CRE made all the difference in managing logistics and affairs in the Ivory Coast. Charles Janson and Robin Dunbar have been supportive scientific supervisors. Last, but certainly not least, Ronald Noë, by believing in the project and providing the opportunity to work in the TMP, has made the project come to fruition.

References

Boesch, C. and Boesch-Achermann, H. (2000). *The Chimpanzees of the Taï Forest: Behavioral Ecology and Evolution.* Oxford: Oxford University Press.

Boshoff, A. F., Palmer, N. G., Vernon, C. J. and Avery, G. (1994). Comparison of the diet of crowned eagles in the savanna and forest biomes of south-east South Africa. *South African Journal of Wildlife Research*, **24**, 26−31.

Brown, L. H. (1971). On the relationship of the crowned eagle *Stephanoaetus coronatus* and some of its prey animals. *Ibis*, **113**, 240−3.

Brown, L. H., Gargett, V. and Steyn, P. (1977). Breeding success in some African eagles related to theories about sibling aggression and its effects. *Ostrich*, **48**, 65−71.

Brown, L. H., Urban, E. K. and Newman, K. (1982). *The Birds of Africa*, Vol 1. New York: Academy Press.

Bshary, R. and Noë, R. (1997). Red colobus and Diana monkeys provide mutual protection against predators. *Animal Behaviour*, **54**, 1461–74.

Cordeiro, N. (1992). Behavior of blue monkeys (*Cercopithecus mitis*) in the presence of crowned eagles (*Stephanoaetus coronatus*). *Folia Primatologica*, **59**, 203–7.

Cowlishaw, G. C. (1994). Vulnerability to predation in baboon populations. *Behaviour*, **131**, 293–304.

Daneel, A. B. C. (1979). Prey size and hunting methods of the crowned eagle. *Ostrich*, **50**, 120–1.

Gautier-Hion, A. and Tutin, C. E. G. (1988). Simultaneous attack by adult males of a polyspecific troop of monkeys against a crowned eagle. *Folia Primatologica*, **51**, 149–51.

Herbinger, I., Boesch, C. and Rothe, H. (2001). Territory characteristics among three neighboring chimpanzee communities in the Taï National Park, Côte d'Ivoire. *International Journal of Primatology*, **22**, 143–67.

Isbell, L. A. and Young, T. P. (1993). Human presence reduces predation in a free-ranging vervet monkey population in Kenya. *Animal Behaviour*, **45**, 1233–5.

Jenny, D. (1996). Spatial organisation of leopards *Panthera pardus* in Taï National Park, Ivory Coast: is rainforest a "tropical haven?" *Journal of Zoology, London*, **240**, 427–40.

Krause, J. and Godin, J. G. (1994). Shoal choice in the banded killifish (*Fundulus diaphanous, Teleostei, Cyprinodontidae*): effects of predation risk, fish size, species composition and size of shoals. *Ethology*, **98**, 128–36.

Leland, L. and Struhsaker, T. T. (1993). Teamwork tactics. *Natural History*, **102**, 42–7.

Lima, S. L. (1987). Vigilance while feeding and its relation to the risk of predation. *Journal of Theoretical Biology*, **124**, 303–16.

Lindstrom, A. (1989). Finch flock size and risk of hawk predation at a migratory stopover site. *The Auk*, **106**, 225–32.

Maisels, F. G., Gautier, J. P., Cruikshank, A. and Bosefe, J. P. (1993). Attacks by crowned hawk-eagles (*Stephanoaetus coronatus*) on monkeys in Zaire. *Folia Primatologica*, **61**, 157–9.

Malan, G. and Shultz, S. (2002). Nest-site selection of the crowned eagle in the forests of KwaZulu-Natal, South Africa and Taï, Ivory Coast. *Journal of Raptor Research*, **36**, 300–8.

Manly, B. F. J., McDonald, L. L. and Thomas, D. L. (1993). *Resource Selection by Animals: Statistical Design and Analysis for Field Studies*. London: Chapman.

McGraw, W. S., Cooke, C. and Shultz, S. (2006). Primate remains from African crowned eagle (*Stephanoaetus coronatus*) nests in Ivory Coast's Taï forest: implications for primate predation and early hominid taphonomy in South Africa. *American Journal of Physical Anthropology*, **131**, 151–165.

Mitani, J. C., Sanders, W. J., Lwanga, J. S. and Windfelder, T. L. (2001). Predatory behavior of crowned hawk-eagles (*Stephanoaetus coronatus*) in Kibale National Park, Uganda. *Behavioral Ecology and Sociobiology*, **49**, 187–95.

Murdoch, W. W. (1969). Switching in generalised predators: experiments on predator specificity and stability of prey populations. *Ecological Monographs*, **39**, 335–54.

Noë, R. and Bshary, R. (1997). The formation of red colobus-Diana monkey associations under predation pressure from chimpanzees. *Proceedings of the Royal Society London*, B, **264**, 253–9.

Orians, G. H. and Pearson, N. E. (1978). On the theory of central place foraging. In *Analysis of Ecological Systems*, ed. B. J. Horn, G. R. Stairs and R. D. Mitchell pp. 155–77. Columbus: Ohio State University Press.

Seyfarth, R. M., Cheney, D. L. and Marler, P. (1980). Monkey responses to three different alarm calls: evidence for predator classification and semantic communication. *Science*, **210**, 801–3.

Shultz, S. (2001). Notes on interactions between monkeys and African crowned eagles in Taï National Park, Ivory Coast. *Folia Primatologica*, **72**, 248–50.

Shultz, S. (2002). Population density, breeding chronology and diet of crowned eagles *Stephanoaetus coronatus* in Taï National Park, Ivory Coast. *Ibis*, **144**, 135–8.

Shultz, S., Faurie, C. and Noë, R. (2003). Behavioral responses of Diana monkeys to male-long distance calls: changes in ranging, association patterns, and activity. *Behavioral Ecology and Sociobiology*, **53**, 238–45.

Shultz, S. and Noë, R. (2002). The consequences of crowned eagle central-place foraging on predation risk in monkeys. *Proceedings of the Royal Society of London*, B, **269**, 1797–802.

Shultz, S., Noë, R., McGraw, W. S. and Dunbar, R. I. M. D. (2004). A community-level evaluation of the impact of prey behavioral and ecological characteristics on predator diet composition. *Proceedings of the Royal Society of London*, B, **271**, 725–32.

Skorupa, J. (1989). Crowned eagles *Stephanoaetus coronatus* in rainforest: observations on breeding chronology and diet at a nest in Uganda. *Ibis*, **131**, 294–8.

Steyn, P. (1982). *Birds of Prey of Southern Africa*. Cape Town: David Philip.

Struhsaker, T. T. (1997). *Ecology of an African Rainforest: Logging in Kibale Forest and the Conflict Between Conservation and Exploitation*. Gainesville, Fla: University of Florida Press.

Struhsaker, T. and Leakey, M. (1990). Prey selectivity by crowned hawk-eagles on monkeys in the Kibale Forest, Uganda. *Behavioral Ecology and Sociobiology*, **26**, 435–44.

Treves, A. (2000). Theory and method in studies of vigilance and aggregation. *Animal Behaviour*, **60**, 711–22.

Vernon, C. J. (1984). The breeding periodicity of the crowned eagle. In *Proceedings of the 2nd symposium on African predatory birds*, ed. J. M. Mendelsohn and C. W. Sapsford, Durban, Natal Bird Club, pp. 127–38.

Zuberbühler, K. (2000). Interspecies semantic communication in two forest primates. *Proceedings of the Royal Society of London*, B, **267**, 713–18.

Zuberbühler, K. and Jenny, D. (2002). Leopard predation and primate evolution. *Journal of Human Evolution*, **43**, 873–6.

Zuberbühler, K., Jenny, D. and Bshary, R. (1999). The predator deterrence function of alarm calls. *Ethology*, **105**, 477–90.

Zuberbühler, K., Noë, R. and Seyfarth, R. M. (1997). Diana monkey long-distance calls: messages for conspecifics and predators. *Animal Behaviour*, **53**, 589–604.

8 *Monkey alarm calls*

K. Zuberbühler

Introduction

Most primates vocalize when threatened by a predator. These signals, usually termed alarm calls (from old Italian "*all arme*" = "to arms" on the approach of an enemy), are interesting for a number of reasons. First, they are relatively uncomplicated to examine. Alarm calls are typically highly discrete signals, and it is thus often not difficult to study both causes and consequences of this behavior. Second, alarm calls are interesting because they pose a problem for evolutionary theory. Paradoxically, they are often amongst the most prominent and noticeable signals in a species' repertoire. But why should an individual behave conspicuously in the presence of a predator, hereby revealing its presence and location? Finally, alarm calls have obtained some significance for the empirical study of pre-linguistic abilities. Alarm calls are unique because they are well suited for experimental work and so provide a unique tool for empirically accessing the cognitive mechanisms underlying an individual's behavior.

The purpose of this chapter is to review some key findings in the study of primate alarm calling behavior. The first section is devoted to the problem of why primates produce seemingly maladaptive behavior in the presence of a predator. Which evolutionary processes could have provided a selective advantage for individuals to behave in this counter-intuitive way? A second section is concerned with the cognitive processes that underlie call production and perception. What are the mental processes that underlie primate alarm calling behavior in non-human primates and what is their relevance for the evolution of linguistic capacities in humans?

The evolution of alarm calls

Three kinds of hypotheses have been put forward to explain why primates produce conspicuous vocalizations in the presence of a predator. First, alarm calls can provide a selective advantage to the signaller if they increase the survival chances of closely related kin (Maynard Smith 1965). Costly calling is then outweighed by the benefits of increased survival of

Monkeys of the Taï Forest, ed. W. Scott McGraw, Klaus Zuberbühler and Ronald Noë. Published by Cambridge University Press. © Cambridge University Press 2007.

194

recipients that carry a proportion of the caller's genes (the kin selection hypothesis). Second, alarm calling is advantageous if it enhances the competitiveness of the caller in the context of reproduction (the sexual selection hypothesis). Third, alarm calling is beneficial if it decreases the vulnerability of the caller to the predator directly (the individual selection hypothesis). This can be the case if alarm calling affects the behavior of nearby conspecific recipients (the prey "manipulation" hypothesis). For example, individuals responding to an alarm call may create a "cloud of confusion" for the predator from which the caller can directly benefit. Another version of the individual selection hypothesis states that some predators are directly averted by the alarm calls of their prey (the "perception advertisement" hypothesis). This is especially the case for predators that rely on unprepared prey for a successful hunt.

Alarm calls favored by kin selection
Primate alarm calling behavior is audience dependent. For example, vervet monkey females adjust the rate of alarm calling depending on whether or not their own offspring is present (Cheney & Seyfarth 1985). This observation suggests that callers benefit from behaving conspicuously in the presence of predators, probably because it increases the survival chances of their offspring. In most animal species, the empirical evidence for the kin selection hypothesis is strongest for the parent-offspring relation (e.g. Blumstein *et al.* 1997), and alarm calling may thus be conceptualized as a form of parental care. However, kin selection further predicts that individuals should be willing to engage in risky alarm call behavior if this increases the survival chances of closely related non-descendent kin, apart from own offspring. There is some support for this idea although the overall empirical evidence is weak. For example, spider monkeys alter their alarm call behavior as a function of the number of kin in the vicinity (Chapman *et al.* 1990). Similarly, Kloss's gibbons produce long-distance alarm calls that carry into neighboring home ranges, which are often occupied by the callers' close relatives (Tenaza & Tilson 1977), suggesting that non-descendent kin could be the intended recipients of these calls.

Alarm calls favored by sexual selection
Sexual selection is likely to impact on the evolution of primate alarm calls if the signals increase the mating and reproductive success of callers. For example, particularly keen callers may gain social status in a group and find it easier to obtain matings (Zuberbühler 2002a).

Since in non-human primates, this argument will mainly apply for males, and since both males and females produce alarm calls, sexual selection is likely to act as a secondary evolutionary force. Male vervet monkeys (*Cercopithecus aethiops*) give alarm calls at higher rates in the presence of adult females than adult males (Cheney & Seyfarth 1990), suggesting that alarm calls may be part of a mating strategy to enhance the caller's reproductive success. In some forest monkey species, adult male alarm calls are structurally different from those of the adult females (Gautier & Gautier 1977). The male calls are low-pitched, high amplitude signals given in repeated bursts, which carry over long distances through dense tropical forest habitat, sometimes up to one kilometer.

Sexual selection may be in a position to explain the curious differences between male and female alarm calls. According to Darwin (1859), sexual selection "... depends, not on the struggle for existence, but on a struggle between the males for possession of the females; the result is not death of the unsuccessful competitor, but few or no offspring." Most forest guenons live in polygynous social systems, with one adult male and several adult females. Male–male competition is high in these species, and males are notorious targets of sexual selection, leading to conspicuous male traits, such as loud vocalizations that are useful in male-male competition or that females find attractive (Clutton-Brock & Albon 1979, Anderson 1994). In Diana monkeys, males typically control a group of females for some years, and mate with them until replaced by another male. If females exert control over a male's tenure length, then males should find it beneficial to engage in costly anti-predator behavior if this affects female tolerance towards them (Eckardt & Zuberbühler 2004). It is also relevant to point out that during puberty sudden changes occur in the vocal behavior of male guenons, specifically a drop in pitch and the loss of some of the juvenile vocal repertoire (Gautier & Gautier 1977). Some observations suggest that sub-adult male Diana monkeys go through a phase in which their alarm shows remnants of a female alarm call, but also the first emerging elements of a fully developed male loud call, suggesting that males go through a transition phase when their calls develop from female alarm calls to male loud calls (Zuberbühler 2002a), a further sign that sexual selection has acted secondarily on the structure and usage of male monkey alarm calls.

Alarm calls favored by individual selection

As outlined before, in some cases individuals enjoy direct benefits from producing alarm calls, for instance if alarm calls elicit collective anti-predator behavior in nearby recipients, which confuses or disorients the predator (the confusion effect or "prey manipulation" hypothesis, e.g. Charnov & Krebs 1975). Visibility in the Taï forest is very low, rarely exceeding 20 m, and members of a monkey group can only maintain visual contact with a fraction of the group. Empirical work would have to show that callers are more likely to give alarm calls if the group forages in open areas or exposed emergent trees in which mass escape could cause confusion to a predator.

Alarm calls can directly interfere with a predator's hunting tactic, for example if predators depend on unwary prey (the "perception-advertisement" hypothesis, e.g. Bergstrom & Lachmann 2001). Diana monkeys alarm calls appear to function this way when interacting with leopards (Zuberbühler & Jenny, Chapter 5) and possibly crowned eagles (Shultz & Thomsett, Chapter 7). Crowned eagles and leopards reliably elicit high rates of alarm calls in Diana and other monkeys, while two other equally dangerous predators, the chimpanzees and human poachers, never do (Zuberbühler *et al.* 1999). The most likely explanation for these differences is that chimpanzees and humans, but not leopards or eagles, are able to pursue monkeys in the trees, which greatly increases the costs of conspicuous alarm calling. Chapter 5 reviews some studies on leopard hunting behavior and how it is affected by primate alarm calling. Briefly, forest leopards hunt monkeys by approaching unwary groups and hiding in their vicinity, presumably to wait for individuals to descend to the ground. Once the monkeys detect a hiding leopard, they produce alarm calls at very high rates, which typically causes the leopard to move on and leave the area, showing that alarm calls function to advertise perception and deter stalking leopards.

The cognitive bases of alarm calls

Pioneering fieldwork with East African vervet monkeys has shown that some primate alarm calls function to refer to external events, demonstrating that basic referential abilities are a natural component of primate cognition (Struhsaker 1967, Seyfarth *et al.* 1980). Vervet monkeys produce acoustically distinct vocalizations in response to a number of predators. Call use is not random, but individuals produce particular call types to certain classes of predators. Recipients respond to these calls as they would the actual predators, not unlike humans responding to speech utterances. Eagle alarm calls, for example, may cause individuals to look into the sky and run for cover, whereas snake alarm calls may cause them to adopt a bipedal vigilance posture. These findings have led to the hypothesis that some primate calls convey meaning in the same way as human speech sounds. Because calls have the same effect on recipients as the corresponding external events they have therefore been termed referential, or semantic, signals (Seyfarth *et al.* 1980).

Do primate alarm calls qualify as semantic signals?

Philosophers have argued that gaining access to the mental processes that steer another organism's behavior is an unachievable endeavor, the other-mind-problem (Nagel 1995). Drawing analogies with human mental states, derived from introspection, appear to be the only solution, an inevitably anthropocentric approach of questionable value (Nagel 1974). Its hazards are particularly apparent when species with a considerably less complicated nervous system produce seemingly complex behavior. For example, honeybees can notify each other about the presence and location of distant food sources using symbolic dance language (von Frisch 1973). Incoming worker bees produce a specific dance motion that describes the geographic location of a previously encountered food source accurately enough for others to go out and find the food source. Hence, from a functional perspective, there does not seem to be much difference between human linguistic behavior and honeybee dance language, although the associated mental processes might be profoundly different. For example, the bees' communication system is possibly rooted in a simple form of perceptual processing, in which recipients respond directly to the physical dimensions of a signal, rather than any associated mental entities. Human language processing appears to be different because speech sounds refer to connected cognitive structures shared by the signaller and recipient (e.g. Yates & Tule 1979).

The function of semantic knowledge: predator-specific alarm calls in guenons

Diana monkeys and Campbell's monkeys live in small groups of one single adult male and several adult females with their offspring (Uster & Zuberbühler 2001, Wolters & Zuberbühler 2003, Buzzard & Eckardt, Chapter 2). The two species frequently form associations, and some groups spend up to 90 per cent of their time in association with one another despite high levels of feeding competition (Wolters & Zuberbühler 2003). In both species, males and females differ in their vocal repertoires. Males only produce a small number of conspicuous loud calls in response to predators and other disturbances, and sometimes also without any apparent reason. Females and juveniles of both species produce contact calls that can vary in their acoustic fine structure, often also as a function of context

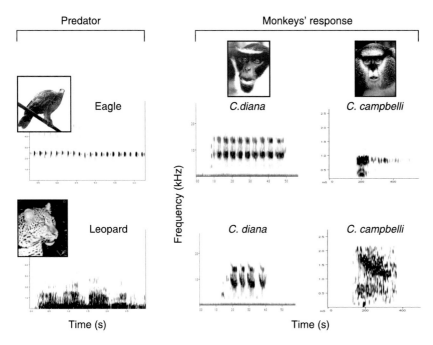

Figure 8.1. Alarm calls produced by male Diana monkeys and Campbell's monkeys in response to leopards and crowned eagles. (Data from Zuberbühler *et al.* 1997, Zuberbühler 2001.)

(Uster & Zuberbühler 2001, Lemasson *et al.* 2004). Other calls are given in the context of agonistic interactions, but these calls have not been studied systematically (Zuberbühler *et al.* 1997). Diana monkeys also produce alert calls in response to a variety of stimuli, although there is a possibility that there are consistent context-dependent differences. Again, no systematic data are available for this variation. Finally, individuals produce acoustically distinct alarm calls to crowned eagles and leopards, which are relatively easy to distinguish by ear and are given specifically to these two predators (see Figure 8.1). Alarm calls are also given to chimpanzees, humans, and all sorts of unknown or unspecific disturbances, although no systematic studies have been conducted for these contexts. A striking pattern is that alarm calls to chimpanzees and humans, if they occur, are never part of a group activity. Although these calls can be quite loud they are typically given by a single animal only, and only at low rates. For some unknown reason these alarm calls never elicit vocal responses in other group members. Diana monkeys also appear to have a call for remarkable things on the ground, such as a large reptile or an unfamiliar human observer. The acoustic structures of these calls have not yet been described.

To investigate whether the alarm calls of these monkeys denote particular predator types a series of playback experiments was conducted. Wild groups were sought throughout a roughly 50 km^2 study area east of the C.R.E. research station in the Taï forest. The home range of both monkey species is normally less than 1 km^2, suggesting that the study area must have contained at least 50 different Diana and Campbell's monkey groups. Once a group was located, usually at a distance of 100 m or so, the experimenter slowly and silently approached to about 50 m and set up the playback and recording equipment without being detected by the monkeys. Then, a short recording was played back to them in order to simulate the presence of a predator, either a leopard or a crowned eagle. Playback stimuli were of the following types: a 15-s recording of leopard growls; a series of male Diana monkey or Campbell's monkey leopard alarm calls; a 15-s recording of crowned eagle shrieks; or a series of male Diana monkey or Campbell's monkey eagle alarm calls. The vocal response of all group members was recorded on cassette tape. No group was tested more than once on a particular playback stimulus. Figure 8.2 summarizes the average call rates per group, not including the adult males.

Female Diana monkeys responded to predator vocalizations and male alarm calls by giving their own acoustically distinct alarm calls. These vocal responses were highly selective in the sense that playbacks of eagle shrieks, male Diana monkey eagle alarm calls, or male Campbell's monkey eagle alarm calls all elicited one type of predator-specific alarm call from females – the eagle alarm call. In contrast, playback of leopard growls, Diana males' leopard alarm calls, or Campbell's males' leopard alarm calls all elicited an acoustically different alarm call – the females' leopard alarm call. Hence, the main organizing principle in the responses of the female Diana monkeys to the six different playback stimuli was the predator type indicated by the playback stimulus, rather than its acoustic properties or the type of signaller. Figure 8.2 further shows that Diana monkeys produced a large number of alert calls (given to all sorts of disturbances) and contact calls (given as part of anti-phonal calling behavior) in the presence of both predators.

The content of semantic knowledge: mental representations of predator types

As outlined before, the bee dance language example has raised some concerns about the cognitive processes underlying alarm call perception. Do the monkeys really respond to their alarm calls as a result of a mental representation of the predator class or do they simply react to the acoustic

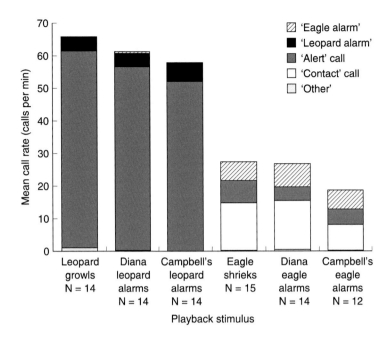

Figure 8.2. Vocal responses of female Diana monkeys to playbacks of predator vocalizations or alarm call series of male Campbell's or Diana monkeys. Vertical bars represent mean number of the different calls given by various Diana monkey groups tested in these experiments. These calls were given by the adult females and subadult individuals, but not by the adult male of each group (Data from Zuberbühler *et al.* 1997, Zuberbühler 2000c).

features of their calls? Under field conditions, the choice of experimental techniques to further investigate this problem is limited. Habituation-dishabituation procedures, originally developed for pre-linguistic children (Eimas *et al.* 1971), have been most successful (e.g. Cheney & Seyfarth 1988). The following experiments have applied one form of this procedure, the prime–probe technique, to primates living in undisturbed natural conditions in order to investigate cognitive processes underlying alarm call behavior. The prime–probe technique does not contain a habituation phase, but individuals are only exposed once to the critical information. The effects of this manipulation are then tested on the animals' subsequent response to a probe stimulus.

In a first experiment, the playback speaker was positioned in the vicinity of wild Diana monkey groups. Monkeys were first primed with a stimulus simulating the presence of a predator (either the calls of a crowned eagle or a leopard, or the eagle or leopard alarm calls given

BASELINE

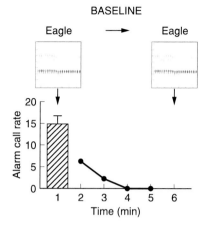

TEST

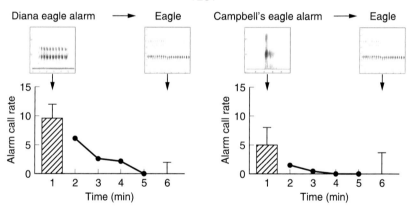

CONTROL

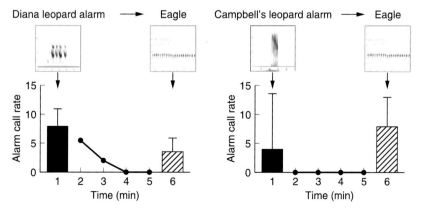

by a monkey, respectively). After a short period of silence a probe stimulus was presented, simulating the presence of the corresponding or non-corresponding predator. This way it was possible to show whether subjects were able to take into account the semantic information conveyed by the priming stimulus. In the baseline condition, the semantic and acoustic information were repeated because subjects heard the same stimulus twice. For example, subjects heard eagle shrieks followed by eagle shrieks again. The prediction was the subjects responded strongly to prime and weakly to the probe stimulus. In the test condition, subjects heard the eagle (or leopard) alarm calls of another monkey followed by vocalizations of the corresponding predator. For example, subjects heard Diana eagle alarm calls followed by eagle shrieks. In this condition, the semantic features remained the same whereas the acoustic features changed. If subjects only attended to the acoustic features of a signal they were expected to respond strongly to both prime and probe stimulus. However, if they attended to the semantic-conceptual information conveyed by the calls then they should respond strongly to the prime and weakly to the probe stimulus. Finally, in the control condition both the acoustic and the semantic features changed from priming to probe stimulus. For example, subjects heard Diana eagle alarm calls followed by leopard growls. Since both the acoustic and the semantic features were different, subjects were expected to produce many predator-specific alarm calls to both the prime and the probe stimuli.

Results were consistent with the idea that Diana monkeys processed the semantic features of these calls. Both eagle shrieks and leopard growls, two very powerful stimuli, lost their effectiveness in eliciting alarm calls when subjects were primed first with the corresponding male alarm calls. Figure 8.3 illustrates the response to eagles. It further shows that it did not matter to the monkeys whether the alarm calls presented as priming stimuli were given by a conspecific Diana male or by a heterospecific Campbell's male. Although the alarm calls of the two species differed in their acoustic structure, the basic finding remained the same: the monkeys ceased to respond to a predator if they were previously warned of its presence by

Figure 8.3. Results of the prime—probe experiments using eagle shrieks as probe stimuli. (Data from Zuberbühler *et al.* 1999, Zuberbühler 2000c.) Histograms represent the median number of eagle alarm calls (hatched) or leopard alarm calls (solid) given in the first minute after a playback stimulus. Error bars represent the third quartile. The points connected by lines between them represent the median alarm call rate during the five-minute period of silence in between two playback stimuli. Using leopard growls as a probe stimulus yielded analogous results.

a semantically corresponding alarm call, regardless of its species' origin or acoustic features. Presumably, Diana monkeys assumed the presence of a particular predator type when hearing other monkeys' alarm calls and were not surprised to hear the predator's vocalizations a few minutes later, suggesting that call processing involved mental representations of the different predator classes.

The acquisition of semantic knowledge: effects of experience

As mentioned earlier, leopards and chimpanzees prey on Taï monkeys. Although both cause substantial mortality in the monkey populations the two predators differ fundamentally in their hunting tactic. The monkeys, in turn, use different anti-predator strategies to defend themselves (Zuberbühler & Jenny, Chapter 5). Leopards follow a sit-and-wait tactic and depend on unprepared prey, and monkeys respond with conspicuous alarm calling behavior. Taï chimpanzees have sophisticated climbing skills that allow them to hunt efficiently for monkeys, mainly red colobus (*Procolobus badius*, Boesch & Boesch 1989). Chimpanzees capture Diana monkeys less frequently, but individuals are nevertheless exposed to many predation events due to their high association rates with red colobus monkeys (Wachter *et al.* 1997). The typical response of all Taï monkeys to an approaching chimpanzee party is to silently hide in the canopy.

Yet, the monkeys' choice of an appropriate anti-predator strategy to chimpanzees is complicated by the fact that chimpanzees themselves occasionally also fall prey to leopards (Zuberbühler & Jenny 2002; Chapter 5). When encountering a leopard, chimpanzees give loud and conspicuous alarm screams, the "SOS" screams (Goodall 1986). Hearing chimpanzee alarm screams thus, not only signals the presence of a group of chimpanzees, it also indicates the presence of a leopard. Thus, if the Diana monkeys understand the meaning of the chimpanzee alarm screams, they will be forced to choose between two mutually exclusive anti-predator strategies.

To investigate whether Diana monkeys understand the meaning of chimpanzee alarm screams, tape recordings of chimpanzee social screams, chimpanzee alarm screams, or leopard growls were played back to different groups. Results showed that the monkeys differed in their vocal responses to the three stimuli: both males and females remained silent after hearing chimpanzee social screams but gave loud and conspicuous alarm calls after hearing leopard growls and chimpanzee alarm screams. Statistical analyses revealed that both adult males and females were significantly more likely to respond with alarm calls to a playback with

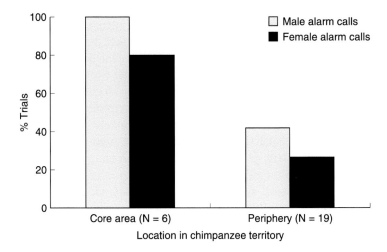

Figure 8.4. Relationship between a Diana monkey group's tendency to respond with leopard alarm calls to chimpanzee alarm screams and their location within the resident chimpanzee territory. (Data from Zuberbühler 2000b.)

chimpanzee alarm screams than to chimpanzee social screams. Males and females were also significantly more likely to respond with alarm calls to a playback of leopard growls than to chimpanzee alarm screams. The latter finding was the result of high variation in responses to chimpanzee alarm screams. In the majority of groups of Diana monkeys, both males and females responded by giving their own leopard alarm calls when hearing chimpanzee alarm screams. In some groups, however, individuals adopted a cryptic response after hearing chimpanzee alarm screams, thus showing no evidence that they perceived the chimpanzee alarm screams as a sign of leopard presence. It has been found that chimpanzees only use a relatively small proportion of their large territory regularly (Herbinger *et al.* 2001). Diana groups living in the periphery of a chimpanzee territory, therefore, encounter chimpanzees much less frequently than core area groups. Interestingly, results showed that peripheral groups were significantly less likely to respond with leopard alarm calls to the chimpanzee alarm calls than groups living in the core area (Figure 8.4).

Did Diana groups living in the periphery of a chimpanzee territory remain silent because they did not understand the meaning of the chimpanzee alarm screams? To investigate this hypothesis the following playback experiment was conducted. As with the previous one, it consisted of three different conditions: a baseline, a test and a control. In each trial, a Diana monkey group heard two playback stimuli, a prime and a probe, separated by an interval of five minutes of silence. Across conditions, priming and probe stimuli varied with respect to their acoustic and referential (semantic) resemblance. In the baseline condition, individuals first heard the growls of a leopard followed after five minutes by playback of a second set of growls. As predicted male Diana monkeys produced significantly fewer leopard alarm calls to the second set of growls than to the first. In the control condition, the monkeys first heard playback of chimpanzee social screams, indicating the presence of chimpanzees, followed by leopard growls, indicating the presence of a leopard. Here, individuals were unable to predict the presence of a leopard from the chimpanzee screams. As predicted, males responded strongly to subsequent leopard growls, suggesting that they were surprised to hear a leopard. In the test condition, finally, monkeys heard the chimpanzee alarm screams to a leopard, followed by leopard growls. Results showed that males who responded with leopard alarm calls to the prime ("conspicuous males") subsequently remained silent to leopard growls, hereby resembling the behavior of males tested in the baseline condition. Only one out of the 14 conspicuous males tested responded with alarm calls to leopard growls, even though this stimulus is normally highly effective in eliciting alarm calls. The response of these males was not different from males tested in the baseline condition, but it differed significantly from the response of the control males. In contrast, the males who remained silent to the leopard alarm calls ("cryptic males"), responded with alarm calls to the leopard growls, resembling the males subjected to the control condition, but it differed from the males subjected to the baseline condition. Figure 8.5 illustrates these findings.

In sum, in some groups individuals behaved as if they recognized that the chimpanzee alarm screams could signal the presence of a leopard. Groups living in the core area of the resident chimpanzee community were more likely to do so than peripheral groups. The prime–probe experiment suggested that these individuals were able to understand the factors underlying the production of chimpanzee alarm calls, because priming with both chimpanzee alarm calls and leopard growls had similar effects on the monkeys' responses to the probe.

One common feature of the playback experiments discussed so far was the highly specific referential properties of the stimuli used, in which acoustically distinct utterances appeared to be associated with one particular type of predator. Theoretically, these relations do not require much cognitive flexibility on behalf of the recipient, since the same utterance will always refer to the same external event. However, in their natural habitats monkeys encounter numerous signals that have more

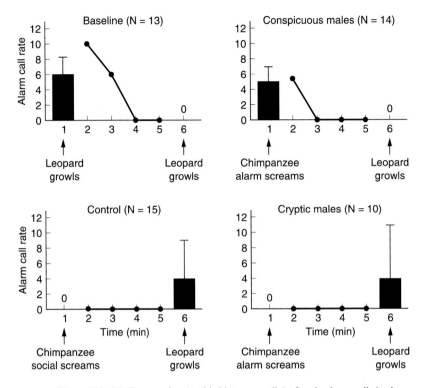

Figure 8.5. Median number (+ third inter-quartile) of male alarm calls in the baseline, test and control condition. (Data from Zuberbühler 2000b.)

vague and ambiguous referents, which require the monkeys to take into account additional information before being able to respond adaptively. The predator alarm calls of the ground-dwelling crested guinea fowl (*Guttera pulcheri*) provide a good example. Chimpanzees do not hunt guinea fowl, but leopards and human poachers may take them. These birds forage in large groups and, when chased, produce conspicuously loud and rattling sounding alarm calls (Seavy 2001) regardless of the type of predator present.

Observations have indicated that Diana monkeys respond to guinea fowl alarm calls as if a leopard were present, suggesting that they may not understand that several predator types can cause the birds to give alarm calls. To address this issue, the following playback was conducted. First, the monkeys' responses to alarm calls of crested guinea fowl were compared with their responses to alarm calls of the helmeted guinea fowl, *Numida meleabris*, a closely related species that does not occur in the forest. Diana monkeys responded to crested guinea fowl alarm calls

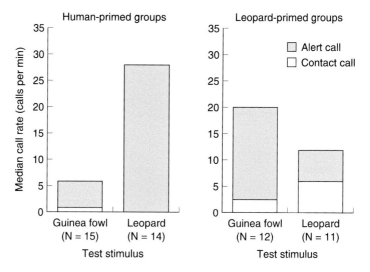

Figure 8.6. Diana monkey responses to guinea fowl alarm calls and leopard growls after having been primed with either human speech (left) or leopard growls (right). (Data from Zuberbühler 2000a.)

as if a leopard were present but did not show much response to the alarm calls of the helmeted guinea fowl (Zuberbühler 2000a). Results were thus consistent with the idea that Diana monkeys treated guinea fowl alarm calls as indicative of the presence of a leopard, even though these calls are also given to humans chasing them. To test whether Diana monkeys were able to take this fact into account, different groups were primed either with leopard growls or human speech to simulate the presence of either of these two predators. After a period of silence, recordings of guinea fowl alarm calls were played to the monkeys from the same location. If the monkeys were able to link the guinea fowl alarm calls to the presence of a specific predator, then their response to the alarm calls should vary depending on the predator vocalizations used as a priming stimulus.

Results showed that the priming stimuli affected the way Diana monkeys responded to guinea fowl alarm calls (Figure 8.6). When primed with human speech, Diana monkeys remained mostly quiet to subsequent guinea fowl alarm calls, a behavioral pattern that was not found in groups that were primed with leopard growls. This difference did not arise because human-primed groups behaved cryptically to any subsequent stimulus. When leopard growls were played to human-primed groups, the monkeys responded by giving many alarm calls. Monkeys responded with a significantly lower call rate to guinea fowl alarm calls when primed with human speech than after being primed with leopard growls. The number of leopard alarm calls produced, however, did not separate the two groups: in only one out of 15 cases did monkeys (male or female) produce leopard alarm calls to guinea fowl alarm calls after being primed with human speech. Similarly, groups primed with leopard growls produced leopard alarm calls to guinea fowls in only 3 out of 12 cases. No significant differences were found in the number of female leopard alarm calls or male leopard alarm calls to guinea fowl alarm calls between human-primed and leopard-primed groups. This paralleled the behavior of control groups primed and re-tested with leopard growls, where females produced leopard alarm calls in only 1 out of 11 cases to the second leopard playback. In sum, these data suggested that when responding to guinea fowl alarm calls, the Diana monkeys responded to the most likely cause for which the birds gave the calls, rather than the calls themselves.

The flexibility of semantic knowledge: complex utterances and call combinations

In contrast with human language, animal communication is often described as event-bound with no evidence for originality or creativity (Ghazanfar & Hauser 1999). It is a hallmark of a human language that it consists of a collection of syntactic rules that specify how words are put together into phrases and sentences, the source of its creativity, which allow speakers to construct an infinite number of previously unheard messages. Children learn to extract, and without any specific instructions, the syntactic rules of their native language. Although combinatorial properties have been found in animal communication systems, most analyses have focussed on the structural rules that underlie the system (e.g. Hailman & Ficken 1987) rather than how it may affect the calls' meanings.

A recent field study suggested that, as recipients, non-human primates may possess some of the cognitive capacities required to extract and interpret the meaning from call sequences (Zuberbühler 2002b). When faced with general disturbances, such as the thundering sound of a falling tree, male Campbell's monkeys normally produce a pair of brief and low-pitched "boom" vocalizations before giving a series of alarm calls. These boom calls are invariably given in pairs separated by some seconds of silence and typically precede an alarm call series by about 30 seconds. "Boom" introduced alarm call series are given to a number of disturbances, such as a falling tree or large breaking branch, the far-away alarm calls of a neighboring group, or a distant predator. Common to these contexts is the lack of immediate danger, unlike when callers are surprised by a close predator. When hearing "boom" introduced Campbell's alarm calls, Diana monkeys do not respond with their own alarm calls, which contrasts sharply to their vocal response to normal, that is "boom" free, Campbell's alarm calls (Figures 8.2 and 8.3). These observations have led to the hypothesis that the booms selectively affect the meaning of subsequent alarm calls.

The following playback experiments were conducted to address the problem. In two baseline conditions, different Diana monkey groups heard a series of five male Campbell's monkey alarm calls given to a crowned eagle or a leopard. Subjects were expected to respond strongly, i.e. to give many eagle or leopard alarm calls, as in the previous experiments (Figure 8.2). In the two test conditions, different Diana monkey groups heard playbacks of the exact same Campbell's alarm call series, but this time two "booms" were artificially added 25 seconds before the alarm calls. If Diana monkeys understood that the "booms" acted to modify the semantic specificity of subsequent alarm calls, then they should give significantly fewer predator-specific alarm calls in the test conditions compared to the baseline conditions. Results of this experiment replicated the natural observations. Playbacks of Campbell's eagle alarm calls caused the Diana monkeys to give their own eagle alarm calls, while playbacks of Campbell's leopard alarm calls caused them to give leopard alarm calls (Figure 8.2).

Playback of booms alone did not cause any noticeable change in Diana monkey vocal behavior, but they had a significant effect on how the monkeys responded to subsequent Campbell's alarm calls (Figure 8.7). If "booms" preceded playbacks of Campbell's leopard alarms, subjects no longer responded with leopard alarm calls. An analogous pattern was found if Campbell's eagle alarm calls were used as playback stimuli. Both the adult male and female Diana monkeys produced their own acoustically distinct eagle alarm calls to this stimulus. If "booms" preceded the Campbell's eagle alarms, however, subjects did not respond with eagle or leopard alarm calls. If Diana monkey alarm calls were used as playback stimuli instead of Campbell's alarm calls, then the "booms" of the Campbell's monkeys no longer had an effect on the Diana monkeys' alarm call behavior. The booms, in other words, affected the way the Diana monkeys interpreted the meaning of subsequent Campbell's alarm calls. They seemed to indicate to nearby listeners that whatever message followed did not require any anti-predator response. Judging from the Diana monkeys' response to these playback stimuli, therefore,

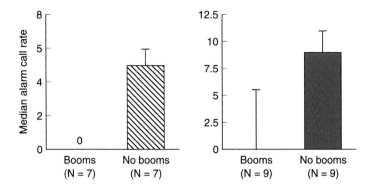

Figure 8.7. Median alarm call responses of male Diana monkeys from different groups to the different playback conditions (median call rates + third quartile during the first minute after beginning of a playback. Black: leopard alarm calls; hatched: eagle alarm calls). (Data from Zuberbühler 2002.)

the booms modified the meaning of the subsequent alarm call series and transformed them from highly specific predator labels, requiring immediate anti-predator responses, into more general signals of disturbance that did not require any direct responses.

More recent work on free-ranging putty-nosed monkeys has reported similar findings. Males produce two types of loud calls, the hacks and pyows to their predators, which include crowned eagles and leopards (Eckardt & Zuberbühler 2004). The two call types did not function as referential signals, however, because both call types occurred within alarm calling sequences irrespective of predator category (Arnold & Zuberbühler 2006a, 2006b). Nevertheless, call production was not random, but there were striking regularities in the patterning of the calls given in response to the two predators. Call sequences to eagles were structurally different from call sequences to leopards, suggesting that the sequences, rather than the individual calls, functioned as carriers of semantic content and provided the basis for selecting appropriate anti-predator responses (Arnold & Zuberbühler 2006a, 2006b).

The encoding of semantic information

The previous sections have focussed on the cognitive capacities of call recipients and emphasized the flexibility involved in call perception. Comparatively less is known about the mechanisms underlying call production, both in terms of call structure and call use. Currently, the available evidence suggests that, as signallers, non-human primates are substantially more limited in their flexibility than are humans

(Tomasello & Zuberbühler 2002). First, primates appear to have only limited control over their articulators. Cross-fostering experiments (Owren *et al.* 1993), ontogenetic data (e.g. Hammerschmidt *et al.* 2001), and hybridization studies (e.g. Geissmann 1984) all suggest that primates are unable to generate relevant vocal variability, much in contrast to what has been found in some other groups of mammals (Janik *et al.* 1994). However, some social variables have lasting effects on the acoustic structure of some primate calls (Marshall *et al.* 1999, Snowdon & Elowson 1999). Campbell's monkeys, for example, produce acoustically distinct contact call variants, which they can or cannot share with other group members. For each individual, the use of shared variants changes over time, as do the social relationships amongst the individuals (Lemasson & Hausberger 2004). A playback experiment has provided evidence that acoustic fine structure of Campbell's monkey contact calls is perceptionally salient and socially meaningful to these individuals (Lemasson *et al.* 2005).

Sound production mechanisms

Several factors account for the fact that there is generally little acoustic variation in primate vocal repertoires compared to human speech sounds. According to classic theory, acoustic variation can result from two sources: the larynx and the vocal tract (Fant 1960). Many animal calls are the result of changes at the level of the larynx. This leads to changes in the fundamental frequency of the call. In some human languages changes in the fundamental frequency can be carriers of meaning, such as in Mandarin Chinese. A second more relevant source of variation concerns the vocal tract and its various articulators. In human speech the tongue plays a central role during articulation and it is particularly remarkable how little is known about the phylogenetic origins of enhanced tongue motor control in humans. In non-human primates, limited articulation has been described, mainly in rhesus monkeys (Hauser *et al.* 1993, Hauser & Ybarra 1994). Coo and grunt calls are the product of active vocal tract filtering, but acoustic differences have not been linked to particular external events (Rendall *et al.* 1998). The effects of vocal tract filtering can only be assessed with special analysis techniques that separate the effect of the glottal source from that of the vocal tract (e.g. Owren & Bernacki 1998).

A number of studies have examined the vocal production mechanisms underlying Diana monkey alarm calling behavior (Riede & Zuberbühler 2003a, 2003b, Riede *et al.* 2005). Male Diana monkey alarm calls are remarkably low-pitched (fundamental frequency 33 to 120 Hz), comparable to the human pulse register (fundamental frequency 10 to 90 Hz; Henton & Bladon 1988). Non-linear phenomena are virtually absent in these calls and pulses are not interrupted by any other vibration modes of the vocal folds. The pulsed phonation in male Diana monkey alarm calls, therefore, appear to be a special adaptation to deliver a robust broadband signal at the level of the source for subsequent vocal tract filtering.

Diana monkey leopard and eagle alarm calls possess formant characteristics (Riede & Zuberbühler 2003b). The two alarm call types differ most prominently in the downward modulation of the first formant at the beginning of each syllable, with leopard alarm calls exhibiting a downward modulation that is three-fold stronger than eagle alarm calls, providing further evidence that non-human primates can rely on the formant modulations to convey information about external events (Owren 1990a, 1990b).

In sum, non-human primates are able to perform some basic vocal tract changes and the resulting acoustic variation can function to encode important events in the environment, a defining feature of human speech production. Although humans clearly show a number of speech-related adaptations, such as a lowered larynx and a high tongue and lip motility, non-human primates are not fundamentally impaired by any anatomical constraints, suggesting that the reason for their relatively simple vocal behavior must be neural.

Call use

As with call structure, non-human primates' flexibility in call use is rather limited. Many primate calls appear to be used in adult-like contexts from early on (Seyfarth & Cheney 1997). For example, infant vervet monkeys give eagle alarm calls to various moving things in the sky (Seyfarth & Cheney 1980). Similarly, young pigtail macaques (*Macaca nemestrina*) are less precise when using agonistic screams than are adults (Gouzoules & Gouzoules 1989). However, some flexibility of use appears to persist even in adults, but the topic is not well researched. For example, Japanese macaques show population-level differences in their use of food and contact calls (Green 1975, Sakura 1989). In the case of the alarm calls it is of interest what kinds of mental processes underlie call production, specifically, whether monkeys actually encode the predator type when producing alarm calls. Research on California ground squirrels, *Spermophilus beecheyi*, has shown that this species possess two acoustically different alarm calls, the "whistle" and the "chatter-chat." However, these alarms appear to be given in response to the degree of threat imposed

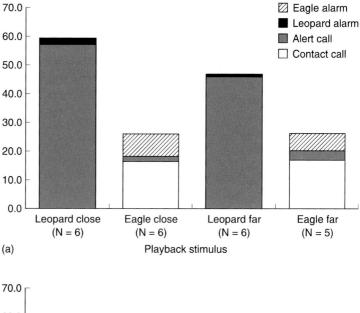

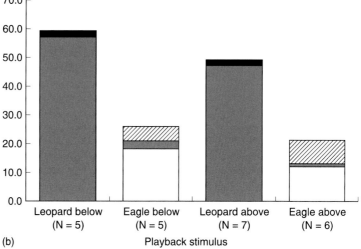

Figure 8.8. Diana monkey responses to playbacks of predator vocalizations presented with varying degrees of threat: (a) distance, (b) direction of attack (Data from Zuberbühler 2000d).

by a predator, rather than its biological class (Owings & Virginia 1978, Owings & Leger 1980). "Whistles" are given whenever a predator arrives suddenly and a response is urgent, while "chatter-chat" alarms are given to predators spotted at a distance. Typically, such predators are mammalian carnivores but squirrels have been observed to give "chatter-chat" alarms to a distant hawk (Leger *et al.* 1980). Domestic chickens, *Gallus domesticus*, provide another interesting example. In this species, males give aerial and ground predator alarm calls (Gyger *et al.* 1987), but individuals give ground alarm calls to many objects moving on the substrate and aerial alarm calls to many objects moving above in free space, regardless of whether or not they are predators, suggesting that chickens respond to the location of a threat. To investigate which aspects the Diana monkeys responded to when giving alarm calls, the presence of a predator was simulated in various ways (Zuberbühler 2000b). A playback speaker was positioned in the vicinity of Diana monkey groups, such that the distance to the group was either "close" or "far" (about 25 or 75 m), the elevation of the speaker was either "below" or "above" the group (about 2 or 30 m off the ground), and the predator was either a "leopard" or an "eagle" (15-s playback of leopard growls or eagle shrieks). Results showed that Diana monkeys consistently responded to predator type, regardless of distance or direction of predator attack (see Figure 8.8). The experiment was replicated with Campbell's monkeys, with comparable results: the predator type was the main determinant of alarm calling behavior in this species as well (Zuberbühler 2001).

Summary

Primate alarm calls have been under selection pressure by all three major evolutionary forces, kin selection, sexual selection and individual selection. Alarm calls have dissuasive effects on the hunting behavior of leopards, suggesting that callers enjoy substantial individual benefits. Alarm calls are also interesting because they provide one of the best windows into the mind of non-human animals. Research has shown that some non-human primates are able to employ elements of their vocal repertoire to label external events, thereby demonstrating that referential abilities are part of primate cognition. The field studies reviewed in this chapter have identified various cognitive capacities, including the ability to take into account semantic, combinatorial and pragmatic cues, suggesting that primates possess a number of capacities that may also be involved in human language processing. As signallers, non-human primates are curiously limited, however. A more complete understanding of the semantic capacities of non-human primates will require further research involving other primate species, particularly the three colobines, the sooty mangabeys and the chimpanzees (e.g. Slocombe & Zuberbühler 2005a, 2005b).

References

Anderson, M. (1994). *Sexual Selection*. Princeton, NJ: Princeton University Press.

Arnold, K. and Zuberbühler, K. (2006a). The alarm calling system of putty-nosed monkeys, *Cercopithecus nictitans martini*. *Animal Behaviour*, **72**, 643−53.

Arnold, K. and Zuberbühler, K. (2006b). Language evolution: semantic combinations in primate calls. *Nature*, **441**, 303.

Bergstrom, C. T. and Lachmann, M. (2001). Alarm calls as costly signals of antipredators vigilance: the watchful babbler game. *Animal Behaviour*, **61**, 535−43.

Blumstein, D. T., Steinmetz, J., Armitage, K. B. and Daniel, J. C. (1997). Alarm calling in yellow-bellied marmots: 2. The importance of direct fitness. *Animal Behaviour*, **53**, 173−84.

Boesch, C. and Boesch, H. (1989). Hunting behavior of wild chimpanzees in the Taï National Park. *American Journal of Physical Anthropology*, **78**, 547−73.

Chapman, C. A., Chapman, L. J. and Lefebvre, L. (1990). Spider monkey alarm calls: honest advertisement of warning kin. *Animal Behaviour*, **39**, 197−8.

Charnov, E. L. and Krebs, J. R. (1975). Evolution of alarm calls: altruism or manipulation? *American Naturalist*, **109**, 107−12.

Cheney, D. L. and Seyfarth, R. M. (1985). Vervet monkey alarm calls: manipulation through shared information? *Behaviour*, **94**, 739−51.

Cheney, D. L. and Seyfarth, R. M. (1988). Assessment of meaning and the detection of unreliable signals by vervet monkeys. *Animal Behaviour*, **36**, 477−86.

Cheney, D. L. and Seyfarth, R. M. (1990). Attending to behaviour versus attending to knowledge: examining monkeys' attribution of mental states. *Animal Behaviour*, **40**, 742−53.

Clutton-Brock, T. M. and Albon, S. D. (1979). The roaring of red deer and the evolution of honest advertisement. *Behaviour*, **69**, 145−70.

Eckardt, W. and Zuberbühler, K. (2004). Cooperation and competition in two forest monkeys. *Behavioral Ecology*, **15**, 400−11.

Eimas, P. D., Siqueland, P., Jusczyk, P. and Vigorito, J. (1971). Speech perception in infants. *Science*, **212**, 303−6.

Fant, G. (1960). *Acoustic Theory of Speech Production*. The Hague: Mouton.

Gautier, J.-P. and Gautier, A. (1977). Communication in Old World monkeys. In *How Animals Communicate*, ed. T. A. Sebeok Bloomington, IN: Indiana University Press, pp. 890−964.

Geissmann, T. (1984). Inheritance of song parameters in the gibbon song, analysed in 2 hybrid gibbons (*Hylobates pileatus* x *H. lar*). *Folia Primatologica*, **24**, 216−35.

Ghazanfar, A. A. and Hauser, M. D. (1999). The neuroethology of primate vocal communication: substrates for the evolution of speech. *Trends in Cognitive Sciences*, **3**, 377−85.

Goodall, J. (1986). *The Chimpanzees of Gombe: Patterns of Behaviour*. Cambridge, MA: Harvard University Press.

Gouzoules, H. and Gouzoules, S. (1989). Design features and developmental modification of pigtail macaque, *Macaca nemestrina*, agonistic screams. *Animal Behaviour*, **32**, 182–93.

Green, S. (1975). Variation of vocal pattern with social situation in the Japanese monkey (*Macaca fuscata*): a field study. In *Primate Behavior Developments in Field and Laboratory Research*, ed. L. A. Rosenblum. New York: Academic Press, pp. 1–102.

Gyger, M., Marler, P. and Pickert, R. (1987). Semantics of an avian alarm call system: the male domestic fowl, *G. domesticus*. *Behaviour*, **102**, 15–40.

Hailman, J. P. and Ficken, M. S. (1987). Combinatorial animal communication with computable syntax: chick-a-dee calling qualifies as "language" by structural linguistics. *Animal Behaviour*, **34**, 1899–901.

Hammerschmidt, K., Freudenstein, T. and Jurgens, U. (2001). Vocal development in squirrel monkeys. *Behaviour*, **138**, 1179–204.

Hauser, M. D., Evans, C. S. and Marler, P. (1993). The role of articulation in the production of rhesus monkey, *Macaca mulatta*, vocalizations. *Animal Behaviour*, **45**, 423–33.

Hauser, M. D. and Ybarra, M. S. (1994). The role of lip configuration in monkey vocalizations – experiments using Xylocaine as a nerve block. *Brain and Language*, **46**, 232–44.

Henton, C. and Bladon, A. (1988). Creak as a sociophonetic marker. In *Language, Speech, and Mind: Studies in Honour of Victoria A. Fromkin*, ed. L. M. Hyman and C. N. Li. London: Routledge, pp. 3–29.

Herbinger, I., Boesch, C. and Rothe, H. (2001). Territory characteristics among three neighbouring chimpanzee communities in the Taï National Park, Côte d'Ivoire. *International Journal of Primatology*, **22**, 143–67.

Janik, V. M., Dehnhardt, G. and Todt, D. (1994). Signature whistle variations in a bottlenosed dolphin, *Tursiops truncatus*. *Behavioral Ecology and Sociobiology*, **35**, 243–8.

Leger, D. W., Owings, D. H. and Gelfand, D. L. (1980). Single note vocalizations of California ground squirrels: graded signals and situation – specificity of predator and socially evoked calls. *Zeitschrift für Tierpsychologie*, **52**, 227–46.

Lemasson, A. and Hausberger, M. (2004). Patterns of vocal sharing and social dynamics in a captive group of Campbell's monkeys (*Cercopithecus campbelli campbelli*). *Journal of Comparative Psychology*, **118**, 347–59.

Lemasson, A., Hausberger, M. and Zuberbühler, K. (2005). Socially meaningful vocal plasticity in adult Campbell's monkeys (*Cercopithecus campbelli*). *Journal of Comparative Psychology*, **112**, 220–9.

Lemasson, A., Richard, J. P. and Hausberger, M. (2004). A new methodological approach to context analysis of call production. *Bioacoustics – the International Journal of Animal Sound and Its Recording*, **14**, 111–25.

Marshall, A., Wrangham, R. and Clark Arcadi, A. (1999). Does learning affect the structure of vocalizations in chimpanzees? *Animal Behaviour*, **58**, 825–30.

Maynard-Smith, J. (1965). The evolution of alarm calls. *American Naturalist*, **99**, 59–63.

Nagel, T. (1974). What is it like to be a bat? *Philosophical Review*, **83**, 435−50.

Nagel, T. (1995). *Other Minds*. Oxford: Oxford University Press.

Owings, D. H. and Leger, D. W. (1980). Chatter vocalizations of California ground squirrels: predator- and social-role specificity. *Zeitschrift für Tierpsychologie*, **54**, 163−84.

Owings, D. H. and Virginia, R. A. (1978). Alarm calls of California ground squirrels (*Spermophilus beecheyi*). *Zeitschrift für Tierpsychologie*, **46**, 58−70.

Owren, M. J. (1990a). Acoustic classification of alarm calls by vervet monkeys (*Cercopithecus aethiops*) and humans (*Homo sapiens*). 1. Natural calls. *Journal of Comparative Psychology*, **104**, 20−8.

Owren, M. J. (1990b). Acoustic classification of alarm calls by vervet monkeys (*Cercopithecus aethiops*) and humans (*Homo sapiens*). 2. Synthetic calls. *Journal of Comparative Psychology*, **104**, 29−40.

Owren, M. J. and Bernacki, R. H. (1998). Applying Linear Predictive Coding (LPC) to frequency-spectrum analysis of animal acoustic signals. In *Animals Acoustic Communication*, ed. S. L. Hopp, M. J. Owren and C. S. Evans. Berlin: Springer, pp. 129−62.

Owren, M. J., Dieter, J. A., Seyfarth, R. M. and Cheney, D. L. (1993). Vocalizations of rhesus (*Macaca mulatta*) and Japanese (*M. fuscata*) macaques cross-fostered between species show evidence of only limited modification. *Developmental Psychobiology*, **26**, 389−406.

Rendall, D., Owren, M. J. and Rodman, R. S. (1998). The role of vocal tract filtering in identity cueing in rhesus monkey (*Macaca mulatta*) vocalizations. *Journal of the Acoustical Society of America*, **103**, 602−14.

Riede, T., Bronson, E., Hatzikirou, B. and Zuberbühler, K. (2005). The production mechanisms of Diana monkey alarm calls: morphological data and a model. *Journal of Human Evolution*, **48**, 85−96.

Riede, T., Bronson, E., Hatzikirou, B. and Zuberbühler, K. (2006). Multiple discontinuities in nonhuman vocal tracts − A response to Lieberman (2006). *Journal of Human Evolution*, **50**, 222.

Riede, T. and Zuberbühler, K. (2003a). Pulse register phonation in Diana monkey alarm calls. *Journal of the Acoustical Society of America*, **113**, 2919−26.

Riede, T. and Zuberbühler, K. (2003b). The relationship between acoustic structure and semantic information in Diana monkey alarm calls. *Journal of the Acoustical Society of America*, **114**, 1132−42.

Sakura, O. (1989). Variability in contact calls between troops of Japanese macaques: a possible case of neutral evolution of animal culture. *Animal Behaviour*, **38**, 900−2.

Seavy, E. A. (2001). Associations of crested guinea fowl *Guttera pucherani* and monkeys in Kibale national park, Uganda. *Ibis*, **143**, 310−12.

Seyfarth, R. M. and Cheney, D. L. (1980). The ontogeny of vervet monkey alarm calling behavior: a preliminary report. *Zeitschrift für Tierpsychologie*, **54**, 37−56.

Seyfarth, R. and Cheney, D. L. (1997). Behavioral mechanisms underlying vocal communication in nonhuman primates. *Animal Learning and Behavior*, **25**, 249−67.

Seyfarth, R. M., Cheney, D. L. and Marler, P. (1980). Monkey responses to three different alarm calls: evidence of predator classification and semantic communication. *Science*, **210**, 801–3.

Slocombe, K. and Zuberbühler, K. (2005a). Functionally referential communication in a chimpanzee. *Current Biology*, **15**, 1779–84.

Slocombe, K. E. and Zuberbühler, K. (2005b). Agonistic screams in wild chimpanzees (*Pan troglodytes schweinfurthii*) vary as a function of social role. *Journal of Comparative Psychology*, **119**, 67–77.

Snowdon, C. T. and Elowson, A. M. (1999). Pygmy marmosets modify call structure when paired. *Ethology*, **105**, 893–908.

Struhsaker, T. T. (1967). Auditory communication among vervet monkeys (*Cercopithecus aethiops*). In *Social Communication Among Primates*, ed. S. A. Altmann. Chicago: University of Chicago Press, pp. 281–324.

Tenaza, R. R. and Tilson, R. L. (1977). Evolution of long-distance alarm calls in Kloss's gibbon. *Nature*, **268**, 233–5.

Tomasello, M. and Zuberbühler, K. (2002). Primate vocal and gestural communication. In *The Cognitive Animal*, ed. M. Bekoff, C. Allen and G. M. Burghardt. Cambridge, MA: Massachusetts Institute of Technology Press, pp. 293–9.

Uster, D. and Zuberbühler, K. (2001). The functional significance of Diana monkey "clear" calls. *Behaviour*, **138**, 741–56.

von Frisch, K. (1973). Decoding the language of the bee. In *Nobel Lecture*.

Wachter, B., Schabel, M. and Noë, R. (1997). Diet overlap and polyspecific associations of red colobus and Diana monkeys in the Taï National Park, Ivory Coast. *Ethology*, **103**, 514–26.

Wolters, S. and Zuberbühler, K. (2003). Mixed-species associations of Diana and Campbell's monkeys: the costs and benefits of a forest phenomenon. *Behaviour*, **140**, 371–85.

Yates, J. and Tule, N. (1979). Perceiving surprising words in an unattended auditory channel. *Quarterly Journal of Experimental Psychology*, **31**, 281–6.

Zuberbühler, K. (2000a). Causal cognition in a non-human primate: field playback experiments with Diana monkeys. *Cognition*, **76**, 195–207.

Zuberbühler, K. (2000b). Causal knowledge of predators' behaviour in wild Diana monkeys. *Animal Behaviour*, **59**, 209–20.

Zuberbühler, K. (2000c). Interspecific semantic communication in two forest monkeys. *Proceedings of the Royal Society of London, B*, **267**, 713–18.

Zuberbühler, K. (2000d). Referential labelling in Diana monkeys. *Animal Behaviour*, **59**, 917–27.

Zuberbühler, K. (2001). Predator-specific alarm calls in Campbell's guenons. *Behavioral Ecology and Sociobiology*, **50**, 414–22.

Zuberbühler, K. (2002a). Effects of natural and sexual selection on the evolution of guenon loud calls. In *The Guenons: Diversity and Adaptation in African Monkeys*, ed. M. E. Glenn and M. Cords. New York: Plenum, pp. 289–306.

Zuberbühler, K. (2002b). A syntactic rule in forest monkey communication. *Animal Behaviour*, **63**, 293–9.

Zuberbühler, K., Cheney, D. L. and Seyfarth, R. M. (1999). Conceptual semantics in a nonhuman primate. *Journal of Comparative Psychology*, **113**, 33–42.

Zuberbühler, K. and Jenny, D. (2002). Leopard predation and primate evolution. *Journal of Human Evolution*, **43**, 873–86.

Zuberbühler, K., Jenny, D. and Bshary, R. (1999). The predator deterrence function of primate alarm calls. *Ethology*, **105**, 477–90.

III *Habitat use*

9 *Positional behavior and habitat use of Taï forest monkeys*

W. S. McGraw

Introduction

Locomotion and posture are important elements of behavioral ecology. Since primates are mobile and most are arboreal, knowing where, why, and how they position themselves in the forest canopy provides a better understanding of many aspects of primate life. Establishing predictive relationships between positional behavior and additional aspects of primate biology not only helps explain the behavior of living animals, but also allows us to infer the behavior of fossil taxa more reliably. Although there have been many studies of one or a few species at single sites, studies of positional behavior of whole (or nearly whole) primate communities are rare. Notable exceptions include the landmark study of Fleagle and Mittermeier (1980) on monkeys in Surinam and the study by Gebo and Chapman (1995a, 1995b) on monkeys in Uganda's Kibale Forest. In an effort to add to the body of comparative data, I began studies on the positional behavior of seven Taï Forest monkeys in 1993. The first study of Taï monkeys by Galat and Galat-Luong (1985) contained basic data on habitat use, however these authors were not primarily concerned with positional behavior.

The monkeys at Taï provide a potentially stringent test of what determines – or at least co-varies with – locomotion and posture. Unlike the Surinam monkeys studied by Fleagle and Mittermeier (1980), the Taï cercopithecids do not represent a great diversity of locomotor adaptations or as great a range of body size (Schultz 1970, Oates *et al.* 1990, Fleagle 1999). With the exception of *Cercocebus atys*, all the Taï monkeys have traditionally been classified as generalized arboreal quadrupeds that lack more specialized adaptations such as hindlimb modifications for vertical clinging, claws, prehensile tails, or highly mobile shoulder joints. The more restricted range of body sizes represented by the seven species provides an opportunity to more rigorously examine the relationship between mass and a variety of positional and

Monkeys of the Taï Forest, ed. W. Scott McGraw, Klaus Zuberbühler and Ronald Noë.
Published by Cambridge University Press. © Cambridge University Press 2007.

habitat variables. Body weights of adult females are *Colobus polykomos* King or Western black and white colobus (8.3 kg), *Procolobus badius* red colobus (8.2 kg), *Procolobus verus* olive colobus (4.2 kg), *Cercopithecus diana* Diana monkey (3.9 kg), *Cercopithecus campbelli* Campbell's monkey (2.7 kg), *Cercopithecus petaurista* lesser spot-nosed monkey (2.9 kg) and *Cercocebus atys* sooty mangabey (6.2 kg) (Oates *et al.* 1990). At the time this chapter was written, data on the positional behavior of *Cercopithecus nictitans* greater spot-nosed monkey had not yet been collected.

This chapter summarizes and synthesizes results from positional studies conducted on the Taï monkeys since 1993 (McGraw 1996a, 1996b, 1998a, 1998b, 1998c, 2000). In addition to providing descriptions of the locomotion, posture and habitat use of each species, I address several specific questions:

1.   What is the relationship between locomotor behavior, body size, canopy use, maintenance activities, support size, and inter-membral index (length of forelimb compared to length of hindlimb)?
2.   Are the relationships between these variables similar to those observed in other primate communities?
3.   Does locomotor behavior significantly change in different habitat types?
4.   Are there correlates to postural behavior?

As the following discussion illustrates, some questions cannot be answered satisfactorily and have in fact spawned many additional ones. There is clearly still much to learn and this chapter will hopefully promote additional research on the positional behavior of the primates at Taï and in other tropical forests.

Methodology

Data were collected from 1993−94 and over shorter periods in 1996, 1998 and 2000. I used an instantaneous time point sampling regime to quantify monkey positional behavior and habitat use. Data on adult females were collected at three-minute intervals; at each instant the behavior of one animal was recorded and this individual was not sampled again until at least 15 minutes had passed. The 15-minute time interval between same-individual samples was designed to avoid dependency of data; a test of independence was performed on all data sets and they were found not to be temporally auto-correlated (see McGraw 1996a, 1996b, 1998a). Females were chosen as study subjects since there are more adult females than males in all study groups. I was therefore able to sample more individuals since

I did not need to wait long to find another independent individual. At every three minute time point I recorded the following data (see Table 9.1 for definitions): (1) time, (2) location within grid, (3) species sampled, (4) general activity, (5) positional behavior, (6) forest stratum, (7) absolute height, and (8) support type. Every effort was made to sample a species for an entire week before rotating to a new species. By rotating species continually throughout the study I minimized seasonal effects on the data.

A major goal of this project was to determine whether monkeys alter their locomotion in structurally different habitats. In order to assess the effects of habitat heterogeneity on positional behavior, I undertook an architectural analysis of the forest within the main study grid. The methodological details can be found in McGraw (1996b) and are discussed only briefly here. The analysis consisted of two parts: (1) determining differences in the relative number of stems (trunks of all sizes) in two forest types and (2) quantifying differences in the availability of different sized supports (boughs, branches and twigs) at regular height intervals in each of the two forest types. Stem density was quantified by sampling a 50 m × 20 m rectangular portion of each 100 m × 100 m grid cell within the main study grid (see Figure 1 in McGraw 1996b). The orientation of the rectangle was alternated in adjacent cells in order to reduce dependency between cells. All trees greater than 2 m in height and with a DBH greater than 4 cm were counted and marked. For each qualifying tree, its DBH was recorded and its height measured using a rangefinder. The circumference of multi-stemmed trees was determined by summing the areas of all stems: the area of each stem was determined using each stem's circumference to figure its radius, calculating the radius of the total circle, and determining the circle's circumference using the equation $2\pi r$. The height of multi-stemmed trees was recorded as the highest point of any stem.

The second phase of the architectural analysis involved sampling the vertical canopy. The spatial distribution of supports at different heights was assessed at six points on a template superimposed on each grid cell (see Figure 2 in McGraw 1996b). The orientation of this sampling template was alternated in adjacent cells. All sampling points were at least 20 m apart to help ensure that sampling points were independent and that I avoided sampling the same tree crown more than once. At each sampling point, I aimed the rangefinder skyward, perpendicular to the ground, and recorded (1) the height of the highest substrate and (2) the diameter of that substrate in the plane directly overhead. This substrate then served as a focal substrate around which a sampling field was designated. All supports in the same plane and within three meters of the focal support were counted and their diameters estimated using a calibrated

Table 9.1. *Definitions*

General Activities	
Traveling	Directed, usually uninterrupted movement between major food sources and/or sleeping sites.
Foraging	Movement during feeding usually confined to single or contiguous trees.
Feeding	Stationary behavior involved during the acquisition and/or processing of food items.
Resting	Stationary behavior involving no other recognizable behavior.
Social	Stationary behavior involving grooming.
Positional Behaviors	
Walking	slow, pronograde quadrupedal locomotion.
Running	fast gaits including bounding and galloping.
Climbing	vertical or semi-vertical ascent in which the arms reach above the head and pull the animal up while the hindlimbs alternately push the body up.
Leaping	progression between discontinuous supports using primarily rapid extension of the hindlimbs for propulsion; landing includes both hind limbs and forelimbs.
Arm swinging	locomotion involving forelimb suspension: bimanualism.
Sit	Ischial callosities and thighs bear the majority of body weight regardless of the position of the limbs.
Quadrupedal stand	Stationary pronograde position.
Lie	Reclining postures of all types including prone lie, supine lie, sprawl.
Stand − forelimb suspend	A modification of the posture defined by Hunt *et al.* (1996: 371) who state "more than half of the body weight (is) supported by the hindlimbs, but there is significant support from the forelimb oriented in an forelimb-suspend pattern." I modify this definition to include those occasions when the forelimb is used to reach food or simply aid in stabilizing the trunk in an upright position. The abducted forelimb need not be elevated above the shoulder (see Figure 5 in McGraw 1998c).
Forest Strata	
Ground	Forest floor.
Stratum 1	Shrub and sapling layer up to five meters in height.
Stratum 2	Under story: small trees and half grown canopy trees usually between five and 15/20 meters.
Stratum 3	Main canopy: closed canopy dominated by larger, more horizontal supports. Divided further into lower main canopy usually between 15 and 23 m (3−) and upper main canopy usually between 24 and 40 meters (3+).
Stratum 4	Emergent layer: largest trees that punctuate the main, continuous canopy; usually greater than 40 m.
Support Types	
Bough	large supports usually greater than 10 cm in diameter in which grasping with hands and feet is not possible.
Branch	medium sized supports between 2 and 10 cm in diameter permitting grasping of hands and feet.
Twig	small flexible terminal branches usually less than 2 cm in diameter.

spotting scope. I then sampled in similar fashion in descending 10 m intervals until the lowest supports were counted. Therefore, the number of canopy classes (10 m samples) is determined by the height of the forest being sampled. When all grid cells were sampled, I tabulated the number of supports in each category per height in each forest type using the size criteria mentioned above (see Table 9.1). I used a three-factor log linear model with two forest types, five height classes and three support types to analyze differences in the relative abundance of different sized supports within the grid.

Results
Canopy use
The frequency with which each monkey species uses each canopy layer during overall locomotion (combined travel and foraging), during travel and during foraging is summarized in Table 9.2. The Taï monkeys share the canopy by using different levels or by avoiding species that prefer the same levels. The seven species fall into one of four general categories (1) high canopy dwellers, (2) canopy generalists, (3) under-story specialists, and (4) ground dwellers. *Colobus polykomos* and *Procolobus badius* are best described as high canopy monkeys. These two similarly sized colobines rarely use the sapling and ground layers; approximately 80 per cent of their overall activity occurs at levels above the under story. Competition over canopy space is minimized by these species avoiding each other and by foraging for different foods (Korstjens 2001). The third Taï colobine, *Procolobus verus*, uses a relatively narrow forest zone spending fully 60 per cent of its time in the under story (stratum 2). Olive colobus and *Cercopithecus diana* are almost always found in association (Korstjens & Schippers 2003, Korstjens & Noë 2004), possibly as an evolved strategy of predator avoidance (Oates & Whitesides 1990). Olive colobus maintain an association with Diana monkeys throughout a shared horizontal range but do not follow Diana monkeys throughout the latter's entire vertical range.

Stratum overlap among the guenon species during certain activities can be considerable and when the species are in association, aggressive encounters are not uncommon (McGraw *et al.* 2002, Buzzard – Chapter 2). Nevertheless, there are some marked differences between congenerics. *Cercopithecus campbelli* and *C. petaurista* are under story specialists that rarely move in the high canopy or emergent layers. Over 90 per cent of Campbell's monkey movement and nearly 80 per cent of lesser spot-nosed monkey movement occurs below the main canopy. Strata use during movement by these guenons is generally similar and the primary difference

Table 9.2. *Taï monkey strata use*

	4	3+	3−	2	1	Ground	N
Procolobus badius							
Overall	3.6	35.8	38.4	21.9	0.26	0.1	4196
Travel	2	36.6	44.3	16.9	0.1	−	797
Forage	2.8	28.1	37.8	30.5	0.6	0.2	669
Colobus polykomos							
Overall	8.9	43.4	27.9	18.9	0.68	0.2	3538
Travel	0.7	32.3	43.5	23.2	−	0.2	538
Forage	4.7	32.6	26.8	33.4	1.8	0.5	380
Procolobus verus							
Overall	0.06	4.3	19.9	61.1	11.9	2.8	1595
Travel	−	4.3	23.9	61.4	4.9	5.6	306
Forage	0.5	1.9	12.9	67.3	11.9	5.5	202
Cercopithecus diana							
Overall	1.5	18.7	29.7	44.1	4.4	1.7	3539
Travel	1.2	13.6	38.4	43.4	1.7	1.7	242
Forage	0.9	19.2	30.1	44.2	3.3	2.3	1311
Cercopithecus campbelli							
Overall	−	1.11	4.3	57.8	21.6	15.2	1437
Travel	−	5.2	5.2	48.9	8.3	32.3	96
Forage	−	1	5	55.8	20.4	17.8	500
Cercopithecus petaurista							
Overall	−	2.8	19	68.7	8.9	0.6	2042
Travel	−	0	32.6	59	5.6	2.8	144
Forage	−	3.3	17.1	68.2	9.8	1.7	368
Cercocebus atys							
Overall	−	0.6	2.4	10.6	19.1	67.24	1343
Travel	−	1.5	3.6	5.8	2.9	85.5	137
Forage	−	0.6	1.5	9.7	16.7	71.4	329

Data under each column are the percentages of locomotor time overall, during travel and during foraging spent in each stratum
4 (emergent layer); 3+ (upper main canopy); 3− (lower main canopy); 2 (under story); 1(shrub layer)

involves use of the ground: Campbell's monkey's moves on the forest floor 15 per cent of the time. *Cercopithecus diana* makes relatively uniform use of all canopy layers spending approximately 50 per cent in the main canopy and above and 50 per cent in the under story and below. In so doing, the Diana monkey is quite unlike all other arboreal monkeys at Taï that show a strong preference for one or two forest layers. It has been argued that the less restricted use of the canopy makes Diana monkeys better watchmen for both terrestrial and arboreal predators (Noë & Bshary 1997). The ability to monitor a wider area of forest makes Diana monkeys a desirable partner in terms of added vigilance. This may explain the high association rate between this species and red colobus (Noë & Bshary 1997), the latter of which are frequently preyed upon by chimpanzees (Boesch 1994). Finally, *Cercocebus atys* is the most divergent species spending nearly two-thirds of its time on the ground.

Comparing forest layers used during travel versus foraging reveals a few trends, though these are not necessarily statistically significant (see Table 9.2) (see McGraw 1998a for significance values). All monkeys use the low main canopy layer (stratum 3−) more for traveling than for foraging. This is most likely due to the fact that substrates preferred for travel by arboreal quadrupeds − large, horizontal supports − are more abundant in the main canopy layers (McGraw 1996b). All species use the under story levels (stratum 1 and 2) more for foraging than travel. No consistent pattern exists in use of the high main canopy (stratum 3+): some species use it more for travel, other species do so less often. Three of the four species that frequent the emergent layers do so more often during foraging than travel.

Most monkeys use different forest levels for travel and foraging. The exception is the Diana monkey which uses strata in similar fashion during both activities. This is surprising since supports used for travel tend to be located in different parts of trees or at different heights than those used for foraging. That Diana monkeys are able to make consistent use of the entire canopy during such different activities is further evidence of this monkey's ability to be an effective sentinel; associated species can rely on Diana monkeys to effectively move through and monitor the range of strata the same way during travel and foraging. In contrast, canopy use in the other guenons is strongly influenced by the nature of movement: *Cercopithecus campbelli* uses the ground almost twice as much during travel compared to foraging while *C. petaurista* uses the main canopy nearly twice as much for travel compared to foraging.

Locomotion and support use

Previous studies have demonstrated that the size of supports used during arboreal activities is correlated with body size (Napier 1962, Fleagle *et al.* 1981, Cant 1986, Garber 1991, Doran 1992, 1993). The frequency that each monkey used each major support type overall and during travel and foraging is listed in Table 9.3. The terrestrial sooty mangabey is clearly distinct from the other species in its preferred support − the forest floor − and I have not included *Cercocebus atys* with the arboreal monkeys in the comparisons of support size and body size. Nevertheless, it is important to note that although Taï mangabeys travel (80.4 per cent) and forage

Table 9.3. *Support use during overall positional behavior (locomotion and posture) and during travel and foraging*

	Ground	Bough	Branch	Twig	Other	N
Colobus polykomos						
Overall	0.2	39.1	37.22	23.29	0.23	3413
Travel	0.2	52.6	23.6	23.2	0.4	538
Forage	0.5	27.4	36.6	34.2	1.3	380
Procolobus badius						
Overall	0.1	43.8	33.4	22.2	0.55	4156
Travel	–	48.9	24.1	25.8	1.1	797
Forage	0.3	26.3	33.6	38.4	1.3	669
Procolobus verus						
Overall	2.07	24.2	46.1	25.9	1.75	1589
Travel	5.8	31.9	39.3	20.4	2.5	306
Forage	–	11.4	38.1	44.4	1.4	202
Cercopithecus diana						
Overall	1.33	17.7	37.1	42.7	1.19	3461
Travel	1.7	43.8	24.8	29.3	0.4	242
Forage	2.1	17.3	36.8	42.3	1.45	1311
Cercopithecus campbelli						
Overall	7.93	13.9	41.5	28.9	7.73	1434
Travel	28.1	23.9	36.5	6.3	5.2	96
Forage	13.2	10.6	38.6	32.8	4.8	500
Cercopithecus petaurista						
Overall	0.58	10.3	49.1	39.2	0.78	2042
Travel	4.2	31.9	55.6	8.3	–	144
Forage	1.1	8.2	44.6	45.6	0.5	368
Cercocebus atys						
Overall	46.84	10.65	17.6	7.67	17.27	1320
Travel	80.4	5.8	3.6	2.9	7.2	137
Forage	65.7	8.5	17	2.12	6.7	329

(65.7 per cent) primarily on the forest floor, less than half of all feeding occurs on the terrestrial substrate. This is because sooty mangabeys routinely take food obtained from the ground into small saplings, low tree branches or on to tree falls to feed (McGraw 1998a, 1998c). This behavior is in addition to acquiring food from low levels of trees themselves.

Red colobus and black and white colobus, the two largest arboreal monkeys, are the most frequent users of the largest supports (boughs) overall, during travel and during foraging. The smaller olive colobus uses boughs much less often, preferring medium and small sized supports in the low canopy and under story levels during most activities. Of the three guenon species, *Cercopithecus diana* uses boughs most often overall and during travel and foraging. This monkey strongly prefers large supports

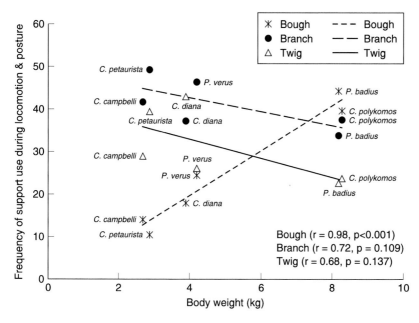

Figure 9.1. The relationship between body weight of adult females and use of each support type (bough, branch, twig) during all positional activities.

during directed travel along arboreal pathways. Although *Cercopithecus petaurista* and *C. campbelli* are the smallest monkeys at Taï, they use the smallest support type (twigs) far less frequently than do larger species. The primary difference between these guenons is the frequency that *C. campbelli* uses the forest floor and "other" supports (tree falls, lianas, etc.). Indeed, the most striking characteristic of Campbell's monkey locomotion is the frequency with which it drops to the forest floor for rapid, long distance travel.

Figures 9.1–9.3 depict the relationship between body size, support type, and maintenance activity. For overall support use, bough use is significantly associated with body size (see Figure 9.1): larger monkeys use the largest supports more often than do smaller monkeys ($r = 0.98$, $p < 0.001$). This relationship is also true for bough use during travel ($r = 0.88$, $p < 0.05$) (see Figure 9.2) and during foraging ($r = 0.956$, $p < 0.05$) (see Figure 9.3). There are no other statistically significant relationships between body mass and use of branches and twigs during any activity. Taken together, these data suggest that employing body size to predict use of supports other than the largest (boughs) is difficult.

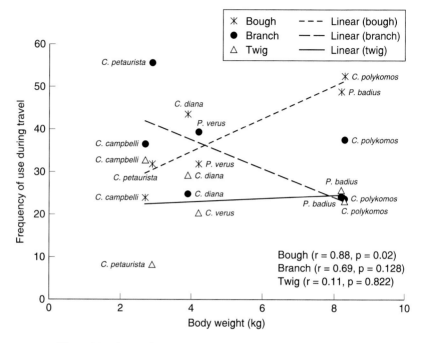

Figure 9.2. The relationship between body weight of adult females and use of each support type (bough, branch, twig) during travel.

With regard to general activities and support size, all monkeys use boughs most often during travel and with one exception (*C. campbelli*) least often during feeding. This fits expectations for arboreal quadrupeds that prefer to travel on the most stable supports. The smallest supports are used most often during feeding and foraging by all species, however the two smallest monkeys use twigs much less frequently during foraging than do larger monkeys. This can be explained by differences in the foraging strategies of guenons and colobines (McGraw 2000). *Cercopithecus petaurista* and *C. campbelli* frequently employ a cryptic style to search for insects and leaves along branches and boughs nearer the trunk (Buzzard 2003). Searching for insects and leaves along larger supports decreases the amount of time one would predict small frugivorous monkeys would forage on twigs in the periphery of tree crowns. On the other hand, the larger colobines use twigs more than body size alone predicts because of their habit of distributing weight across and amid multiple twigs in the periphery of tree crowns while sitting and manually gathering leaves. During travel, the larger colobines use more twigs than expected because of the frequency with which they leap between

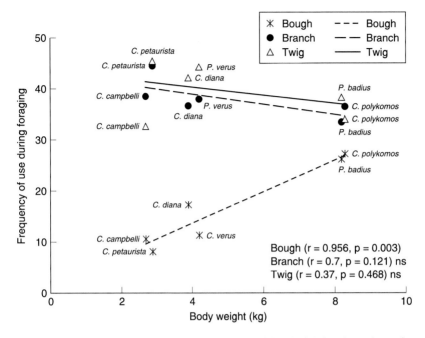

Figure 9.3. The relationship between body weight of adult females and use of each support type (bough, branch, twig) during foraging.

canopy gaps that are fringed by small supports. *Procolobus badius* and *Colobus polykomos* routinely accomplish leaps of great distances by flinging themselves between the terminal supports of one tree across a canopy gap and into the terminal supports of adjacent trees. Landing forces are better dampened by touching down amid multiple, flexible twigs. The smaller guenons do not leap as far and are less constrained by body size and the resulting forces of touchdown. Indeed, a number of authors have remarked that when large colobines leap, they aim for a general area of tree crown (twigs) whereas smaller monkeys are much more precise in their leaps and frequently land on relatively larger supports (Struhsaker 1975, Napier 1976).

Locomotion, body size, and limb length
The locomotor profile of each monkey is listed in Table 9.4. The dominant behavior for all monkeys during all activities is walking. Running is less common, accounting for less than 10 per cent of overall locomotion in most species. The three colobines all leap more than the cercopithecines and the smallest colobine (*Procolobus verus*) leaps most often.

Table 9.4. *Locomotor profiles of Taï monkeys*

	Arm swing	Climb	Leap	Run	Walk	N
Colobus polykomos						
Overall locomotion	–	14.3	14.5	29.4	41.8	918
Travel	–	8	17.6	31.8	42.6	538
Forage	–	24.2	13.3	12	50.5	380
Procolobus badius						
Overall locomotion	3.9	17	17.8	8.2	53.1	1466
Travel	2.9	12.2	20.8	9.3	54.8	797
Forage	5	21.6	15	5	53.4	669
Procolobus verus						
Overall locomotion	–	12	20.4	22.4	45.2	508
Travel	–	6.3	25	29.6	39.1	306
Forage	–	20.8	13.5	10.5	54.5	202
Cercopithecus diana						
Overall locomotion	0.1	19.4	10.4	10.8	59.3	1553
Travel	–	6.2	16.7	23.8	53.3	242
Forage	0.1	21	9.4	7.9	61.6	1311
Cercopithecus campbelli						
Overall locomotion	–	14.5	5.2	7.7	72.6	596
Travel	–	3.1	5.2	23	68.8	96
Forage	–	16.4	5.2	4.8	73.5	500
Cercopithecus petaurista						
Overall locomotion	–	18.8	10.1	9.8	61.3	512
Travel	–	6.9	5.6	26.4	61.1	144
Forage	–	23.4	11.4	2.7	62.5	368
Cercocebus atys						
Overall locomotion	–	12.5	1.02	5.7	80.7	466
Travel	–	2.9	0.72	5.1	91.3	137
Forage	–	16.5	1.2	0.9	81.4	329

Data under each column are the percentages of locomotor time overall, during travel, and during foraging that each behavior was observed

Overall rates of climbing for the seven species are quite similar, ranging between 12 per cent and 19.4 per cent. Bi-manual locomotion was observed rarely in two species – *Procolobus badius* and *Cercopithecus diana*.

Various authors have argued that there are predictable relationships between leaping, climbing and body size (Fleagle *et al.* 1981, Cant 1987, 1992, Cannon & Leighton 1994). When different sized primates use the same strata, larger monkeys should be able to "bridge" more gaps between supports while smaller monkeys must negotiate the same gaps by leaping across discontinuities. Small-bodied primates should leap more than their larger counterparts. In terms of vertical movement, larger monkeys are expected to show higher climbing frequencies since they are more likely to

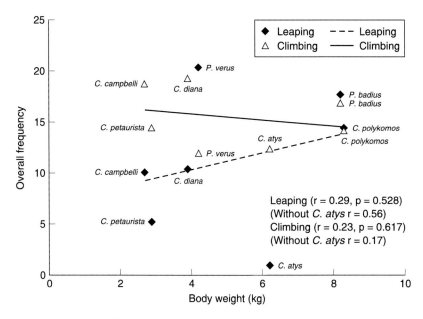

Figure 9.4. The relationship between body weight of adult females and overall frequencies of leaping and climbing.

encounter fewer supports on which they can be quadrupedal (Fleagle & Mittermeier 1980).

Among the Taï monkeys (see Figure 9.4), there is little association between body size, climbing, and leaping (Leaping $r = 0.29$, $p = 0.528$; climbing $r = -0.23$, $p = 0.617$). Even subtracting the terrestrial *Cercocebus atys* does little to strengthen these relationships ($r = 0.56$, $p = -0.17$, respectively). It is difficult to explain why the Taï monkeys do not conform to the patterns seen in most other primates, however the only other study of a group of sympatric cercopithecids (Gebo & Chapman 1995a) yielded similar results: when all primates are considered, body size is not a good predictor of leaping or climbing frequencies. Comparing body size within subfamilies also fails to yield a consistent pattern: among colobines, the most frequent leaper was the smallest (*Procolobus verus*) while the most frequent leaper among arboreal cercopithecines was the largest (*Cercopithecus diana*). A stronger relationship exists with phylogeny: all colobines leap more than do cercopithecines (Jungers *et al.* 1998). The fact that body size is not closely associated with locomotor behavior in these primates means we must be cautious when using modern analogs to infer the behavior of extinct primates.

Table 9.5. *Inter-membral indices*[a] *of Taï forest monkeys*

Colobus polykomos	81.6
Procolobus badius	90.5
Procolobus verus	80.1
Cercopithecus diana	80.8
Cercopithecus petaurista	80
Cercopithecus campbelli	84.6
Cercocebus atys	84

[a]Indices calculated from specimens collected in the Taï forest

It is widely recognized that leaping frequency is associated with relative limb length (Ankel-Simons 2000). The inter-membral index (humerus length + radius length/femur length + tibia length × 100) of each species is presented in Table 9.5. Primates with relatively longer hindlimbs tend to be more active leapers than those with shorter hindlimbs (Fleagle 1999). This relationship is not found among the seven Taï cercopithecids: there is no obvious relationship between leaping frequency and intermembral index (r = 0.022, p = 0.962). This is also true when the Taï guenons (McGraw 2002) and colobines (McGraw 1996a) are treated separately. The fact that the most active leaper − olive colobus − has a low inter-membral index offers some consolation, however it is possible that leap length rather than frequency exerts as strong a selective pressure on limb length. Although I did not collect data on the size of gaps crossed, my impression is that *Colobus polykomos* and *Procolobus badius*, two monkeys usually found in higher reaches of the canopy, are capable of leaping the greatest horizontal distances. For large monkeys living high in the canopy, the ability to cross gaps to escape predators may be the most important selective pressure, an argument put forth by Gebo *et al.* (1994).

Locomotion and maintenance activities

Various studies have shown that the locomotor profiles of primates differ during travel compared to foraging (Rose 1977, Susman 1984, Hunt 1992). This is because during travel, "the endpoints are more general ... the animals appear to chose regular pathways ... that permit long uninterrupted bouts of locomotion on relatively stable supports. By contrast, the endpoints of locomotor bouts during feeding are more precise and to a large extent predetermined by the location of the food sources on the relatively small terminal branches (Fleagle & Mittermeier 1980)." One therefore expects higher frequencies of climbing and lower frequencies of leaping and running during foraging since monkeys must negotiate the small, slender, terminal branches in the periphery of tree crowns where food can be concentrated. In contrast, travel is more likely to occur on larger, more horizontal pathways that facilitate running. Leaping should occur more often during travel as monkeys encounter large gaps between adjacent tree crowns while moving between more distant endpoints (Cannon & Leighton 1994).

The locomotion of the Taï monkeys during travel and foraging conforms to many of these predictions (see Table 9.4). All species climb more during foraging than during travel. Reaching food in tree peripheries routinely involves quadrumanous clambering or climbing because the most abundant support in the fringes of crowns − twigs − does not easily accommodate quadrupedalism in mid-large sized monkeys. For most Taï species, the energetically costly (and often dangerous) act of leaping is performed more often during travel when monkeys are using more direct routes to pass between distant endpoints (Cant 1992). All species run more during travel and walking is the least variable behavior in terms of general maintenance activities.

Locomotion and habitat structure

A major goal of locomotor field studies is to better understand the precise relationship between behavior and the structural milieu in which it takes place. Locomotor behavior has evolved to solve problems of habitat (Fleagle 1976, 1979, 1980, Cant 1992, Cannon & Leighton 1994) and habitual behaviors are reflected in primate anatomy (Stern & Larson 2001, Larson & Stern 2002). There is a functional chain between where an animal lives (habitat) how it moves (locomotor behavior) and how it is built (anatomy). Understanding the relationships of these variables in living primates allows us to more reliably interpret the probable behavior of fossil species (Dagosto 1994, Garber & Pruetz 1995, Kowalewski *et al.* 2002). The accuracy of these predictions depends on the strength of relationship between anatomy and behavior. An important issue currently being debated concerns the flexibility of locomotor behavior: is it conservative or does it fluctuate significantly in different environments? What range of behavior is represented by a morphological complex? What time frame (days, months, years) is the relevant unit for assessing locomotor variability? These are critical questions for paleontologists because if positional behavior differs significantly in different habitats, then our ability to consistently infer the positional behavior of fossils is weakened.

One way to assess the variability of positional behavior is to compare locomotion between habitat types. To this end, I sampled the behavior of five monkeys in two forest types within the main study grid. My principal

Table 9.6. *Distribution of supports in two forest types*

Height	Bough	Branch	Twig	Total
Forest type A				
>40 m		29	85	114
31–40 m	58	200	429	687
21–30 m	127	264	582	973
11–20 m	108	295	626	1029
0–10 m	26	374	607	1007
Total	319	1162	2329	3810
Forest type B				
>40 m		7	48	55
31–40 m	13	156	314	483
21–30 m	101	427	748	1276
11–20 m	35	503	1526	2064
0–10 m	13	827	1031	1871
Total	162	1920	3667	5749

Data are summed frequencies of each support type at each 10 m interval. Three factor (forest type, height, support type) log-linear model of support differences: Interaction: [G(Williams) = 70.328 (significant)] (critical χ^2 value (0.05) = 15.507)

goal was to determine if frequencies of major locomotor activities changed significantly in structurally different forests. In order to do this, I first undertook an architectural analysis of forest within the main study grid. The methods are discussed briefly above and in McGraw (1996a, 1996b).

Based on the forest survey, the main study grid can be divided into two general forest types corresponding to varying degrees of disturbance. Forest type A – undisturbed forest – contains fewer total stems (7,786/18,000 m^2 sampled) then did disturbed forest type B (9,068/18,000 m^2 sampled). There were 2,435 trees and 5,531 saplings in undisturbed forest and 2,885 trees and 6,183 saplings in disturbed forest. Areas of disturbance within the study area resulted from two large trails, an abandoned logging road, and the fact that a number of large trees were selectively removed from the forest approximately ten years prior to the study.

Results of the canopy survey further distinguish the two areas of forest (see Table 9.6). The three factor log-linear model indicates that there are significantly different numbers of each support type at each forest level; the forest types differ not only in the relative number of supports at each height interval but also in the types of supports within each height interval. Generally speaking, disturbed forest (type B) is denser at lower levels with greater numbers of twigs and branches and fewer boughs. Undisturbed forest is more open: there are fewer overall supports but more

Table 9.7. *Locomotor profiles of five species in two forest types*

	Arm swing	Climb	Leap	Run	Walk	N	G-value	χ^2 value
Colobus polykomos						*918*	*7.02*	*7.815*
Forest A (undisturbed)	0	13.3	9.2	32.6	44.9			
Forest B (disturbed)	0	14.4	15.1	29.2	41.4			
Procolobus badius						*1466*	*4.63*	*9.488*
Forest A (undisturbed)	4.9	17.6	16.9	6.9	53.6			
Forest B (disturbed)	3.3	16.7	18.4	8.9	52.8			
Procolobus verus						*507*	*6.73*	*7.815*
Forest A (undisturbed)	0	11.7	21	21	46.4			
Forest B (disturbed)	0	12.3	19.9	24	44.1			
Cercopithecus diana						*1553*	*3.44*	*7.815*
Forest A (undisturbed)	0	18.1	9.7	11.8	60.4			
Forest B (disturbed)	0	20.6	11.1	9.9	58.4			
Cercopithecus campbelli						*596*	*4.15*	*7.815*
Forest A (undisturbed)	0	14.1	5	7.8	73			
Forest B (disturbed)	0	15.1	5.4	7.5	72			

Data under each column are the percentage of total locomotor time that each locomotor behavior was observed. Test of conditional independence: G(Williams) = 659.547. Three-factor interaction: G(Williams) = 26.055 (ns)

large (boughs) supports. Comparison of support density between 11 and 20 meters across forest types illustrates this point: undisturbed forest has almost three times the number of boughs despite having half as many total supports.

The overall locomotor frequencies of five species sampled in each forest type are presented in Table 9.7. The results show that although the major behavioral differences *between* species are still readily apparent (test for conditional independence, G value = 659.547), the overall loco-motor profiles of the five species *across* forest types do not change significantly (Williams-corrected G values are all $< \chi^2$). Closer examina-tion of the behavior of *Colobus polykomos*, *Procolobus badius*, and *Cercopithecus diana* indicates that inter-forest locomotion is constant even when maintenance activities are controlled for (McGraw 1998c). The results of this analysis are presented in Table 9.8 (from McGraw 1998c) and indicate that the overall locomotor profile of each of the three species did not differ significantly between habitats during travel and foraging.

A detailed discussion of the support use in both forest types can be found in McGraw (1996b, 1998c). In general, when overall locomotion (travel and foraging combined) is considered, there is no statistical difference in the supports selected by each species across forest types within each

Table 9.8. *Locomotion in two forest types (from McGraw 1998c)*

	Undisturbed forest		Disturbed forest	
Behavior	Traveling	Foraging	Traveling	Foraging
Colobus polykomos				
Arm swing	–	–	–	–
Climb	7.9	21.9	6.9	23.6
Leap	7.9	12.5	16.5	13
Run	41.3	15.6	40	14.4
Walk	42.9	50	36.6	49
Procolobus badius				
Arm swing	3.7	6.2	2.5	4.2
Climb	12	24.5	12.3	19.8
Leap	19.4	14	21.6	15.6
Run	8.7	3.5	9.7	5.9
Walk	56.2	51.8	53.9	54.5
Cercopithecus diana				
Arm swing	–	–	–	0.1
Climb	2.8	20.5	9.1	21.6
Leap	15.7	8.6	17.4	10
Run	27.8	8	20.5	7.8
Walk	53.7	62.9	53	60.5

For each species, the values in the traveling and foraging columns are the percentages of locomotor time spent engaged in the five locomotor behaviors
There is no significant difference in locomotion between forest types during traveling and during foraging for any species. G(Williams) for each species during these activities are *Colobus polykomos* (3.7, 0.08), *Procolobus badius* (1.7, 5.2), and *Cercopithecus diana* (5.5, 1.2) respectively

height interval. That is, despite differences in the relative availability of each support type, most monkeys use each support with the same frequency in each habitat. Patterns of support use across the different forest types during travel and foraging are more complex. Support use of *C. polykomos* did not differ in both forest types regardless of activity (see Table 9.3, McGraw 1998c). Support use of *P. badius* did not differ during travel but this species chose different supports during foraging in each forest type (see Table 9.4, McGraw 1998c). *Cercopithecus diana* showed the greatest variation and chose different supports during travel and foraging in each forest type (see Table 9.5, McGraw 1998c).

Taken together, this analysis suggests that locomotor equivalence is maintained through the selection of similar supports in each forest type. The notion that locomotion in these cercopithecids is conservative with respect to support size and availability is in line with other studies (e.g. Doran & Hunt 1994, Garber & Pruetz 1995, Kowalewski *et al.* 2002) that found conservative positional behavior in platyrrhines and apes (but see Dagosto 1994, Gebo & Chapman 1995a, Dagosto & Gebo 1998). This is encouraging news for those interested in inferring the behavior of fossil species based on the relationship between anatomy and behavior in living species.

Posture

Posture has traditionally been a neglected area of positional behavior (Fleagle 1980, McGraw 1998c) since most biologists assume that locomotor behaviors exert a greater selective pressure on a primate's postcranial anatomy. While perhaps true, most primates spend far more time engaged in non-locomotor aspects of positional behavior and many postural behaviors – even among generalized arboreal quadrupeds – may still exert significant forces on the limbs (Rose 1974). For these reasons, investigating postural differences between species can be a rewarding approach towards understanding broader aspects of behavior.

The frequencies with which each species engaged in each posture and during three general maintenance activities are summarized in Table 9.9. Postural differences fall along subfamily lines and seem to be related to foraging techniques, the distribution of preferred food items and the energetics of digestion. Virtually all feeding by the three colobine species is done from sitting positions while the four cercopithecine species feed often while standing or, in the case of the Diana monkey, using the stand/forelimb-climb posture (Hunt *et al.* 1996) (see Figure 1.15). This fundamental difference is best explained by the spatial distribution of preferred food items. Colobines – as predominantly leaf eating monkeys – are more able to position themselves amid a food patch and feed banquet style until all food items (e.g. leaves or flowers) within arms reach have been harvested. Cercopithecines rely more on fruit and insects that tend to be more patchily distributed and, in the case of the latter, mobile. Rather than sacrifice time and energy by continually rising from a sitting position, cercopithecines often feed standing up while scanning the canopy for their next feeding location. This tendency to feed "on the go" is also reflected in the time colobines and cercopithecines devote to feeding and foraging. Both groups spend similar amounts of time feeding, however moving between less ubiquitously distributed food items such as fruit and insects is a more time intensive activity and this is reflected in the greater time all cercopithecines must devote to foraging than do colobines (see Table 9.10).

The second major difference between subfamilies concerns rates of reclining (sprawl and lie); colobines recline frequently whereas

Table 9.9. *Postural profiles of Taï monkeys*

	Sit	Stand	Sprawl	Lie	St/F-S	Other	N
Colobus polykomos							
Overall	89.5	0.19	2.9	6.3	0.84	0.27	2495
Resting	82.8	0	6.4	10.8	0	0	1191
Social	80.1	0	0	19.9	0	0	181
Feeding	97.3	0.41	0	0	1.8	0.49	1123
Procolobus badius							
Overall	87	1.4	4.2	6.1	0.55	0.75	2690
Resting	80.4	0	9.1	10.5	0	0	1248
Social	72.5	13.5	0	13.2	0	0.8	244
Feeding	97.3	0.17	0	0	1.2	1.33	1198
Procolobus verus							
Overall	90.7	1.3	3.8	3.7	0.28	0.22	1081
Resting	86.9	0	7.4	5.7	0	0	557
Social	81	9.5	0	7.6	0	1.9	103
Feeding	98.3	0.72	0	0	0.72	0.26	421
Cercopithecus diana							
Overall	59.6	22.1	0.16	0.77	15.2	2.17	1908
Resting	87.3	8.7	0.5	3.5	0	0	403
Social	86.5	10.8	0	0	0	2.7	37
Feeding	51.5	25.9	0	0	19.9	2.7	1468
Cercopithecus petaurista							
Overall	81.9	12.9	0	< 1	3.1	1.3	1530
Resting	91.5	7.9	0	0	0	< 1	662
Social	75	11.4	0	13.6	0	0	88
Feeding	74.6	17.4	0	0	6.2	1.8	780
Cercopithecus campbelli							
Overall	70.9	21.9	0.16	0.77	4.8	1.47	838
Resting	97.2	2.4	0	0	0	0.4	290
Social	92.5	0	0	7.5	0	0	40
Feeding	54.1	34.8	0	0	7.9	3.2	508
Cercocebus atys							
Overall	80.9	15.7	0	0	1.9	1.5	854
Resting	98.4	1.2	0	0	0	0.4	248
Social	89.6	10.3	0	0	0	0.1	87
Feeding	71.1	23.5	0	0	3.1	2.3	519

Data under each column are the percentages of time spent in each postural category overall and during resting, social activities and feeding. St/F-S = Stand/Forelimb suspend

cercopithecines do so rarely. This difference is most likely related to the colobines need for longer periods of inactivity in order to digest leaves; all colobines rest a significantly greater amount of time than do cercopithecines (see Table 9.10).

The morphological correlates of different postural behaviors in quadrupedal monkeys have not been extensively explored (McGraw 1998c).

Table 9.10. *Overall activity budgets (from McGraw 1996a)*

	Traveling	Foraging	Resting	Social	Feeding	N
Colobus polykomos	15.1	10.8	33.9	5.3	34.9	3538
Procolobus badius	18.9	15.8	29.9	6.3	29.1	4196
Procolobus verus	19.1	12.7	35	6.7	26.5	1595
Cercopithecus diana	28.5	28.3	8.8	1.2	33.2	3461
Cercopithecus petaurista	7.1	17.9	32.5	4.3	38.2	2042
Cercopithecus campbelli	6.7	34.8	20.2	2.8	35.5	1434
Cercocebus atys	10.3	24.5	18.5	7.9	38.8	1343

Rose (1974) found that colobines had larger ischial callosities than do cercopithecines which is expected given the much higher rate of sitting observed in the former. It is possible that the more mobile shoulder and hindlimb complexes of colobines (Napier 1963, Morbeck 1979) facilitates reclining postures such as sprawl that involve sustained periods of shoulder abduction and hip extension (see Figure 9.7 in McGraw 1998c).

Supports used during postural behavior
The frequencies with which each species used each major support category during overall posture and during individual maintenance activities are summarized in Table 9.11. Bough use is significantly associated with body weight; larger monkeys use boughs more often than do smaller monkeys during overall posture (see Figure 9.5). During feeding, use of branches (r = 0.89, p = 0.115) and boughs (r = 0.811, p = 0.05) both increase with body size (see Figure 9.6). Twig use tends to decrease with body size during feeding postures but this is not significant (r = −0.72, p = 0.108) most likely because of the comparatively low frequency that the smallest monkey − *C. campbelli* − uses this support type. *C. campbelli* spends more time foraging/feeding on the ground and "other" supports than would be predicted from body size alone. This cryptic monkey frequently feeds on fallen trees and large supports looking for insects as it scans the under story (McGraw 2002).

During rest periods, large arboreal monkeys prefer boughs, small monkeys prefer branches, and twigs are generally avoided. Sooty mangabeys only occasionally rest directly on the ground, preferring instead the branches of low trees or, more commonly, tree falls. For the seven species, bough and branch use during rest are significantly correlated with body size, but in opposite directions (see Figure 9.7); as body size increases bough use increases while branch use decreases. The relationship between supports used during social behavior and body size is similar to

Table 9.11. *Support use during posture*

	Ground	Bough	Branch	Twig	Other	N
Colobus polykomos						
Overall posture	0.16	38.1	40.1	21.6	0.04	2495
Resting	0	57.4	36.6	6	0	1191
Social	0	74	24.9	1.1	0	181
Feeding	0.16	13.9	46	39.9	0.04	1123
Procolobus badius						
Overall posture	0.17	47.1	35.6	17.1	0.08	2690
Resting	0	67.6	29	3.3	0.1	1248
Social	0	76.6	22.5	0.41	0.49	244
Feeding	0.17	18.8	46.4	34.6	0.03	1198
Procolobus verus						
Overall posture	0.95	24.2	49.8	23.8	1.25	1081
Resting	0	35.2	55.1	7.4	2.3	557
Social	0	42.9	46.7	5.7	4.7	103
Feeding	0.95	5.7	42.4	50.5	0.45	421
Cercopithecus diana						
Overall posture	0.89	15.7	34.5	48.1	0.81	1908
Resting	0	31.3	60.7	7.2	0.8	403
Social	0	62.2	29.7	8.1	0	37
Feeding	0.89	8.6	33.5	55.9	1.11	1468
Cercopithecus petaurista						
Overall posture	<1	10.3	49.1	39.2	<1	1530
Resting	0	12.7	69.2	17.2	<1	662
Social	2.3	25	51.1	19.3	2.3	88
Feeding	0	3.8	32.7	62.7	<1	780
Cercopithecus campbelli						
Overall posture	4.1	14.8	43.6	28.3	9.2	838
Resting	0	20.3	59.3	10.3	10.1	290
Social	0	25	45	2.5	27.5	40
Feeding	4.1	10.8	34.6	42.1	8.4	508
Cercocebus atys						
Overall posture	33.2	12.4	20.3	5.6	28.5	854
Resting	6.5	23.4	16.1	0	54	248
Social	19.5	13.8	2.3	0	64.4	87
Feeding	48.4	7	24.2	9.3	11.1	519

Data under each column are the percentages of postural time overall, during resting, during social behavior and during feeding that each support type was used

that of resting and body size (see Figure 9.8); with increasing body size, bough use increases while branch use decreases.

Future work

Studies of primate positional behavior are relatively uncommon and the number of species for which we have quantitative data on free-ranging

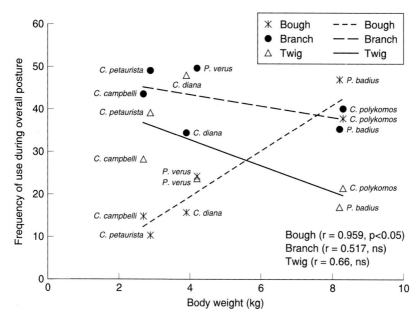

Figure 9.5. The relationship between body weight of adult females and use of each support type (bough, branch, twig) during overall posture (combined feeding, resting and social behavior).

populations is few (Dagosto & Gebo 1998). As this review highlights, there are many unanswered questions and additional investigations of the relationship between locomotion, posture, ecology, and anatomy are sorely needed. For example, since it appears that size is not always a good predictor of locomotor behavior, one priority should be to identify reasons why the relationship between mass and locomotor behavior is strong in platyrrhines (Fleagle & Mittermeier 1980) but not in cercopithecids (Gebo & Chapman 1995a, McGraw 1998b). Additional data on the positional behavior of *Cercopithecus nictitans* could shed light on this problem since this is one of the largest arboreal guenons; studies of this species are currently under way (Bitty and McGraw in press). The problem of locomotor variation in different habitat types is an important one for those interested in more precisely associating locomotor tendencies with morphological complexes. To this end, it would be useful to collect positional data on these species at other sites within Taï National Park or elsewhere in West Africa. In the meantime, we should begin identifying the anatomical correlates to differences in positional behavior between the Taï species. Over the last fifteen years, our research team has collected a large

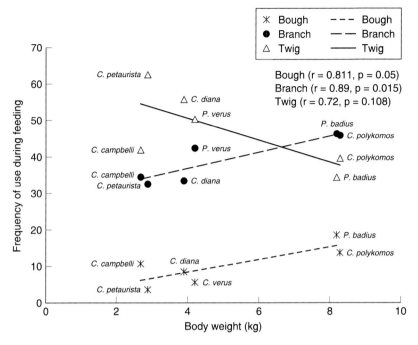

Figure 9.6. The relationship between body weight of adult females and use of each support type (bough, branch, twig) during feeding.

number of monkey skeletons and functional studies on Taï cercopithecid postcrania are now underway. Each of these areas of investigation should lead to a broader understanding of extant primate biology and provide for more reliable reconstructions of the behavior of extinct species.

Species summaries
Colobus polykomos is the largest arboreal monkey at Taï. This monkey prefers the main canopy for most activities and is the species most frequently found in the emergent layer. It prefers to travel on boughs and feed primarily on branches and twigs. Black and white colobus are excellent leapers. Although walking is the principal means of locomotion, running is common. *C. polykomos* is the only Taï monkey that bounds, a behavior seen in other black and white colobus species which involves simultaneous touchdown of the forelimbs followed by simultaneous touchdown of the hindlimbs (see Figure 2 in McGraw 1998b). During foraging, *C. polykomos* is the most frequent climber. Sitting is the dominant posture during all non-locomotor activities and various forms of reclining are used frequently. Boughs and branches are the preferred

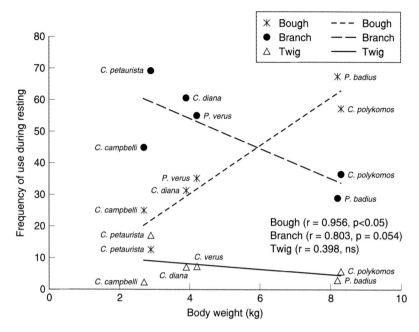

Figure 9.7. The relationship between body weight of adult females and use of each support type (bough, branch, twig) during resting.

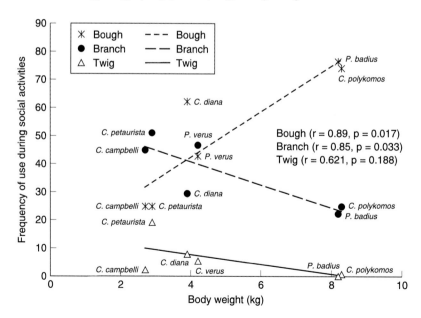

Figure 9.8. The relationship between body weight of adult females and use of each support type (bough, branch, twig) during social activities.

supports for feeding and they are used even more often during periods of rest and social behavior. Black and white colobus spend the least time foraging of all species.

Procolobus badius at Taï prefer the main canopy and are occasionally found in the emergent layers. Red colobus prefer to travel on boughs while branches and twigs are used most often during foraging. Walking comprises just over 50 per cent of locomotion during travel and foraging and, in contrast to the other two Taï colobines, red colobus do not run often. Sitting is the dominant postural behavior during all activities and various forms of reclining are common during rest and social activities. Red colobus prefer to rest and engage in social activities on boughs and feed on branches and twigs.

Procolobus verus is a cryptic monkey that strongly prefers the under story layer for all activities and is rarely found in the high main canopy. Locomotion during feeding takes place primarily on branches and twigs while branches and boughs are preferred for travel. The olive colobus does run occasionally and is the most frequent leaper at Taï. Like its congenerics, the dominant posture during all non-locomotor activities is sitting; over 98 per cent of all feeding is done from this position. *P. verus* reclines during resting and social periods though not as frequently as *P. badius*, and *Colobus polykomos*. Branches are the preferred supports for all postural activities except feeding; over 50 per cent of feeding occurs on twigs.

Cercopithecus diana is the most active and agile monkey in the forest. It can be found in all forest layers – including the ground – though it spends the largest portion of time in the under story (stratum 2). Diana monkeys prefer to travel on boughs and forage on twigs and branches. Diana monkeys are good leapers and this behavior is performed at all forest levels. Sitting is the dominant posture during all maintenance activities however just over 50 per cent of all feeding is done from a sitting position. Diana monkeys employ transitional postures, such as quadrupedal stand and stand-forelimb climb, during feeding that allow quick, efficient movement to the next feeding site. The frequent, continual movement throughout the canopy layers probably increases their sentinel abilities. Diana monkeys prefer to feed on twigs, rest on branches and engage in social activities on boughs. This monkey has the most uniform activity budget devoting similar amounts of time to feeding, foraging and traveling.

Cercopithecus campbelli is a cryptic monkey that spends 90 per cent of its time in the forest under story (stratum 2) and below (McGraw 2002). Most locomotion tends to be slow, cautious and deliberate. The exception is when *C. campbelli* runs along the forest floor during long distance travel. During arboreal travel, *C. campbelli* prefers medium sized supports while most foraging is carried out on branches and twigs. This monkey does not leap frequently. Sitting is the dominant posture overall, during resting and during social activities. Over 30 per cent of all feeding is performed while standing. The most striking characteristic of support use during posture is the frequency that *C. campbelli* uses "other" supports; this monkey is often found on tree falls, tree buttresses, lianas, etc.

Cercopithecus petaurista prefers the under story and low main canopy layers. This monkey generally feeds and forages at lower heights than it travels. Branches are preferred supports during travel; branches and twigs are used in similar frequencies during foraging. *C. petaurista* runs frequently during travel and climbs frequently during foraging. Leaping is uncommon. Sitting is the dominant posture during all activities. *C. petaurista* stands while feeding 50 per cent less often than does *C. campbelli*. Branches are preferred for all postural activities except feeding. Feeding takes place primarily on twigs.

Cercocebus atys is the only terrestrial monkey at Taï; 85 per cent of travel and 71 per cent of foraging occurs on the forest floor. Low forest levels are used for foraging and resting. Although a significant amount of food is obtained from the ground (Fleagle & McGraw 2002), sooty mangabeys routinely feed on supports above or on (e.g. tree falls) the forest floor. A significant amount of climbing occurs during foraging/ feeding. Sooty mangabeys are not frequent leapers but they are capable of performing long leaps to cross large gaps in the canopy. Postural behavior is typically cercopithecine (McGraw 1998c); sitting dominates, feeding is frequently done while standing and reclining postures are rare. *C. atys* uses a wide variety of supports during postural activities including the ground, tree falls, lianas, boughs, and branches.

Acknowledgments
The following individuals helped during different phases of this project: Ronald and Bettie Noe, John Fleagle, Randall Susman, Charles Janson, Patricia Wright, Diane Doran, Dan Gebo, and the skilled assistants of the Taï Monkey Project. Funding was generously provided by the National Science Foundation, Leakey Foundation, American Society of Primatology, Conservation International, Primate Conservation Incorporated, New York Zoological Society, SUNY at Stony Brook, and the Ohio State University. Many thanks to the Centre Suisse de Recherche Scientifique and its director, Dr. Olivier Girardin, for logistical

and research support. For permission to work in the Ivory Coast, I thank the Ministere d'Enseinment Superieur et Recherche Scientifique, the Ministere d'Agriculture et Resources Animales de Cote d'Ivoire, and the Directorate of the Instutute d'Ecologie Tropicale. The comments of Ronald Noë, John Cant, and Suzanne Shultz significantly improved this chapter.

References

Ankel-Simons, F. (2000). *Primate Anatomy: An Introduction*. New York: Academic Press.

Bitty, E. A. and McGraw, W. S. (in press). Locomotion and habitat use of stampflii's putty-nosed monkey (*Cercopithecus nictitans stampflii*) in the Taï National Park, Ivory Coast. *American Journal of Physical Anthropology*.

Boesch, C. (1994). Chimpanzees-red colobus monkeys: a predator-prey system. *Animal Behaviour*, **47**, 1135–48.

Buzzard, P. J. (2003). Ecological partitioning in Taï forest guenons: *Cercopithecus campbelli, C. petaurista, C. diana*. *American Journal of Physical Anthropology*, Supp **36**, 73.

Cannon, C. H. and Leighton, M. (1994). Comparative locomotor ecology of gibbons and macaques: selection of canopy elements for crossing gaps. *American Journal of Physical Anthropology*, **93**, 505–24.

Cant, J. G. H. (1986). Locomotion and feeding postures of spider and howling monkeys: field study and evolutionary interpretations. *Folia Primatologica*, **46**, 1–14.

Cant, J. G. H. (1987). Effects of sexual dimorphism in body size on feeding postural behavior of Sumatran orangutans (*Pongo pygmaeus*). *American Journal of Physical Anthropology*, **74**, 143–8.

Cant, J. G. H. (1992). Positional behavior and body size of arboreal primates: a theoretical framework for field studies and an illustration of its application. *American Journal of Physical Anthropology*, **88**, 273–83.

Dagosto, M. (1994). Testing positional behavior of Malagasy lemurs: a randomization approach. *American Journal of Physical Anthropology*, **94**, 189–202.

Dagosto, M. and Gebo, D. L. (1998). Methodological issues in studying positional behavior: meeting Ripley's challenge. In *Primate Locomotion: Recent Advances*, ed. E. Stasser, J. Fleagle, A. Rosenberger and H. McHenry, New York: Plenum Press, pp. 5–30.

Doran, D. M. (1992). The ontogeny of chimpanzee and pygmy chimpanzee locomotor behaviors: a case study of paedomorphism and its behavioral contexts. *Journal of Human Evolution*, **23**, 139–58.

Doran, D. M. (1993). Sex differences in adult chimpanzee positional behavior: the influence of body size on locomotion and posture. *American Journal of Physical Anthropology*, **91**, 99–116.

Doran, D. M. and Hunt, K. D. (1994). Comparative locomotor behavior of chimpanzees and bonobos: species and habitat differences. In *Chimpanzee Cultures*, ed. R. W. Wrangham, W. C. McGrew and P. Heltne, Cambridge MA: Harvard University Press, pp. 93–108.

Fleagle, J. G. (1976). Locomotion and posture of the Malayan siamang and implications for hominoid evolution. *Folia Primatologica*, **26**, 245–69.

Fleagle, J. G. (1979). Primate positional behavior and anatomy: naturalistic and experimental approaches. In *Environment, Behavior, and Morphology: Dynamic Interactions in Primates*, ed. M. E. Morbeck, H. Preuschoft, N. Gomberg. New York: Gustav Fischer, pp. 313–25.

Fleagle, J. G. (1980). Locomotion and posture. In *Malayan Forest Primates*, ed. D. Chivers. Plenum Pres: New York, pp. 191–207.

Fleagle, J. G. (1999). *Primate Adaptation and Evolution*. Academic Press: New York.

Fleagle, J. G. and McGraw, W. S. (2002). Skeletal and dental morphology of African papionins: unmasking a cryptic clade. *Journal of Human Evolution*, **42**, 267–92.

Fleagle, J. G. and Mittermeier, R. A. (1980). Locomotor behavior, body size and comparative ecology of seven Surinam monkeys. *American Journal of Physical Anthropology*, **52**, 301–14.

Fleagle, J. G., Mittermeier, R. A. and Skopec, A. (1981). Differential habitat use by *Cebus apella* and *Saimiri sciureus* in Central Surinam. *Primates*, **22**, 361–7.

Galat, G. and Galat Luong, A. (1985). La communuaté de primates diurnes de la forêt de Taï, Côte d'Ivoire. *Revue de Ecologie (Terre Vie)*, **40**, 3–32.

Garber, P. A. (1991). A comparative study of positional behavior in three species of tamarin monkeys. *Primates*, **32**, 219–30.

Garber, P. A. and Pruetz, J. D. (1995). Positional behavior in moustached tamarin monkeys; effects of habitat on locomotor variability and locomotor stability. *Journal of Human Evolution*, **28**, 411–26.

Gebo, D. L. and Chapman, C. A. (1995a). Positional behavior in five sympatric Old World monkeys. *American Journal of Physical Anthropology*, **97**, 49–76.

Gebo, D. L. and Chapman, C. A. (1995b). Habitat, annual, and seasonal effects on positional behavior in red colobus monkeys. *American Journal of Physical Anthropology*, **96**, 73–82.

Gebo, D. L., Chapman, C. A., Chapman, L. J. and Lambert, J. (1994). Locomotor response to predator threat in red colobus monkeys. *Primates*, **35**, 219–23.

Hunt, K. D. (1992). Positional behavior of *Pan troglodytes* in the Mahale mountains and Gombe Stream National Parks, Tanzania. *American Journal of Physical Anthropology*, **87**, 83–106.

Hunt, K. D., Cant, J. G. H., Gebo, D. L., Rose, M. D., Walker, S. E. and Youlatos, D. (1996). Standardized descriptions of primate locomotor and postural modes. *Primates*, **37**, 363–87.

Jungers, W. L., Burr, D. B. and Cole, M. S. (1998). Body size and scaling of long bone geometry, bone strength, and positional behavior in cercopithecoid primates. In *Primate Locomotion: Recent Advances*, ed. E. Strasser, J. Fleagle, A. Rosenberger and H. McHenry. New York: Plenum Press, pp. 309–30.

Korstjens, A. H. (2001). *The Mob, the Secret Sorority, and the Phantoms: An analysis of the socio-ecological strategies of the three colobines of Taï*. Ph.D. Thesis, University of Utrecht, Netherlands.

Korstjens, A. and Noe, R. (2004). Mating system of an exceptional primate, the olive colobus *(Procolobus verus)*. *American Journal of Primatology*, **62**, 261–73.

Korstjens, A. H. and Schippers, E. P. (2003). Dispersal patterns among olive colobus in Taï National Park. *International Journal of Primatology*, **24**(3), 515–39.

Kowalewski, M., Alvarez, C., Pereyra, D., Violi, E. and Zunino, G. (2002). A preliminary study of positional behavior in *Alouatta caraya* in northern Argentina. *American Journal of Physical Anthropology*, Supp. **34**, 97.

Larson, S. G. and Stern, J. T. (2002). Forearm rotation and the "Origin of the Hominoid Lifestyle:" response to Sarmiento. *American Journal of Physical Anthropology*, **119**, 95.

McGraw, W. S. (1996a). *The positional behavior and habitat use of six sympatric monkeys in the Taï forest, Ivory Coast*. Ph.D. thesis. SUNY at Stony Brook, USA.

McGraw, W. S. (1996b). Cercopithecid locomotion, support use, and support availability in the Taï Forest, Ivory Coast. *American Journal of Physical Anthropology*, **100**, 507–22.

McGraw, W. S. (1998a). Comparative locomotion and habitat use of six monkeys in the Taï Forest, Ivory Coast. *American Journal of Physical Anthropology*, **105**, 493–510.

McGraw, W. S. (1998b). Locomotion, support use, maintenance activities, and habitat structure: the case of the Taï Forest cercopithecids. In *Primate Locomotion: Recent Advances*, ed. E. Strasser, J. Fleagle, A. Rosenberger and H. McHenry. Plenum Press: New York, pp. 79–94.

McGraw, W. S. (1998c). Posture and support use of Old World monkeys (*Cercopithecidae*): the influence of foraging strategies, activity patterns, and the spatial distribution of preferred food items. *American Journal of Primatology*, **46**, 229–50.

McGraw, W. S. (2000). Positional behavior of *Cercopithecus petaurista*. *International Journal of Primatology*, **21**, 157–82.

McGraw, W. S. (2002). Positional diversity in the guenons. In *The Guenons: Diversity and Adaptation in African Monkeys*, ed. M. Glenn, M. Cords. New York: Kluwer Academic Press, pp. 113–31.

McGraw, W. S., Plavcan, J. M. and Adachi, K. (2002). Female adult *Cercopithecus diana* killed by other females: evidence for the use of canine teeth as weapons by females. *International Journal of Primatology*, **23**(6), 1301–8.

Morbeck, M. E. (1979). Forelimb use and positional adaptations in *Colobus guereza*: integration of behavioral, ecological, and anatomical data. In *Environment, Behavior, and Morphology: Dynamic Interactions in Primates*, ed. M. E. Morbeck, H. Preuschoft and N. Gomberg. New York: Gustav Fischer, pp. 95–117.

Napier, J. R. (1962). Monkeys and their habitats. *New Scientist*, **295**, 88–92.

Napier, J. R. (1963). Brachiation and brachiators. *Symposia of the Zoological Society of London*, **10**, 183–95.

Napier, J. R. (1976). Primate locomotion. In *Oxford Biology Readers*, ed. J. J. Head. Oxford: Oxford University Press, pp. 1–16.

Noë, R. and Bshary, R. (1997). The formation of red colobus-Diana monkey associations under predation pressure from chimpanzees. *Proceedings of the Royal Society of London*, **64**, 253−9.

Oates, J. F. and Whitesides, G. H. (1990). Association between olive colobus (*Procolobus verus*), Diana guenons (*Cercopithecus diana*), and other forest monkeys in Sierra Leone. *American Journal of Primatology*, **21**, 129−46.

Oates, J. F., Whitesides, G. H., Davies, A. G. *et al.* (1990). Determinants of variation in tropical forest primate biomass: new evidence from West Africa. *Ecology*, **71**, 328−43.

Rose, M. D. (1974). Postural adaptations in New and Old World Monkeys. In *Primate Locomotion*, ed. F. A. Jenkins. Academic Press: New York, pp. 201−22.

Rose, M. D. (1977). Interspecific play between free ranging guerezas (*Colobus guereza*) and vervet monkeys (*Cercopithecus aethiops*). *Primates*, **18**, 957−64.

Schultz, A. H. (1970). The comparative uniformity of the cercopithecoidea. In *Old World Monkeys*, ed. J. R. Napier and P. H. Napier. Academic Press: New York, pp. 39−51.

Stern, J. T. and Larson, S. G. (2001). Telemetered electromyography of supinators and pronators of the forearm in gibbons and chimpanzees: implications for the fundamental positional adaptation of hominoids. *American Journal of Physical Anthropology*, **115**, 253−68.

Struhsaker, T. T. (1975). *The Red Colobus Monkey*. Chicago: University of Chicago Press.

Susman, R. L. (1984). The locomotor behavior of *Pan paniscus* in the Lomako Forest. In *The Pygmy Chimpanzee*, ed. R. L. Susman. Plenum Press: New York, pp. 369−93.

IV *Conservation*

10 Can monkey behavior be used as an indicator for poaching pressure? A case study of the Diana guenon (Cercopithecus diana) and the western red colobus (Procolobus badius) in the Taï National Park, Côte d'Ivoire

I. Koné and J. Refisch

Introduction

Market and subsistence hunting often result in the unsustainable exploitation of game in tropical forests, even when these forests are still intact (Oates *et al.* 2000). Poaching of wild animals is currently the major threat to the long-term conservation of the Taï National Park, Côte d'Ivoire (Hoppe-Dominik 1995, 1997, P.A.C.P.N.T. 1997). Previous studies of the effects of poaching in the Taï National Park have focused on differences in animal density and abundance between poached areas and non-poached areas (Hoppe-Dominik 1995, 1997, Refisch & Koné 2005) and on the harvest of game species within and around the park (Caspary *et al.* 2001). However, hunting not only affects number but also behavior (Verdade 1996). Knowledge of the impact of poaching on wildlife behavior can provide reliable indicators for the spatial and temporal development of poaching.

In addition to human hunters there are three major predators of monkeys in the Taï National Park: crowned hawk eagles (*Stephanoaetus coronatus*), leopards (*Panthera pardus*), and chimpanzees (*Pan troglodytes verus*). Leopards and crowned hawk eagles are typical ambush predators whose hunting success is considerably reduced if they are detected before the final strike (Hoppe-Dominik 1984, Klump & Shalter 1984, Zuberbühler *et al.* 1999b). In contrast, chimpanzees and humans are

Monkeys of the Taï Forest, ed. W. Scott McGraw, Klaus Zuberbühler and Ronald Noë.
Published by Cambridge University Press. © Cambridge University Press 2007.

pursuit hunters and search for prey using acoustic cues (Gautier-Hion & Tutin 1988, Boesch & Boesch 1989). Once they have located a group, both chimpanzee and human predators reach prey in the canopy by co-operative hunting or with shotguns, respectively. Monkeys adjust their anti-predator behavior to the hunting techniques used by their predators. In fact, monkeys react to the presence of ambush predators such as crowned hawk eagles and leopards with vocalizations and approach the predators, probably for the purpose of mobbing (Boesch & Boesch 1989, Dind 1995, Zuberbühler *et al.* 1999b, Shultz 2001). In contrast, in the presence of pursuit hunters (chimpanzees and humans) monkeys use temporary cryptic behavior (Boesch & Boesch 1989, Zuberbühler *et al.* 1997). Chimpanzee hunting differs from human hunting in several respects. In addition to the use of shotguns by humans, four major differences can be found between the hunting techniques of the Taï chimpanzees and those of poachers:

1.   Taï chimpanzees frequently hunt in relatively large groups (Boesch & Boesch 1989, Boesch 1994) while poachers most often search for prey alone or in small groups rarely exceeding two persons (personal communication with villagers, Koné & Refisch)
2.   Hunting chimpanzees regularly engage in loud displays and vocalizations which inform the monkeys of their presence (Kummer & Noë 1994) while poachers remain silent during hunting expeditions
3.   Taï chimpanzees generally do not kill more than one monkey per attack (Boesch & Boesch 1989) while poachers can kill several monkeys during one hunt (Bshary 2001)
4.   Poachers in the Taï region often imitate animal calls to feign the presence of either leopards or crowned hawk eagles which cause the monkeys to react with vocalization and approach (Boesch & Boesch 1989, Dind 1995, Zuberbühler *et al.* 1999b, Shultz 2001). Bshary (2001) documented that monkeys in heavily poached areas of the Taï National Park are rarely fooled by humans feigning the presence of either leopards or crowned hawk eagles and remain cryptic.

Since monkeys behave differently in response to different predators, our aim in this study was to document the specific behaviors displayed by the monkeys in response to human hunting. In particular, our goal was to determine whether suites of behaviors could be used to discriminate between monkeys living in poached areas and monkeys living in non-poached areas. We focused on the Diana guenon (*Cercopithecus diana*) and the western red colobus (*Procolobus badius*) because (1) they are relatively abundant in the Taï National Park (Galat-Luong & Galat 1982, Hoppe-Dominik 1995), (2) they are heavily poached in that park (P.A.C.P.N.T. 1997), and (3) their behavior under the pressure of non-human predation is well documented (Boesch & Boesch 1989, Adachi 1995, McGraw & Noë 1995, Bshary & Noë 1997a, 1997b, Höner *et al.* 1997, Noë & Bshary 1997, Wachter *et al.* 1997, Zuberbühler *et al.* 1997, 1999a, 1999b, Zuberbühler 2000a, 2000b). This study is among the first to emphasize the impact of poaching on primate behavior in the Taï National Park. The results should contribute to a more complete understanding of the effects of human hunting on primate populations, and its implications for conservation.

Methods
Study sites
We observed *P. badius* and *C. diana* in a non-poached and a poached site. The non-poached site is situated on the periphery of the Taï Monkey Project (TMP) research site located 500 m east of the Research station of the Centre de Recherche en Ecologie (CRE) in the western part of the Taï National Park. The continuous presence of researchers and assistants at this site since 1991 has efficiently prevented poaching. The monkey groups that we studied near the periphery of the primary research area of the TMP were not habituated to humans even though those monkeys had frequent contact with habituated monkeys and passing observers. The poached area is situated 2 km northeast of the CRE research station. Indicators of poaching activities including gun shells, trails, and hunting camps were regularly recorded at this site. In order to prevent repeated sampling gun shells were removed after recording.

Refisch and Koné (2001) compared the tree species composition and the forest structure in both areas and confirmed that both areas were comparable in terms of species richness and vegetation structure. In addition, both study areas were characterized by a comparable occurrence of non-human predators. Leopards, crowned hawk eagles, and chimpanzees were frequently observed in both areas and chimpanzees were at least partly habituated in non-poached sites (Shultz 2001, personal observation Koné & Refisch). Therefore, we assumed that the predation pressure by each of these non-human predators was comparable for the monkeys in both areas. We also assumed that all sites were characterized by the same climate since the distance between them was relatively short.

Observations and tabulation of data

We followed two groups of *C. diana* and two groups of *P. badius* in poached and non-poached areas respectively from dawn to dusk one to three times per month between October 1997 and December 1998 for a total of 14 months. We observed red colobus groups for 21−25 days for a total of 211−309 hours and Diana groups for 16−24 days for a total of 118−230 hours. When weather conditions were appropriate, observations started at 07:00 and ended at 17:30. Observations consisted of 15-minute group scans made every hour (Altman 1974). Infants and juveniles were ignored in order to reduce biases due to behavioral differences between age classes (Martin & Bateson 1985) and also because poachers usually aim at larger individuals (personal communication with villagers, Koné & Refisch). We gathered information on the following behaviors that can be easily observed in the field and are likely to be involved in defense strategies against predators.

Strata use

The vegetation was divided into six strata (McGraw 1998) and for each individual observed, we noted the stratum used. Stratum 0 is the ground level, stratum 1, (0−5 m), is composed of shrubs and small trees, stratum 2, (5−15 m), is composed of the under story, stratum 3−, (15−24 m), represents the lower canopy, stratum 3+, (24−40 m), represents the upper canopy, and stratum 4 (beyond 40 m) is composed of emergent trees.

Body exposure

We divided the monkey body into eight different parts: head, tail, four limbs, and torso subdivided into two parts. For each individual observed, we noted the number of body parts entirely visible or at least half-visible. Later we distinguished between unexposed monkeys when 1−2 body parts were visible, fairly exposed monkeys when 3−4 body parts were visible, and very exposed monkeys when 5−8 body parts were visible.

Utilization of plant parts to hide

We noted whether each individual observed used foliage and/or branches to avoid being seen by the observer. The utilization of plant parts to hide behind was taken into account despite its apparent correlation with body exposure because during the pilot phase of the study we frequently observed monkeys adopting a "hiding" posture even though they were still in full view. We define hiding posture as a posture in which few body parts of the monkey are visible.

Frequency of different activities

We noted the activity of each individual observed. Four main activities were considered: foraging including the search for food and its ingestion, locomotion including any movement of non-foraging monkeys within the same tree or between trees, resting and social interactions including grooming, playing, and fighting.

Food items consumed

We noted food items consumed during foraging to test whether monkeys change their feeding behavior in relation to the predation risk (Godin 1990). We distinguished eight different food items: ripe fruits, unripe fruit, young leaves, mature leaves, flowers, buds, invertebrates, and unidentified items.

Group cohesion

We noted the number of adult or sub-adult neighbors within a 5 m radius around each individual. Three categories of individuals were defined: isolated individuals that had no neighbor within a 5 m radius, paired individuals that had one adult neighbor, and multiple individuals that had more than one adult or sub-adult neighbor within a 5 m radius.

Individuals' reaction to observer

We noted the reaction of individual monkeys when they detected the observer. Seven types of reaction were distinguished: instantaneous silent fleeing, fleeing after inspection, vocalization after inspection, fleeing after vocalization, inspection, vocalization, and no change in behavior. One might argue that the reaction of individuals sampled earlier in a scan period could influence that of individuals sampled later in the same scan period but we are convinced that it would always be a surprise for any individual monkey to find himself facing a human. Therefore we assumed that the reaction of all individuals sampled during a given scan period would be representative of their perception of humans.

Reaction distance

For each individual that reacted to the presence of the observer, we noted the distance between the observer and the vertical line drawn from the monkey's position. Reaction could be observed from the following three distance categories: less than 10 m ("reacting close to"), up to 20 m ("reacting from fairly far away"), or more than 20 m ("reacting from afar"). For each of these eight behaviors variables were defined, as listed

Table 10.1. *Definition of study variables*

Behavior	Variable
Strata use	Using stratum 0 Using stratum 1 Using stratum 2 Using stratum 3− Using stratum 3+ Using stratum 4
Body exposure	Unexposed body Fairly exposed body Very exposed body
Utilization of plant parts to hide behind	Hiding behind plant parts Not hiding behind plant parts
Activity budget	Foraging Locomotion Resting Socializing
Food items consumed	Eating ripe fruits Eating young leaves Eating mature leaves Eating flowers Eating flower buds Eating invertebrates Eating unidentified items
Cohesion of the group	Being isolated Being paired Being aggregated
Individuals' reaction to the presence of the observer	Silent fleeing Fleeing after inspection Vocalizing after inspection Fleeing after vocalization Inspecting Vocalizing Not changing behavior
Reaction distance	Reacting from close to Reacting from fairly far away Reacting from afar

in Table 10.1. Later, data were tabulated as "one-zero" scores for each of these variables.

Data analyses

For each study group, we calculated the frequency of "1" values for the study variables by season and by period of the day (morning, i.e. 07:00–12:00 and afternoon, i.e. 13:00–17:00). Following Collinet *et al.* (1984), we distinguished four seasons: two dry seasons (November–February, and July–August) and two rainy seasons (March–June, and September–October). Previous studies have shown that hunting by chimpanzees in the Taï National Park is more intense in September (Boesch & Boesch 1989) and hunting by humans in forest zones is more intense during rainy seasons (Caspary 1999). In addition it was demonstrated that monkeys behaved differently in different seasons and different periods of the day (Galat-Luong & Galat 1982).

Percentages of the behavioral variables were calculated before controlling for potential inter-correlations of the behavioral variables. For each species we carried out two steps of principal component analysis (PCA). In the first, all variables were included except those with an extremely low frequency of "1" scores (using strata 0–2 and 4). The result of this analysis enabled us to select the most relevant and least interrelated ones. Those variables with the highest correlation coefficients in absolute value on a given axis were considered relevant (Phillipeau 1986). A second PCA was then carried out with the selected variables to determine the principal components or axes defined by two so-called key variables and correlated variables. The two key variables for a given principal component are the variables with the lowest negative and the highest positive correlation coefficients on that principal component. Positive and negative values of each principal component described two different behaviors. The results of this PCA enabled us to visualize the observations in bivariate plots and identify "behavioral groups" which consist of observations with similar behavioral features. When the positive part of a variable with two categories is correlated with one part of a given axis, the contrasting category is negatively correlated with the opposite part of the axis. Therefore, by examining the correlation among the original variables, we could still determine how the eliminated variables had contributed to the formation of the behavioral groups (James & McCulloch 1990). In the PCAs each study group observed during a given season and a given period of the day represented one observation from the statistical point of view. Hence, sixteen "observations" of poached monkeys and sixteen "observations" of non-poached monkeys were considered for each of the two study species.

After the two PCAs, we applied a discriminant analysis to the selected variables, in which the discriminant functions were calculated thus permitting the classification of observations into either the "poached" or the "non-poached" category. The results enabled us to calculate the percentage of misclassified cases.

Multivariate statistics were calculated with the SAS 6.12 package (SAS Institute Inc., Cary, NC, USA; Leroux-Scribe 1989).

Results
The impact of poaching on the behavior of C. diana

The first three axes of the first PCA used with 32 variables accounted for 26.55 per cent, 13.89 per cent, and 10.49 per cent of the total variance, making a total of 50.92 per cent. It has to be noted however, that in all PCAs we ignored the importance of the subordinate axes since they only explained a marginal proportion of the total variance. Those variables with a correlation coefficient close to the variance expressed by a given axis were considered as relevant to that axis (Phillipeau 1986). Thus variables with a correlation coefficient equalling in absolute value at least 0.25 on axis 1, 0.14 on axis 2, and 0.10 on axis 3 (see Appendix 10.1a) were considered to be relevant. Percentages for each behavioral variable are presented in Appendix 10.2a. When two variables referred to the same behavior, we selected the one with the lowest negative correlation coefficient with another study variable (if any) referring to the same behavior. The reduced list of study variables consisted of the following 16 variables:

- Using stratum 3+
- Unexposed body
- Very exposed body
- Not hiding behind plant parts
- Resting
- Socializing
- Eating mature leaves
- Eating flowers
- Being isolated
- Being paired
- Silent fleeing
- Vocalizing after inspection
- Fleeing after vocalization
- Inspecting
- Reacting from close to
- Reacting from fairly far away.

The second PCA was then carried out with a reduced set of 16 variables. The first three axes of this PCA accounted for 32.06 per cent, 21.08 per cent, and 12.03 per cent of the total variance, with a total of 65.17 per cent. The variables with a correlation coefficient equalling in absolute value at least 0.32 on axis 1, 0.21 on axis 2, and 0.12 on axis 3– were considered

Table 10.2. *Key variables and their correlated variables for each axis of the second principal component analysis of* Cercopithecus diana *behavioral data. Key variables were identified by the highest correlation coefficient on the positive and negative side in absolute value. The positive correlation of a variable with either of the key variables was considered as significant when it was equal or larger than 0.30*

Axis	Side	Key variable	Positively correlated variables
Axis 1	positive	Very exposed body	Not hiding behind plant parts, Being paired
	negative	Unexposed body	Using stratum 3+, Being isolated
Axis 2	positive	Inspecting	Eating mature leaves, Vocalizing after inspection
	negative	Silent fleeing	Fleeing after vocalization
Axis 3	positive	Resting	Using stratum 3−, Vocalizing after inspection
	negative	Silent fleeing	

to be relevant. Table 10.2 lists the variables that contributed most to the formation of each of these axes and the relevant variables that had a significant positive correlation with these key variables.

The variables that form each axis express particular behavioral patterns. Positive and negative sides of each axis thus described two different behavioral patterns.

Axis 1 was formed by a set of variables that described hiding behavior. The positive side of axis 1 described the behavior of Diana monkeys that did not hide (*very exposed body*) and did not use plant parts to form a barrier between them and ground predators (*not hiding behind plant parts*). These monkeys frequently used the main canopy (*using stratum 3* negatively correlated with *using stratum 3+*) and were frequently found with a neighbor within a 5 m radius (*being paired*). The negative side of axis 1 described the behavior of Diana monkeys that frequently hide (*unexposed body*) behind plant parts (*hiding behind plant parts* negatively correlated with *not hiding behind plant parts*). These monkeys frequently used the upper canopy (*using stratum 3+*) and had no neighbor in their vicinity (*being isolated*).

Axis 2 was formed by a set of variables that described fleeing behavior. The positive side of axis 2 characterized the Diana monkeys that showed an inspection behavior and did not flee (*inspecting*) when they had detected the observer. Another reaction of these monkeys to the presence of the observer consisted in performing alarm calls after an inspection (*vocalizing after inspection*). They frequently consumed mature leaves (*eating mature leaves*).

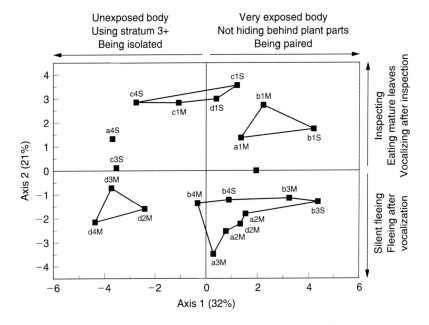

Figure 10.1. Plot of *Cercopithecus diana* observations on the first two axes of the principal component analysis. The four distinct behavioral groups are characterized by presence or absence of poaching pressure. Code for observations: a, b: monkeys in non-poached areas; c, d: monkeys in poached areas; 1 and 3: rainy seasons; 2 and 4: dry seasons; M: morning; S: afternoon.

The negative side of axis 2 characterized the Diana monkeys that fled away immediately after the observer had been detected (*silent fleeing*). These monkeys often gave alarm calls before fleeing (*fleeing after vocalization*).

Axis 3 consisted of a set of variables that described activity patterns of monkeys. The positive side of axis 3 characterized the Diana monkeys that spent most of their time resting (*resting*) in high strata (*using stratum 3+*). When the observer was detected, these monkeys reacted with alarm calls after inspection (*vocalizing after inspection*).

The negative side of axis 3 characterized Diana monkeys that fled instantaneously when they detected the observer (*silent fleeing*).

The results of the first two axes of the second PCA (see Figure 10.1) demonstrated that "observations" of *C. diana* can be classified into four distinct behavioral groups on the basis of the selected variables. The groups were correlated with the presence or absence of poaching pressure. No behavioral group was found to be correlated with seasons or periods of the day. The results of the other two binary combinations of the first three

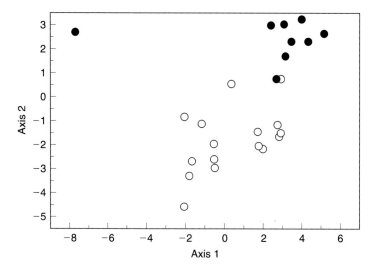

Figure 10.2. Plot of *Cercopithecus diana* observations on the first two axes of the discriminant analysis. Observations from poached area (●) and from non-poached areas (○) are correctly classified with a precision of 79 per cent.

axes of the second PCA also demonstrated that the different behavioral groups were only correlated with the presence or absence of poaching pressure.

The conclusion that there were different behavioral groups depending on poached or non-poached status was supported by the Discriminant Analysis (see Figure 10.2) when "observations" of *C. diana* were classified into the two distinct categories "poached" and "non-poached." The results showed that 76.92 per cent of classification into the first category and 81.25 per cent of classification into the second one were correct. The rate of misclassification was estimated at 21 per cent.

The impact of poaching on the behavior of **P. badius**

The first three axes of the first PCA used with 32 variables accounted for 19.83 per cent, 16.98 per cent, and 11.48 per cent of the total variance, with a total of 48.29 per cent. We considered those variables with a correlation coefficient equalling in absolute value at least 0.19 on axis 1, 0.15 on axis 2, and 0.11 on axis 3 (see Appendix 10.1b) to be relevant. Percentages for each behavioral variable are presented in Appendix 10.2b).

The reduced list of study variables is composed of 15 variables:

- Using stratum 3−
- Using stratum 3+

- Hiding behind plant parts
- Unexposed body
- Very exposed body
- Not hiding behind plant parts
- Eating unripe fruits
- Eating mature leaves
- Eating flowers
- Being isolated
- Being paired
- Fleeing
- Vocalizing
- Reacting from fairly far away
- Reacting from afar.

The second PCA was carried out with these 15 variables. The first three axes of the PCA account for 27.26 per cent, 21.46 per cent, and 17.82 per cent of the total variance, with a total of 66.54 per cent. Variables with a correlation coefficient equalling in absolute value at least 0.27 on axis 1, 0.22 on axis 2, and 0.18 on axis 3 were considered to be relevant. Table 10.3 lists the variables that contributed the most to the formation of the first three axes and the relevant variables that had a significant positive correlation with these key variables.

Axis 1 was formed by a set of variables that expressed hiding behavior.

The positive side of axis 1 described the behavior of monkeys that hide (*unexposed body*) behind plant parts (*hiding behind plant parts*). These monkeys were most often isolated within the group as no other individual is found in their close vicinity (*being isolated*).

The negative side of axis 1 described the behavior of monkeys that did not hide (*very exposed body*). No plant part formed a barrier between them and the observer (*not hiding behind plant parts*) and these monkeys were most frequently found in pairs (*being paired*).

Axis 2 was formed by a set of variables that expressed strata use. The positive side of axis 2 described the behavior of red colobus that used the upper canopy (*using stratum 3+*) from where plant parts formed a barrier between them and the observer (*hiding behind plant parts*). These monkeys fed upon mature leaves (*eating mature leaves*) and were frequently found in pairs (*being paired*). The negative side of axis 2 described the behavior of red colobus that used the lower canopy (*using stratum 3−*) where plant parts did not form a barrier between them and the observer (*not hiding behind plant parts*).

Table 10.3. *Key variables and their correlated variables for each axis of the second principal component analysis of* Procolobus badius *behavioral data. Key variables were identified by the highest correlation coefficient on the positive and negative side in absolute value. The positive correlation of a variable with either of the key variables was considered as significant when it was equal or larger than 0.30*

Axis	Side	Key variable	Positively correlated variable
Axis 1	positive	Unexposed body	Hiding behind plant parts, Being isolated
	negative	Very exposed body	Not hiding behind plant parts, Being paired
Axis 2	positive	Using stratum 3+	Hiding behind plant parts, Eating mature leaves, Being paired
	negative	Using stratum 3−	Not hiding behind plant parts
Axis 3	positive	Vocalizing	Reacting from far away
	negative	Silent fleeing	Eating unripe fruits

Axis 3 was formed by a set of variables that expressed the reaction to the presence of the observer. The positive side of axis 3 characterized red colobus that performed alarm calls when they had detected the observer (*vocalizing*). This type of reaction was produced when the observer was far from the monkeys (*reacting from afar*). The negative side of axis 3 characterized red colobus that fled instantaneously when they had detected the observer (*silent fleeing*). This type of reaction was displayed when the observer was fairly far away (*reacting from fairly far away* negatively correlated with *reacting from afar*). Unripe fruits were frequently consumed (*eating unripe fruits*).

The results of axes 1 and 3 of the second PCA (see Figure 10.3) demonstrated that three behavioral groups could be defined for *P. badius*, but these groups were essentially correlated with the season of observation and not with poaching pressure or period of day. Each behavioral group consisted of poached and non-poached monkeys observed at different periods of day but during the same season. The results of the other two binary combinations of the first three axes of the second PCA also demonstrated no correlation of the different behavioral groups with the presence or absence of poaching pressure.

We used a discriminant analysis to test whether the observations could be classified into the categories "poached" and "non-poached" as we did for *C. diana*. No clear differences could be found between poached and non-poached *P. badius* individuals (see Figure 10.4). The proportion of correctly classified *P. badius* in the non-poached category

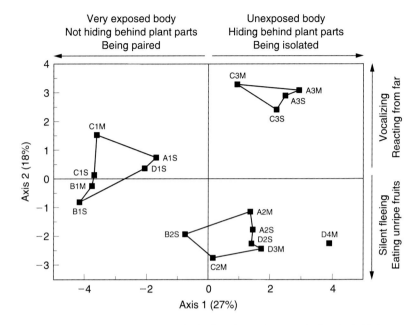

Figure 10.3. Plot of *Procolobus badius* observations on the first and third axes of the principal component analysis. The three distinct behavioral groups are essentially characterized by the season but neither by the presence or absence of poaching pressure nor by the period of the day. Code for observations: a, b: monkeys in non-poached areas; c, d: monkeys in poached areas; 1 and 3: rainy seasons; 2 and 4: dry seasons; M: morning; S: afternoon.

was 43.75 per cent while that in the non-poached category was 25 per cent. The rate of misclassification was estimated at 66 per cent.

Discussion
Diana monkeys' responsiveness to poaching pressure
In the case of *C. diana*, the behavioral groups obtained from the PCAs were correlated with the presence or absence of poaching pressure. This suggests that the study variables reliably establish a behavioral difference between poached and non-poached Diana monkeys. Using these variables, the error in classifying observed Diana monkeys as either "poached" or "non-poached" was estimated at 21 per cent. An error rate of this magnitude is not surprising in a study of primate behavior, which is influenced by innumerable factors. We can conclude that even with this error rate the behavior observed in *C. diana* clearly reflected a reaction to poaching pressure. Differences observed between poached and non-poached monkeys were not a result of differences between the sites.

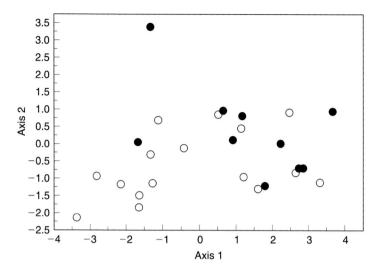

Figure 10.4. Plot of *Procolobus badius* observations on the first two axes of the discriminant analysis. Observations from poached area (•) and from non-poached areas (○) cannot be precisely classified with the study of behavioral data.

In the presence of humans, *C. diana* in poached areas hide systematically, exhibit inspection behavior, and rest in high strata (see Figure 10.1). Being successfully hidden is the best way to escape from human predators since the latter use weapons and can reach prey from afar (Hames 1979). The use of plant parts by monkeys as a barrier from humans offers the double advantage of being hidden and protected against an eventual shot. Furthermore, using the upper canopy increases the chance of escaping from a shot (personal communication from villagers, Koné & Refisch). It is also advantageous for monkeys in poached areas to keep a certain distance to the proximate individual since aggregations of many individuals are the preferred targets of poachers. In poached areas it is particularly important for monkeys to be vigilant and they have to react while poachers are still fairly far from them. Spending most time resting in the higher strata is an appropriate behavior in poached areas since all other activity would increase the conspicuousness of monkeys, which is used by humans to spot them (Gautier-Hion & Tutin 1988). This behavior contrasted with the usual behavior of *C. diana* that consists of a high mobility (McGraw 1998) and a preference for the stratum 3– (Kingdon 1997).

C. diana that exhibited an inspection behavior and did not flee certainly remained vulnerable to poachers. But these monkeys exhibited a simultaneous behavior and search for protection against poachers by hiding behind plant parts in high strata (see Figure 10.1).

Western red colobus' lack of responsiveness to poaching pressure?
Results of the discriminant analysis indicated that the variables of this study do not allow the classification of *P. badius* into "poached" or "non-poached" categories. The risk of misclassification was estimated at 66 per cent. If we conclude that *P. badius* did not adjust its behavior to poaching pressure, one should, nevertheless, bear in mind the possible incompleteness of our study variables.

Results of the PCAs demonstrated, however, that *P. badius* behaved differently during the dry and rainy seasons (see Figure 10.3). During the dry seasons *P. badius* most often used the high strata and fled instantaneously when they detected humans. During the rainy seasons *P. badius* were routinely found lower in the canopy and more often gave alarm calls in response to detected humans. The behavioral pattern adopted during the dry seasons would be appropriate in a context of poaching. In fact, *P. badius* using the upper canopy were also protected from eventual poachers by plant parts, and were not found in aggregations of many individuals but in pairs. *P. badius* that fled silently when they had detected humans were vigilant enough while the latter were still fairly far away.

In contrast, the behaviors exhibited during the rainy seasons would not be appropriate in an area where poaching is common. In fact, when using the lower canopy monkeys were still reachable by poachers and were neither hidden nor protected by plant organs. Reacting with alarm calls to the presence of humans in a poached area is not an appropriate behavior. Even if the monkeys vocalized when humans were still far away, that behavior certainly made it easy to spot and follow them. These results suggest that *P. badius* are more vulnerable to poaching during rainy seasons.

Conclusion and recommendations

Over-hunting is seriously threatening animal populations in tropical forests even when the latter are still intact. Whereas previous studies have focussed on the impact of poaching on the abundance of primates, it remains largely unknown whether primates adjust their behavior in poached areas. Because poachers tended to avoid the research area of the Taï Monkey Project it was possible to contrast the behavior of monkeys in non-poached areas with that of their conspecifics in poached areas.

1. Results of this study suggest that in the case of *C. diana* the behavioral variables used in this study can be used to classify monkey populations into two distinct groups: poached and non-poached. Diana monkeys that are frequently exposed to poachers adjust their behavior and become cryptic in the presence of humans and no other factor such as season or time of day appeared to account for the difference. The risk of misclassification was estimated at 21 per cent, which is judged acceptable with respect to the difficulties when considering all sources of variability.

2. We did not find a significant difference in the behavior of the red colobus between the two areas. Although behavioral groups could be defined by our analysis, they were associated with season rather than poaching pressure.

3. This study suggests, therefore, that the behavior of Diana monkeys can be used as an indicator for poaching pressure. However this finding cannot be automatically extrapolated to other monkey species.

We recommend the use of *C. diana* behavior as an indicator for poaching pressure in different regions of the Taï National Park. This is feasible in practice, since bio-monitoring teams would simply have to stop and make a 15-minute group scan sampling whenever they encounter a Diana group during transect counts, which are done routinely. With their results, they should be able to indicate on a map the zones where poaching pressure is high enough to induce behavioral changes in Diana monkeys and those where it is not. Such a map could be used to identify areas of high surveillance priority. Repeated mapping over the years would permit the monitoring of the way poaching pressure evolves in space and time. This is important for assessing the efficiency of conservation strategies.

P. badius are among the most vulnerable monkey species to poaching pressure. In fact, their population size has dramatically decreased in some heavily poached areas of the Taï National Park where other monkey species are still well represented (Hoppe-Dominik 1995, Kone personal observation). Chapman *et al.* (1999) argued that red colobus are the easiest primates to hunt in Africa and one form – *Procolobus badius waldroni* – is possibly extinct (Oates *et al.* 2000). Unless urgent measures are taken to protect the remaining populations of the Taï *P. badius*, these monkeys might face extinction too. This study shows that the behavior of *P. badius* in the rainy seasons could make them more vulnerable to poaching. Protection efforts should thus be particularly reinforced during the rainy seasons.

Acknowledgments
We thank the *Ministère de l'Enseignement Supérieur et de la Recherche Scientifique, the University of Cocody, the Ministère de l'Environnement et du Cadre de Vie, the Projet Autonome pour la Conservation du Parc National de Taï* (P.A.C.P.N.T.), the Directorate of the *Centre Suisse de Recherches Scientifiques* (CSRS), Maria and Jakob Zinsstag and then Simone and Olivier Girardin, and the Centre de Recherche en Ecologie (CRE) for the permission to conduct research in the Taï National Park and for financial, academic, and logistic support. Funding for this research was provided by grants from GTZ via its Flanking Program for Tropical Ecology and the Universitätsvere in Bayreuth, Germany. Further support was provided by the West African Research Association (WARA) through a travel grant awarded to Inza Koné to interact more efficiently with Johannes Refisch in the writing stage of this paper. We want to thank in particular Ronald Noë, the founder of the Taï Monkey Project. We are very grateful to Richard Peho and Appolinaire Ouho for essential help in collecting behavioral data. Scott McGraw, Joanna Lambert, Jennifer Jenkins, Dorothy Newman, Alison Davies, and specially Andres Tschannen made significant contributions to this work through their constructive comments on the manuscript.

References
Adachi, K. (1995). A workshop on "polyspecific associations of rain forest primates" – an introduction to a research project in the Taï National Park. *Primate Research*, **11**, 29–36.

Altman, J. (1974). Observational study of behavior: sampling methods. *Behavior*, **49**, 227–67.

Boesch, C. (1994). Chimpanzees-red colobus monkeys: a predator-prey system. *Behavior*, **47**, 1135–48.

Boesch, C. and Boesch, H. (1989). Hunting behavior of wild chimpanzees in the Taï National Park. *American Journal of Physical Anthropology*, **78**, 547–73.

Bshary, R. (2001). Diana monkeys, *Cercopithecus diana*, adapt their anti-predator response behaviour to human hunting strategies. *Behavioural Ecology and Sociobiology*, **50**, 251–6.

Bshary, R. and Noë, R. (1997a). Anti-predation behaviour of red colobus monkeys in the presence of chimpanzees. *Behavioural Ecology and Sociobiology*, **41**, 321–33.

Bshary, R. and Noë, R. (1997b). Red colobus and Diana monkeys provide mutual protection against predators. *Animal Behaviour*, **54**, 1461–74.

Caspary, H. U. (1999). *Utilisation de la faune sauvage en Côte-d'Ivoire et Afrique de l'Ouest – potentiels et contraintes pour la coopération au développement.* Eschborn: GTZ.

Caspary, H. U., Koné, I., Prouot, C. and De Pauw, M. (2001). La chasse et la filière viande de brousse dans l'espace Taï, Côte d'Ivoire. Tropenbos Côte d'Ivoire séries 2, Wageningen, Netherlands.

Chapman, C. A., Balcomb, S. R., Gillespie, T. R., Skorupa, J. P. and Struhsaker, T. T. (1999). Long term effects of logging on African primate communities: A 28-year comparison from Kibale National Park, Uganda. *Conservation Biology*, **14**(1), 207–17.

Collinet, J., Monteny, B. and Poutaud, B. (1984). Le milieu physique. In *Recherche et Aménagement en Milieu Forestier Tropical Humide*, ed. J. Guillaumet, G. Couturier and H. Dosso. le Projet Taï de Côte d'Ivoire. UNESCO, Paris.

Dind, F. (1995). Etude d'une population cible de leopards (*Panthera pardus*) en forêt tropicale humide (Parc National de Taï, Côte d'Ivoire). Mémoire de Diplôme, Université de Lausanne, Switzerland.

Galat-Luong, A. and Galat, G. (1982). Abondance relative et associations plurispécifiques des primates diurnes du P.N.T., Côte d'Ivoire. Rapport du centre O.R.S.T.O.M. d'Adiopodoumé, Abidjan.

Gautier-Hion, A. and Tutin, C. E. G. (1988). Simultaneous attack by adult males of a polyspecific troop of monkeys against a crowned hawk eagle. *Folia Primatologica*, **51**, 149–51.

Godin, J-G. J. (1990). Diet selection under the risk of predation. In *Behavioural Mechanisms of Food Selection*, ed. R. N. Hughues. NATO ASI series 20. Heidelberg, Berlin, Springer Verlag, 739–69.

Hames, R. B. (1979). A comparison of the efficiencies of the shotgun and the bow in neotropical forest hunting. *Human Ecology*, **7**, 219–52.

Höner, O., Leumann, L. and Noë, R. (1997). Polyspecific associations of red colobus (*Colobus badius*) and Diana monkey (*Cercopithecus diana*) groups in the Taï National Park, Ivory Coast. *Primates*, **38**, 281–91.

Hoppe-Dominik, B. (1984). Etude du spectre des proies de la panthère, *Panthera pardus*, dans le Parc National de Taï en Côte d'Ivoire. *Mammalia*, **48** (4), 477–87.

Hoppe-Dominik, B. (1995). L'état actuel des effectifs de grands mammifères dans l'ensemble du Parc National de Taï. Direction de la Protection de la Nature, Abidjan.

Hoppe-Dominik, B. (1997). Suivi et analyse des résultats du travail de la cellule suivi faune sur l'état actuel des effectifs des grands mammifères dans l'ensemble du Parc National de Taï: proposition et mise en œuvre d'un système plus efficace de surveillance. Rapport établi pour GTZ, Abidjan, San Pédro.

James, F. C. and McCulloch, C. E. (1990). Multivariate analysis in ecology and systematics: panacea or pandora's box. *Annual Review in Ecology and Systematics*, **21**, 129–66.

Kingdon, J. (1997). *The Kingdon Field Guide to African Mammals*. London: Academic Press.

Klump, G. M. and Shalter, M. D. (1984). Acoustic behaviour of birds and mammals in the predator context; I. Factors affecting the structure of alarm signals, II. The functional significance and evolution of alarm signals. *Zeitschrift Tierpsychology*, **66**, 189−226.

Kummer, H. and Noë, R. (1994). *The influence of predation by chimpanzees on the behaviour and the social structure of red colobus monkeys.* Project Report − CSRS, Abidjan.

Leroux-Scribe, C. (1989). Les statistiques. Tome 1: L'analyse des données. Paris: Série les Guides SAS, Edition Europstat.

Martin, P. and Bateson, P. (1985). *Measuring behaviour: an introductory guide*, 2nd edition. Cambridge: Cambridge University Press.

McGraw, W. S. (1998). Comparative locomotion and habitat use of six monkeys in the Taï Forest, Ivory Coast. *American Journal of Physical Anthropology*, **105**, 493−510.

McGraw, W. S. and Noë, R. (1995). The Taï Forest monkey project. *African Primates*, **1** (1), 17−19.

Noë, R. and Bshary, R. (1997). The formation of red colobus-Diana monkey associations under predation pressure from chimpanzees. *Proceedings of the Royal Society of Britain*, **264**, 253−9.

Oates, J. F., Abedi-Lartey, M., McGraw, W. S., Struhsaker, T. T. and Whitesides, G. H. (2000). Extinction of a West African red colobus monkey. *Conservation Biology*, **14**(5), 1526−32.

P.A.C.P.N.T. (1997). Rapport d'évaluation des activités du PACPNT d'Octobre 96 à Juillet 97. San Pédro: PACPNT.

Phillipeau, G. (1986). *Comment interpréter les résultats d'une analyse en composantes principales*. Paris: SESI-ITCF.

Refisch, J. and Koné, I. (2001). Influence du braconnage sur les populations simiennes et effets secondaires sur la végétation: un exemple tiré d'une région forestière de régime pluvieux en Côte d'Ivoire. TÖB série F-IV/4f, Eschborn, Netherlands.

Refisch, J. and Koné, I. (2005). Impact of commercial hunting on monkey populations in the Taï region, Côte d'Ivoire. *Biotropica*, **37**(1), 136−44.

Shultz, S. (2001). Notes on interactions between monkeys and African crowned eagles in the Taï National Park, Ivory Coast. *Folia Primatologica*, **72**, 248−50.

Verdade, L. M. (1996). The influence of hunting pressure on the social behavior of vertebrates. *Revista Brasileria de Biologia*, **56**(1), 1−13.

Wachter, B., Schabel, M. and Noë, R. (1997). Diet overlap and polyspecific associations of red colobus and Diana monkeys in the Taï National Park, Ivory Coast. *Ethology*, **103**, 514−26.

Zuberbühler, K. (2000a). Interspecies semantic communication between two forest monkeys. *Proceedings of the Royal Society of London, Britain*, **267**, 713−18.

Zuberbühler, K. (2000b). Causal knowledge of predators' behaviour in wild Diana monkeys. *Animal Behaviour*, **59**, 209−20.

Zuberbühler, K., Cheney, D. L. and Seyfarth, R. M. (1999a). Conceptual semantics in a non-human primate. *Journal of Comparative Psychology*, **113**, 33−42.

Zuberbühler, K., Jenny, D. and Bshary, R. (1999b). The predator deterrence function of primate alarm calls. *Ethology*, **105**, 477−90.

Zuberbühler, K., Noë, R. and Seyfarth, R. M. (1997). Diana monkey long-distance calls: messages for conspecifics and predators. *Animal Behaviour*, **53**, 589−604.

Appendix 10.1a. *List of the variables contributing most to the formation of the first three axes of the Principal Component Analysis of Diana monkey behavior. Threshold defined the minimum absolute correlation coefficient to the respective axis to allow the variable to enter the list*

Axes	Variables with a positive correlation	Correlation coefficient	Variables with a negative correlation	Correlation coefficient
Axis 1 threshold: 0.20	Using stratum 3−	0.280488	Using stratum 3+	− 0.285796
	Not hiding behind plant parts	0.295551	Hiding behind plant parts	− 0.295551
	Very exposed body	0.305783	Unexposed body	− 0.303723
	Foraging	0.250997	Being isolated	− 0.246807
	Being paired	0.201162		
Axis 2 threshold: 0.13	Resting	0.135960	Socializing	− 0.137373
	Eating flower buds	0.220897	Eating flowers	− 0.216771
	Eating ripe fruits	0.168044	Eating unripe fruits	− 0.303636
	Eating mature leaves	0.344417	Being paired	− 0.149714
	Being isolated	0.139451	Silent fleeing	− 0.212855
	Socializing	0.358073	Fleeing after vocalization	− 0.286895
	Vocalizing after inspection	0.308416	Reacting from close to	− 0.243538
	Reacting from fairly far away	0.221738		
	Reacting from afar	0.153140		
Axis 3 threshold: 0.10	Using stratum 3−	0.149270	Using stratum 3+	− 0.148611
	Unexposed body	0.117040	Fairly exposed body	− 0.303857
	Locomotion	0.314457	Resting	− 0.354757
	Eating flower buds	0.124673	Socializing	− 0.132909
	Eating ripe fruits	0.119721	Eating unripe fruits	− 0.106208
	Being isolated	0.100800	Being paired	− 0.137274
	Silent fleeing	0.368403	Vocalizing after inspection	− 0.297955
	Fleeing after inspection	0.316769	Vocalizing	− 0.255570
	Reacting from fairly far away	0.225783	Reacting from close to	− 0.234732

Appendix 10.1b. *List of the variables contributing most of the formation of the first three axes of the Principal Component Analysis of red colobus behavior. Threshold defined the minimum absolute correlation coefficient to the respective axis to allow the variable to enter the list*

Axes	Variables with a positive correlation	Correlation coefficient	Variables with a negative correlation	Correlation coefficient
Axis 1	Using stratum 3−	0.255573	Using stratum 3+	− 0.255975
threshold:	Unexposed body	0.226538	Fairly exposed body	− 0.277434
0.19	Socializing	0.195058	Eating mature leaves	− 0.262768
	Eating unidentified items	0.212670	Being paired	− 0.241664
	Vocalizing	0.204459	Socializing	− 0.245054
	Fleeing after vocalization	0.208553	Fleeing after inspection	− 0.220512
Axis 2	Using stratum 3−	0.199445	Using stratum 3+	− 0.196518
threshold:	Not hiding behind plant parts	0.356248	Hiding behind plant parts	− 0.356248
0.15	Fairly exposed body	0.178142	Unexposed body	− 0.325206
	Very exposed body	0.363526	Locomotion	− 0.157608
	Socializing	0.212359	Eating unripe fruits	− 0.165390
	Eating young leaves	0.199279	Being isolated	− 0.213810
	Being paired	0.150139	Fleeing	− 0.152091
	Not changing behavior	0.253006		
Axis 3	Using stratum 3+	0.256004	Using stratum 3−	− 0.253813
threshold:	Not hiding behind plant parts	0.178994	Hiding behind plant parts	− 0.179724
0.11	Locomotion	0.144580	Resting	− 0.175958
	Eating flower buds	0.113822	Eating unripe fruits	− 0.271832
	Eating flowers	0.144526	Eating young leaves	− 0.144804
	Eating ripe fruits	0.161879	Being aggregated	− 0.271878
	Eating mature leaves	0.163367	Fleeing	− 0.362983
	Being isolated	0.145588	Reacting from fairly far away	− 0.194460
	Not changing behavior	0.179088		
	Socializing	0.163517		
	Vocalizing	0.309938		
	Reacting from afar	0.282639		

Appendix 10.2a. *Percentages of each behavioral variable for Diana monkeys*

Area	Group label	Season	Period of the day	Using stratum 0	Using stratum 1	Using stratum 2	Using stratum 3−	Using stratum 3+	Not hiding behind plant parts	Hiding behind plant parts	Unexposed body	Fairly exposed body
Non-poached	A1M	rainy 1	Morning	0.00	0.00	0.00	0.00	100.00	50.98	49.02	17.65	53.92
	A1S	rainy 1	Afternoon	0.00	0.00	0.00	3.31	96.69	48.76	51.24	21.49	55.37
	A2M	dry 1	Morning	0.00	0.00	0.57	19.26	80.17	35.98	64.02	36.54	44.19
	A2S	dry 1	Afternoon	0.00	0.00	0.00	16.29	83.71	36.16	63.84	35.18	48.20
	A3M	rainy 2	Morning	0.00	0.00	0.00	2.24	97.76	45.38	54.62	44.26	43.70
	A3S	rainy 2	Afternoon	0.00	0.00	0.00	0.88	99.12	47.08	52.92	45.03	40.35
	A4M	dry 2	Morning	0.00	0.00	0.00	0.00	100.00	25.26	74.74	17.89	63.16
	A4S	dry 2	Afternoon	0.00	0.00	0.00	0.00	100.00	23.93	76.07	23.93	55.56
	B1M	rainy 1	Morning	0.00	0.00	0.00	18.46	81.54	66.11	33.89	10.06	57.05
	B1S	rainy 1	Afternoon	0.00	0.00	0.00	17.65	82.35	69.80	30.20	10.20	56.07
	B2M	dry 1	Morning	0.00	0.00	0.00	13.84	86.16	43.30	56.70	29.91	50.45
	B2S	dry 1	Afternoon	0.00	0.00	0.00	17.71	82.29	47.57	52.43	23.61	53.13
	B3M	rainy 2	Morning	0.00	0.00	0.00	14.02	85.98	55.76	44.24	34.85	40.07
	B3S	rainy 2	Afternoon	0.00	0.00	0.00	9.77	90.23	57.84	42.16	33.16	44.22
	B4M	dry 2	Morning	0.00	0.00	0.00	1.18	98.82	20.00	80.00	29.42	54.12
	B4S	dry 2	Afternoon	0.00	0.00	0.00	0.00	100.00	27.63	72.37	32.90	50.00
Poached	C1M	rainy 1	Morning	0.00	0.00	0.00	0.00	100.00	51.66	48.34	20.53	54.96
	C1S	rainy 1	Afternoon	0.00	0.00	0.00	0.00	100.00	60.45	39.55	13.43	52.24
	C2M	dry 1	Morning	0.00	0.00	0.00	27.67	72.33	41.26	58.74	31.55	46.60
	C2S	dry 1	Afternoon	0.00	0.00	0.00	22.91	77.09	40.00	60.00	31.64	46.18
	C3M	rainy 2	Morning	0.00	0.00	0.00	11.22	88.78	50.99	49.01	46.38	41.77
	C3S	rainy 2	Afternoon	0.00	0.00	0.00	13.54	86.46	55.62	44.38	44.38	40.34
	C4M	dry 2	Morning	0.00	0.00	0.00	0.00	100.00	26.19	73.81	28.57	57.14

Appendix 10.2a. (cont.)

Area	Group label	Season	Period of the day	Using stratum 0	Using stratum 1	Using stratum 2	Using stratum 3−	Using stratum 3+	Not hiding behind plant parts	Hiding behind plant parts	Unexposed body	Fairly exposed body
C4S	dry 2	Afternoon	0.00	0.00	0.00	0.00	100.00	20.59	79.41	38.35	50.00
D1M	rainy 1	Morning	0.00	0.00	0.00	13.50	86.50	61.96	38.04	19.01	58.89
D1S	rainy 1	Afternoon	0.00	0.00	0.00	4.30	95.70	54.30	45.70	18.28	58.07
D2M	dry 1	Morning	0.00	0.00	0.00	19.02	80.98	38.59	61.41	22.29	58.70
D2S	dry 1	Afternoon	0.00	0.00	0.00	18.90	81.10	41.46	58.54	31.10	52.44
D3M	rainy 2	Morning	0.00	0.00	0.00	5.88	94.12	30.39	69.61	33.33	48.03
D3S	rainy 2	Afternoon	0.00	0.00	0.00	1.18	98.82	35.88	64.12	28.24	47.06
D4M	dry 2	Morning	0.00	0.00	0.00	0.00	100.00	12.50	87.50	46.42	42.86
D4S	dry 2	Afternoon	0.00	0.00	0.00	4.60	95.40	17.24	82.76	44.82	43.68

Area	Group label	Very exposed body	Foraging	Locomotion	Resting	Socializing	Eating flowers	Eating flower buds	Eating ripe fruits	Eating unripe fruits	Eating unidentified items	Eating invertebrates	Eating mature leaves
Non-poached	A1M	33.61	22.41	29.46	44.81	3.31	0.00	5.56	62.96	11.11	0.00	1.85	16.67
	A1S	34.74	24.74	33.16	40.00	2.11	0.00	0.00	36.17	23.40	0.00	.	40.43
	A2M	30.60	21.31	36.61	38.25	3.83	5.41	2.70	2.70	48.65	0.00	2.70	16.22
	A2S	34.06	18.84	33.33	47.10	0.72	23.08	0.00	0.00	61.54	0.00	0.00	3.85
	A3M	25.40	6.35	46.03	44.44	3.18	25.00	0.00	0.00	75.00	0.00	0.00	0.00
	A3S	15.86	2.44	48.78	48.78	0.00	0.00	0.00	0.00	100.00	0.00	0.00	0.00
	A4M	18.30	19.51	43.90	36.59	0.00	0.00	0.00	81.25	6.25	6.25	0.00	6.25
	A4S	20.55	10.96	45.21	43.84	0.00	0.00	12.50	0.00	62.50	0.00	0.00	12.50
	B1M	45.53	21.14	38.21	39.84	0.81	0.00	16.00	24.00	8.00	0.00	4.00	48.00
	B1S	47.88	14.08	33.80	47.89	4.23	0.00	5.00	35.00	0.00	0.00	10.00	50.00
	B2M	22.59	17.42	37.42	42.58	2.59	0.00	0.00	0.00	70.37	0.00	0.00	25.93
	B2S	25.64	14.10	40.38	42.95	2.56	0.00	0.00	0.00	68.18	0.00	18.18	13.64
	B3M	37.28	27.12	42.37	30.51	0.00	0.00	6.25	0.00	25.00	0.00	68.75	0.00
	B3S	36.73	22.45	34.69	42.86	0.00	0.00	0.00	72.73	9.09	0.00	18.18	0.00
	B4M	27.47	20.88	31.87	45.05	2.20	0.00	5.26	0.00	73.68	5.26	15.79	0.00
	B4S	30.31	10.61	42.42	45.45	1.52	0.00	0.00	14.29	71.43	0.00	14.29	0.00
Poached	C1M	28.07	9.36	38.60	52.05	0.00	0.00	7.14	28.57	21.43	0.00	0.00	42.86
	C1S	36.72	11.86	25.42	61.02	1.69	4.76	4.76	19.05	4.76	0.00	0.00	61.90
	C2M	38.22	10.83	30.57	56.05	2.55	0.00	5.88	11.76	17.65	0.00	17.65	23.53
	C2S	28.99	15.94	31.40	52.17	0.48	0.00	0.00	28.13	40.63	0.00	3.13	6.25
	C3M	27.47	7.69	35.16	56.04	1.10	0.00	0.00	0.00	100.00	0.00	0.00	0.00
	C3S	17.65	1.47	44.12	54.41	0.00	0.00	0.00	100.00	0.00	0.00	0.00	0.00
	C4M	14.81	12.96	24.07	61.11	1.85	0.00	0.00	14.29	57.14	0.00	0.00	28.57
	C4S	11.64	4.65	44.19	51.16	0.00	0.00	0.00	0.00	0.00	0.00	0.00	50.00
	D1M	33.33	1.85	50.00	48.15	0.00	0.00	0.00	0.00	0.00	0.00	0.00	0.00
	D1S	50.00	13.73	37.25	49.02	0.00	0.00	0.00	57.14	0.00	0.00	7.14	35.71
	D2M	14.70	8.82	67.65	23.53	0.00	0.00	0.00	0.00	33.33	0.00	0.00	33.33
	D2S	26.15	12.12	37.88	46.97	3.03	0.00	0.00	0.00	100.00	0.00	0.00	0.00
	D3M	20.00	2.00	54.00	44.00	0.00	0.00	0.00	0.00	100.00	0.00	0.00	0.00
	D3S	26.31	0.00	47.37	52.63	0.00	0.00	0.00	0.00	0.00	0.00	0.00	0.00
	D4M	7.89	2.63	55.26	42.11	0.00	0.00	0.00	0.00	0.00	0.00	0.00	0.00
	D4S	40.00	0.00	60.00	40.00	0.00	0.00	0.00	0.00	0.00	0.00	0.00	0.00

Appendix 10.2a. *(cont.)*

Area	Group label	Eating young leaves	Being isolated	Being paired	Being aggregated	Silent fleeing	Not changing behavior	Socializing	Fleeing after inspection	Vocalizing after inspection	Vocalizing	Fleeing after vocalization	Reacting from close to	Reacting from fairly far away	Reacting from afar
Non-poached	A1M	1.85	84.23	14.52	1.24	13.79	3.45	31.03	3.45	27.59	6.90	13.79	35.71	64.29	0.00
	A1S	0.00	88.08	8.29	3.63	33.33	5.56	22.22	0.00	16.67	11.11	11.11	40.00	60.00	0.00
	A2M	21.62	82.07	14.67	3.26	30.00	0.00	5.00	0.00	15.00	20.00	30.00	50.00	50.00	0.00
	A2S	11.54	85.61	11.51	2.88	47.06	0.00	5.88	0.00	5.88	17.65	23.53	47.05	52.93	0.00
	A3M	0.00	88.24	8.82	2.94	16.67	0.00	0.00	0.00	0.00	50.00	33.33	66.67	33.33	0.00
	A3S	0.00	90.36	8.43	1.20	0.00	0.00	0.00	0.00	14.29	71.43	14.29	42.86	57.15	0.00
	A4M	0.00	90.24	8.54	1.22	25.00	0.00	12.50	0.00	50.00	12.50	0.00	50.00	50.00	0.00
	A4S	12.50	93.15	6.85	0.00	25.00	0.00	25.00	0.00	25.00	25.00	0.00	0.00	100.00	0.00
	B1M	0.00	85.37	11.38	3.25	18.75	0.00	37.50	18.75	12.50	6.25	6.25	31.25	68.75	0.00
	B1S	0.00	82.52	15.38	2.10	37.50	0.00	29.17	0.00	16.67	8.33	8.33	54.17	41.67	4.17
	B2M	3.70	86.54	11.54	1.92	35.29	0.00	17.65	5.88	11.76	5.88	23.53	41.18	58.81	0.00
	B2S	0.00	83.95	12.96	3.89	43.75	9.38	3.13	0.00	21.88	6.25	15.63	40.63	56.26	3.13
	B3M	0.00	85.00	13.33	1.67	37.50	0.00	25.00	0.00	0.00	12.50	25.00	62.50	37.50	0.00
	B3S	0.00	79.25	18.87	1.89	33.33	13.33	13.33	6.67	6.67	13.33	13.33	86.67	13.33	0.00
	B4M	0.00	86.81	13.19	0.00	37.50	0.00	6.25	0.00	6.25	43.75	6.25	43.75	56.25	0.00
	B4S	0.00	85.07	10.45	4.48	46.67	13.33	13.33	6.67	0.00	13.33	6.67	53.34	46.67	0.00

Poached														
C1M	0.00	86.44	9.60	3.94	8.33	0.00	25.00	11.11	36.11	5.56	13.89	16.67	83.83	0.00
C1S	4.76	80.90	15.17	3.94	8.33	2.78	30.56	2.78	38.89	2.78	13.89	25.01	75.01	0.00
C2M	23.53	82.39	13.21	4.40	23.81	2.38	2.38	7.14	21.43	11.90	30.95	35.71	54.76	9.52
C2S	21.88	83.25	11.96	4.79	39.13	0.00	2.17	0.00	19.57	13.04	26.09	55.55	42.22	2.22
C3M	0.00	86.81	13.19	0.00	8.33	0.00	4.17	0.00	12.50	70.83	4.17	29.17	70.84	0.00
C3S	0.00	94.37	4.23	1.41	11.11	0.00	11.11	0.00	11.11	66.67	0.00	33.33	66.67	0.00
C4M	0.00	85.19	14.81	0.00	9.09	0.00	18.18	0.00	36.36	36.36	0.00	72.72	27.27	0.00
C4S	50.00	91.11	8.89	0.00	0.00	0.00	25.00	0.00	37.50	37.50	0.00	25.00	75.00	0.00
D1M	0.00	91.38	5.17	3.45	9.09	7.41	45.45	18.18	18.18	9.09	0.00	0.00	90.90	9.09
D1S	0.00	90.57	8.49	0.94	33.33	0.00	29.63	3.70	22.22	0.00	3.70	14.81	70.37	14.81
D2M	33.33	91.43	8.57	0.00	45.45	0.00	9.09	0.00	9.09	0.00	36.36	18.18	81.82	0.00
D2S	0.00	75.76	22.73	1.52	45.45	0.00	9.09	0.00	9.09	0.00	36.36	27.27	72.72	0.00
D3M	0.00	94.12	5.88	0.00	40.00	0.00	10.00	10.00	10.00	20.00	10.00	10.00	90.00	0.00
D3S	0.00	95.65	4.35	0.00	0.00	0.00	16.67	16.67	16.67	50.00	0.00	66.67	33.33	0.00
D4M	0.00	88.10	9.52	2.38	57.14	0.00	0.00	28.57	0.00	14.29	0.00	0.00	100.00	0.00
D4S	0.00	84.62	7.69	7.69	50.00	0.00	0.00	0.00	0.00	0.00	50.00	0.00	100.00	0.00

Appendix 10.2b. *Percentages of each behavioral variable for red colobus*

Area	Group label	Season	Period of the day	Using stratum 0	Using stratum 1	Using stratum 2	Using stratum 3−	Using stratum 3+	Not hiding behind plant parts	Hiding behind plant parts	Unexposed body	Fairly exposed body
Non-poached	A1M	rainy 1	Morning	0.00	0.00	0.00	0.00	100.00	50.98	49.02	17.65	53.92
	A1S	rainy 1	Afternoon	0.00	0.00	0.00	3.31	96.69	48.76	51.24	21.49	55.37
	A2M	dry 1	Morning	0.00	0.00	0.57	19.26	80.17	35.98	64.02	36.54	44.19
	A2S	dry 1	Afternoon	0.00	0.00	0.00	16.29	83.71	36.16	63.84	35.18	48.20
	A3M	rainy 2	Morning	0.00	0.00	0.00	2.24	97.76	45.38	54.62	44.26	43.70
	A3S	rainy 2	Afternoon	0.00	0.00	0.00	0.88	99.12	47.08	52.92	45.03	40.35
	A4M	dry 2	Morning	0.00	0.00	0.00	0.00	100.00	25.26	74.74	17.89	63.16
	A4S	dry 2	Afternoon	0.00	0.00	0.00	0.00	100.00	23.93	76.07	23.93	55.56
	B1M	rainy 1	Morning	0.00	0.00	0.00	18.46	81.54	66.11	33.89	10.06	57.05
	B1S	rainy 1	Afternoon	0.00	0.00	0.00	17.65	82.35	69.80	30.20	10.20	56.07
	B2M	dry 1	Morning	0.00	0.00	0.00	13.84	86.16	43.30	56.70	29.91	50.45
	B2S	dry 1	Afternoon	0.00	0.00	0.00	17.71	82.29	47.57	52.43	23.61	53.13
	B3M	rainy 2	Morning	0.00	0.00	0.00	14.02	85.98	55.76	44.24	34.85	40.07
	B3S	rainy 2	Afternoon	0.00	0.00	0.00	9.77	90.23	57.84	42.16	33.16	44.22
	B4M	dry 2	Morning	0.00	0.00	0.00	1.18	98.82	20.00	80.00	29.42	54.12
	B4S	dry 2	Afternoon	0.00	0.00	0.00	0.00	100.00	27.63	72.37	32.90	50.00

Poached										
C1M	rainy 1	Morning	0.00	0.00	0.00	100.00	51.66	48.34	20.53	54.96
C1S	rainy 1	Afternoon	0.00	0.00	0.00	100.00	60.45	39.55	13.43	52.24
C2M	dry 1	Morning	0.00	0.00	27.67	72.33	41.26	58.74	31.55	46.60
C2S	dry 1	Afternoon	0.00	0.00	22.91	77.09	40.00	60.00	31.64	46.18
C3M	rainy 2	Morning	0.00	0.00	11.22	88.78	50.99	49.01	46.38	41.77
C3S	rainy 2	Afternoon	0.00	0.00	13.54	86.46	55.62	44.38	44.38	40.34
C4M	dry 2	Morning	0.00	0.00	0.00	100.00	26.19	73.81	28.57	57.14
C4S	dry 2	Afternoon	0.00	0.00	0.00	100.00	20.59	79.41	38.35	50.00
D1M	rainy 1	Morning	0.00	0.00	13.50	86.50	61.96	38.04	19.01	58.89
D1S	rainy 1	Afternoon	0.00	0.00	4.30	95.70	54.30	45.70	18.28	58.07
D2M	dry 1	Morning	0.00	0.00	19.02	80.98	38.59	61.41	22.29	58.70
D2S	dry 1	Afternoon	0.00	0.00	18.90	81.10	41.46	58.54	31.10	52.44
D3M	rainy 2	Morning	0.00	0.00	5.88	94.12	30.39	69.61	33.33	48.03
D3S	rainy 2	Afternoon	0.00	0.00	1.18	98.82	35.88	64.12	28.24	47.06
D4M	dry 2	Morning	0.00	0.00	0.00	100.00	12.50	87.50	46.42	42.86
D4S	dry 2	Afternoon	0.00	0.00	4.60	95.40	17.24	82.76	44.82	43.68

Appendix 10.2b. (*cont.*)

Area	Group label	Very exposed body	Foraging	Locomotion	Resting	Socializing	Eating flowers	Eating flower buds	Eating ripe fruits	Eating unripe fruits	Eating unidentified items	Eating mature leaves
Non-poached	A1M	28.43	16.83	18.81	60.40	3.96	0.00	5.56	22.22	44.44	0.00	22.22
	A1S	23.13	9.09	28.10	60.33	2.48	0.00	9.09	27.27	0.00	0.00	54.55
	A2M	19.26	9.92	22.66	60.62	6.80	9.09	12.12	9.09	39.39	0.00	15.15
	A2S	16.61	18.89	25.08	52.12	3.91	0.00	0.00	15.79	45.61	0.00	24.56
	A3M	12.04	16.57	36.52	44.94	1.97	13.79	6.90	25.86	34.48	0.00	10.34
	A3S	14.62	16.08	34.50	45.61	3.80	12.96	20.37	22.22	14.81	0.00	20.37
	A4M	18.94	13.68	44.21	37.89	4.21	0.00	30.77	0.00	7.69	0.00	61.54
	A4S	20.52	15.38	29.91	53.85	0.85	0.00	40.00	0.00	33.33	0.00	20.00
	B1M	32.89	11.74	29.53	54.36	4.36	0.00	11.43	17.14	2.86	0.00	54.29
	B1S	33.73	15.69	24.71	55.69	3.92	0.00	20.51	12.82	5.13	0.00	38.46
	B2M	19.65	18.75	27.23	50.00	4.02	0.00	0.00	0.00	38.10	0.00	9.52
	B2S	23.26	14.58	31.25	50.00	4.17	2.50	2.50	0.00	60.00	0.00	12.50
	B3M	25.07	13.08	24.61	58.57	3.74	0.00	23.81	7.14	28.57	4.76	11.90
	B3S	22.62	16.20	30.08	50.39	3.34	3.28	6.56	26.23	11.48	6.56	14.75
	B4M	16.48	9.41	34.12	55.29	1.18	0.00	0.00	0.00	0.00	0.00	100.00
	B4S	17.11	10.53	19.74	68.42	1.32	12.50	12.50	0.00	25.00	0.00	50.00

Poached

C1M	24.50	9.27	38.41	50.33	1.98	0.00	0.00	0.00	0.00	0.00	100.00
C1S	34.33	15.67	28.36	52.24	3.73	0.00	0.00	0.00	55.00	0.00	45.00
C2M	21.84	12.14	24.76	60.19	2.92	0.00	0.00	20.00	36.00	0.00	20.00
C2S	22.18	18.55	28.00	51.27	2.18	7.84	0.00	3.92	35.29	0.00	11.76
C3M	11.85	8.55	36.18	50.66	4.61	20.00	24.00	12.00	20.00	4.00	20.00
C3S	15.28	14.70	42.94	38.04	3.75	7.84	1.96	11.76	54.90	3.92	9.80
C4M	14.28	16.67	26.19	54.76	2.38	28.57	0.00	0.00	57.14	0.00	14.29
C4S	11.76	5.88	32.35	61.76	0.00	0.00	0.00	0.00	50.00	0.00	50.00
D1M	22.08	5.56	38.89	50.62	4.94	33.33	0.00	0.00	0.00	0.00	11.11
D1S	23.66	11.29	30.65	56.45	1.62	0.00	42.86	4.76	9.52	0.00	19.05
D2M	19.02	10.93	28.96	58.47	1.64	0.00	0.00	0.00	70.00	0.00	15.00
D2S	16.47	13.41	31.10	53.66	1.83	4.55	0.00	4.55	77.27	0.00	13.64
D3M	18.63	2.94	55.88	41.18	0.00	0.00	0.00	0.00	66.67	0.00	0.00
D3S	24.70	11.76	29.41	58.24	0.59	15.00	0.00	0.00	5.00	0.00	40.00
D4M	10.72	7.14	33.93	57.14	1.79	0.00	25.00	0.00	75.00	0.00	0.00
D4S	11.50	8.05	35.63	54.02	2.30	0.00	14.29	42.86	14.29	0.00	28.57

Appendix 10.2b. (cont.)

Area	Group label	Eating young leaves	Being isolated	Being paired	Being aggregated	Silent	Not changing behavior	Inspecting	Fleeing after inspection	Vocalizing after inspection	Vocalizing close to	Fleeing after vocalization	Fleeing from close to away	Fleeing from fairly far	Fleeing from afar
Non-poached	A1M	5.56	77.67	20.39	1.94	50.00	0.00	30.77	3.85	11.54	0.00	3.85	34.62	65.39	0.00
	A1S	9.09	74.38	19.01	6.62	45.45	4.55	13.64	13.64	4.55	4.55	13.64	36.37	63.63	0.00
	A2M	15.15	75.42	16.95	7.62	72.31	0.00	0.00	0.00	0.00	4.62	23.08	46.17	46.17	7.69
	A2S	14.04	75.90	15.31	8.80	78.18	1.82	0.00	7.27	1.82	0.00	10.91	44.44	51.85	3.70
	A3M	8.62	81.11	15.28	3.62	44.33	3.09	3.09	1.03	0.00	30.93	17.53	36.46	36.45	27.08
	A3S	9.26	81.74	15.07	3.00	55.17	8.62	0.00	3.45	0.00	24.14	8.62	37.91	36.19	25.85
	A4M	0.00	71.58	23.16	3.13	71.43	0.00	0.00	0.00	0.00	28.57	0.00	42.86	57.14	0.00
	A4S	6.67	67.80	21.19	11.02	68.75	0.00	0.00	6.25	12.50	6.25	6.25	12.50	87.50	0.00
	B1M	14.29	70.90	18.73	10.37	42.55	10.64	17.02	17.02	4.26	2.13	6.38	55.32	44.68	0.00
	B1S	23.08	71.88	20.70	7.42	47.06	11.76	8.82	20.59	8.82	0.00	2.94	15.15	81.81	3.03
	B2M	52.38	68.00	21.33	10.67	66.67	8.33	4.17	0.00	0.00	8.33	12.50	47.92	49.99	2.08
	B2S	22.50	69.10	21.18	9.73	70.00	8.00	4.00	0.00	4.00	2.00	12.00	32.00	64.00	4.00
	B3M	23.81	74.53	16.15	9.32	50.00	3.85	3.85	1.92	1.92	19.23	19.23	47.05	41.16	11.76
	B3S	31.15	76.92	17.44	5.64	61.67	1.67	0.00	0.00	0.00	21.67	15.00	55.00	45.00	0.00
	B4M	0.00	74.12	20.00	5.88	23.08	0.00	30.77	23.08	15.38	0.00	7.69	69.00	30.77	0.00
	B4S	0.00	63.64	31.17	5.19	58.82	0.00	0.00	11.76	0.00	17.65	11.76	23.53	76.47	0.00

Poached

C1M	0.00	68.21	25.17	6.62	28.57	0.00	28.57	19.05	0.00	4.76	19.05	9.52	90.48	0.00
C1S	0.00	69.34	29.20	1.46	46.67	0.00	20.00	33.33	0.00	0.00	0.00	40.00	60.00	0.00
C2M	24.00	71.15	16.35	12.50	78.57	0.00	0.00	0.00	0.00	0.00	21.43	35.71	60.70	3.57
C2S	41.18	70.91	18.55	10.54	57.14	0.00	0.00	0.00	3.57	0.00	39.29	57.14	42.85	0.00
C3M	0.00	66.45	23.78	9.77	41.67	6.25	0.00	2.08	6.25	33.33	10.42	25.00	43.74	31.25
C3S	9.80	74.29	16.38	9.31	47.30	6.76	0.00	0.00	0.00	33.78	12.16	44.74	19.74	35.53
C4M	0.00	72.09	23.26	4.65	33.33	0.00	50.00	66.67	0.00	0.00	0.00	33.33	66.67	0.00
C4S	0.00	82.35	17.65	0.00	50.00	15.15	6.06	0.00	0.00	0.00	0.00	0.00	0.00	0.00
D1M	55.56	73.81	22.02	4.17	51.52	5.41	29.73	15.15	13.50	0.00	12.12	48.48	48.48	3.03
D1S	23.81	72.19	21.93	5.88	40.54	0.00	0.00	8.11	0.00	0.00	2.70	44.45	55.57	0.00
D2M	15.00	69.52	22.46	8.02	66.67	0.00	0.00	5.56	0.00	11.11	16.67	41.67	52.78	5.56
D2S	0.00	75.76	16.97	7.28	81.48	0.00	0.00	3.70	0.00	3.70	11.11	33.33	62.96	3.70
D3M	33.33	75.00	16.67	8.34	93.33	0.00	0.00	6.67	0.00	0.00	0.00	26.67	73.34	0.00
D3S	40.00	73.99	20.81	5.21	73.91	0.00	0.00	17.39	4.35	4.35	0.00	52.17	47.83	0.00
D4M	0.00	75.00	16.07	8.93	100.00	0.00	0.00	0.00	0.00	0.00	0.00	33.33	66.66	0.00
D4S	0.00	83.91	14.94	1.15	61.54	0.00	15.38	0.00	0.00	0.00	23.08	15.38	84.61	0.00

11 *Vulnerability and conservation of the Taï monkey fauna*

W. S. McGraw

Conservation is in crisis. Conventional approaches have not worked. Development and conservation are on an accelerating collision course, and proposed solutions are no more than hopeful improvisations . . . . Instead of confronting uncertainty honestly, all too much of the conservation agenda consists of hollow and confused verbiage that promotes dogma rather than dialogue. Blindness to reality is dangerous.

(Schaller 2000:xv)

What is really needed now is not further unproductive debate on whose method is the best, but agreement on what is most important and collaborative action to ensure that as much as possible is conserved.

(Mittermeier *et al.* 1998)

Introduction

The purpose of this chapter is to identify those monkey species most vulnerable to extirpation based on criteria discussed below: the higher the cumulative score, the greater the risk a species is of disappearing. Although all primates within Taï National Park are threatened by hunting to some degree, the aim here is to indicate the probable order that Taï cercopithecids would be eliminated unless better protection is afforded them in the very near future. It is important to note that for at least two species – *Procolobus badius* and *Cercopithecus diana* – Taï National Park is the only forest in Ivory Coast containing sizable and, perhaps, any populations of these species. For this reason, their status is already vulnerable. It is also important to note that hunting pressure, levels of habitat disturbance and species abundances are not homogeneous throughout the park. We know, for example, that poaching pressure in the eastern region of the park is greatest and that while hunting occurs throughout the forest, it is probably not as significant at three sites in the

Monkeys of the Taï Forest, ed. W. Scott McGraw, Klaus Zuberbühler and Ronald Noë.
Published by Cambridge University Press. © Cambridge University Press 2007.

park's western region: our study area, the chimpanzee study area of Boesch and Boesch-Achermann (2000) and the ecotourism site near Guiroutou (Crockford *et al.* 2004). This means that while a species may still be relatively common in western or southern regions of the park, it could already be significantly reduced or absent in areas to the east. For these reasons, the order that monkeys might be extirpated varies by area and the vulnerability totals refer to the likelihood that a taxon would be reduced throughout the entire park.

Background

Since gaining independence in 1960, Ivory Coast has lost approximately 67 per cent of its original forest cover (Tockman 2002). Today, less than one quarter of the country's primary forest remains, totalling approximately 24,000 km^2. Most exists as unprotected remnants of logging activities or as small, isolated patches designated as forest reserves[1]. While large amounts of timber were removed to supply tropical hardwood markets, the need for cultivable lands greatly accelerated the replacement of old growth forest by both subsistence and commercial farmers. Ivory Coast is the world leader in cocoa production and Africa's largest exporter of coffee, but the country's economic future is in jeopardy due to recent sociopolitical trouble. Once regarded as the most economically and politically sound country in West Africa, Ivory Coast is at a critical point in its history as foreign soldiers maintain a fragile ceasefire between government loyalists and a growing rebel movement.

Of equal concern to the country's prosperity are the long-term effects of agricultural policies. Even if peace were to return soon to Ivory Coast, the country's economy has been significantly weakened. Some authorities believe that unless current commercial farming practices are modified, the resulting changes in soil and climatic conditions will threaten the remaining forest as well as the future of crops responsible for changes in the first place (Tockman 2002). Though deforestation rates have slowed recently (annual change in deforestation rate between 1990 and 2000 = 3.1 per cent), farmers continue to encroach on protected forests and illegal cocoa and coffee plantations are abundant in many of the country's forest reserves (Anonymous www.afrol.com 2002). Sacred forests that have remained untouched for centuries are now being slashed and burned to make way for more arable land. The ability and/or motivation for authorities to control this increasing human pressure is questionable.

---

[1] For a contrasting view of deforestation rates in Ivory Coast see Fairhead and Leach (1998).

During surveys in eastern Ivory Coast (1997–2002), I encountered farmers living and working in fields well inside protected forest reserves. When these squatters were questioned, most responded that they either did not know the forests were protected or had never been asked to leave.

The effect of deforestation and illegal hunting on the country's wildlife has been devastating. The number of elephants in Ivory Coast plummeted from approximately 4,800 in 1984 to less than 600 ten years later (Said *et al.* 1995). In Comoe National Park, densities of the 11 most abundant ungulates declined between 60 and 90 per cent in the last 20 years and one primate – *Cercopithecus nictitans* – has likely been hunted to extirpation there (Fischer *et al.* 1999–2000, Fischer & Linsenmair 2001). Most forest reserves in eastern Ivory Coast contain dwindling populations of animals and the common species are only those able to survive in heavily degraded forests. Due to habitat loss and hunting, the primates of the Ivory Coast are at the highest possible risk of extinction (Cowlishaw 1999). One species endemic to eastern Ivory Coast and western Ghana may have already vanished. Surveys to locate populations of *Procolobus badius waldroni* have recovered little evidence that this monkey survives and if it has disappeared, it will be the first primate taxon to have gone extinct in over 400 years (Oates *et al.* 2000, McGraw & Oates 2002, McGraw 2005). Some authorities believe the potential disappearance of this monkey could signal the beginning of a wave of extinctions across West Africa. The future of forest-dwelling primates in this region of Africa is, at best, grim.

In the midst of this devastation sits an oasis of African biodiversity considered one of the most significant rainforests on the continent. Taï National Park is approximately 4570 km^2 and is the largest block of protected, continuous forest in West Africa. It has remained intact largely due to early recognition of its ecological importance. In 1926, the French colonial government declared the Taï Forest a "Forest and Wildlife Refuge." National Park status was achieved in 1972 at which time commercial logging ceased. The forest was recognized as a Biosphere Reserve under the United Nations Educational, Scientific, and Cultural Organization's (UNESCO's) Man and Biosphere Program in 1978 and in 1982, Taï National Park became a World Heritage Site. The park is situated within a bio-diversity "hotspot" and its preservation has been designated a top conservation priority in Africa (Oates 1996, Hacker *et al.* 1998, Mittermeier *et al.* 1998, Myers *et al.* 2000). Many agencies contribute to varied conservation initiatives in and around the park and for years Taï National Park has enjoyed major financial support from the Ivorian government represented by the Ministère d'Agriculture et des Ressources Animales and from organizations such as World Wildlife

Fund, German Technical Cooperation, German Development Bank, and Tropenbos.

Despite its significance as a wildlife sanctuary, Taï National Park and its inhabitants are faced with serious threats to their survival. Foremost among these is increased pressure from humans. Recent studies suggest that there are well over 500,000 people living in the Taï region with approximately 150,000 living immediately adjacent to the park (Caspary *et al.* 2001). Struhsaker (2001a) estimates the population density around the park increased almost 21 per cent to 170/km^2 between 1990 and 2000. The burgeoning human population affects habitat and wildlife directly by clearing more forested areas for farms, felling trees for firewood and by hunting wildlife. The result is an overexploitation of resources already stressed and direct endangerment of various animals and plants that can survive only within the park. Threats to wildlife dependent on the continued health and integrity of rainforest are obvious and immediate. In neighboring Ghana, a strong correlation was found between wildlife extinction rate and human population density (Brashares *et al.* 2001) and overexploitation by humans is believed to be the major threat to one third of the bird and mammal species facing extinction (Rosser & Mainka 2002). Although Taï National Park is considerably larger than other West African forests whose wildlife has been degraded (e.g. Beier *et al.* 2001), it may be no less secure.

It is illegal to hunt in Ivory Coast yet poaching is rampant throughout the country and widespread in the park (Kone & Refisch, Chapter 10). Caspary *et al.* (2001) estimate the value of the bushmeat trade in Ivory Coast (1996) to be well over 100 million dollars contributing 1.4 per cent to the nation's gross national product. In the Taï region alone there are an estimated 73,000 subsistence hunters, 2,200 semi-professional hunters and 220 professional hunters. Activities of professional hunters are concentrated within the park and annual takeoff is estimated at between 56 and 76 tons, consisting primarily of primates and duikers. Estimated bushmeat harvest for 1996 was over 100,000 tons – over twice that of livestock production (Caspary *et al.* 2001). As the human population escalates, taboos that once protected primates are breaking down. For example, followers of Islam are prohibited from eating monkeys unless they are traveling. Many Muslims relocating from Burkina Faso have settled in the Taï region for over 20 years, yet believe they are still traveling. It appears that laws forbidding the consumption of certain foods can be disregarded when one is on a permanent pilgrimage.

Recently, a great deal of effort has been directed towards exploring the issue of sustainable hunting practices within African rainforests (e.g. Robinson & Bennett 2000). Whether traditional hunting is compatible with the survival of primates in a national park should be a non-issue, yet hunting poses the most immediate risk to the Taï primates today. Refisch and Kone (2005a, 2005b) examined the bushmeat trade within the park and concluded that present rates of monkey harvesting by hunters are 3 to 11 times the maximum reproductive values of the respective species. These authors found that while the majority of commercial hunting is being carried out in the eastern region of the park, hunters are rapidly expanding the areas they hunt which inevitably takes them deeper into the park. Kone and Refisch found that the numbers of Red colobus (*Procolobus badius*) are already greatly reduced in areas of moderate hunting and entirely absent in areas of high poaching. This is also true for black and white colobus (*Colobus polykomos*). Near our study site in the park's western region, hunting is more subsistence-based, but this does not imply that off-take rates are sustainable (Robinson & Bennett 2004). Census data confirm that the density of monkeys drops off dramatically immediately beyond the borders of our primary study area (Kone & Refisch, Chapter 10) and it is probable that some species have already been eliminated from pockets of forest between our study site and the nearest villages.

The effects of hunting on Taï's wildlife are significant. Although primates do not comprise the largest percentage of bushmeat, even moderate hunting can significantly impact primate communities because of the relatively long interbirth interval, long gestation period, and longevity of primates compared to other mammals and birds (Cowlishaw & Dunbar 2000). Some monkey species are at greater risk than others and below I discuss the relative vulnerability of each species based on the following criteria.

Habitat sensitivity

Certain monkey species are able to respond to habitat disturbance better than others (Johns & Skorupa 1987). Two species at Taï, the red colobus and Diana monkey, are generally unable to exploit secondary or degraded primary forest. Even modest perturbations in the structure of primary forest can lead to declines in these species, which is one reason they tend to be found away from the park's edges. Based on a reliance on undisturbed forest, red colobus and Diana monkeys are considered most vulnerable. Other species are better able to survive in disturbed habitats and can be found living near villages in areas of secondary growth. Their ability to adapt to habitats other than pristine rainforest is critical to their future survival.

Habitat sensitivity also encompasses dietary considerations and years of research have shown that even closely related species may respond in very different ways to changes in food availability or dietary regime. In Uganda's Kibale Forest, selective logging removed vital resources from the diet of red colobus while actually increasing the dietary options of sympatric black and white colobus. The result was increased densities of the latter and a dramatic decrease in the red colobus population (Oates & Davies 1994).

Based on studies of the Taï monkeys and work on these and closely related species elsewhere in West Africa (e.g. Fimbel 1994, McGraw *et al.* 1999, Oates *et al.* 2000), I scored the sensitivity of each taxon to habitat disturbance as high (3), moderate (2), or low (1). High values indicate that a species is particularly dependent on undisturbed forest and is not usually found in secondary or degraded forest. Moderate values indicate that a species is able to adapt to disturbed habitat, but that it strongly prefers pristine forest. Low values indicate that a species does well and, in fact may thrive in degraded forest. It is important to note that in some cases, the sensitivity of a taxon to habitat disturbance must be inferred by its absence rather than by careful documentation of its failure to adapt to sub-optimal habitats over time.

Body size

Large-bodied species are more vulnerable to local extinction than are smaller ones (Freese *et al.* 1982, Mittermeier 1987). "Natural" factors contributing to the vulnerability of some large-bodied primates include comparatively low reproductive rates, intrinsic rarity, and hunting preference of non-human predators. From a human poacher's vantage, larger primates are generally easier to hunt and provide a greater return on the ballistic investment. In the Taï region, a single shotgun shell costs 500 FCFA (approximately one dollar). Poached adults of the four smaller monkey species can be purchased for between 1,000 and 2,500 FCFA while adult males of the larger species are sold for approximately 6,000 FCFA (personal observation). Large adult male mangabeys can cost 8,000 FCFA. Given the opportunity to kill a 3 kg guenon or 6.5 kg red colobus, a hunter is likely to choose the monkey yielding the most meat. Based on their adult body weight (Oates *et al.* 1990), I classified each species as large (3), medium-sized (2), or small (1) relative to each other.

Substrate preference

A variety of studies have demonstrated that ground-dwelling primates are more susceptible to hunting than are arboreal primates (Mittermeier 1987).

Primates that move predominantly on the forest floor may be caught in snares intended to capture other terrestrial fauna such as duikers and pigs (Struhsaker 1999)[2]. Although all the Taï monkeys spend at least some time on the ground (McGraw 1998a), overall support preference for each was scored as terrestrial (2) or arboreal (1).

Anti-predator behavior
A major research interest of the Taï Monkey project has been the responses of different species to their naturally occurring predators including chimpanzees, leopards, and eagles (e.g. Noë & Bshary 1997, Zuberbuhler *et al.* 1997, 1999, Shultz 2001, Zuberbuhler & Jenny 2002, Shultz & Noë 2002). Throughout their history, human hunters have likely exerted strong selective pressures on the Taï monkey fauna. There is evidence that although monkeys routinely respond in predictable ways to the presence of unfamiliar humans, some species are able to alter their behavior in response to high hunting pressure (Chapter 10). Other species have been less able to effectively respond to the recent dramatic increase in human pressure coupled with the widespread use of firearms. For example, safety in numbers is ultimately no match for a shotgun. There are major differences in how each of the monkey species reacts to human poachers and the ability to detect and elude human hunters is likely correlated with the number of each species a hunter kills. As discussed below, some monkey species are better at escaping hunters than others. Based on our work in the Taï Forest and studies carried out elsewhere in Ivory Coast and Ghana (McGraw 1998b, Oates *et al.* 2000), I scored the ability of each species to effectively elude human hunters as poor (2) or good (1).

Density (and group size)
As discussed by Cowlishaw and Dunbar (2000), population density (a measure of abundance) is an important factor contributing to a

---

[2] Our habituated study animals have been caught in ground snares set by poachers. In one instance, Friederike Range witnessed an adult male mangabey (approximate age of 6–7 years) become trapped by a cable snare in August 2001. The individual screamed violently while other group members observed from a close distance. After approximately 30 minutes, the group moved on to continue foraging at which time TMP members freed the monkey but could not completely detach the snare from the forelimb. The monkey was not observed for approximately 2–3 weeks but then returned to the group. The monkey lost his rank and assumed a position on the group's periphery but eventually was observed interacting and even copulating with females. One year later, the snare was still attached to the male's forelimb. The most recent report indicates that the male had lost his forelimb but was still managing to move with the rest of the group.

primate's rarity. Population density as a measure of abundance contributes to a species' vulnerability and under natural conditions the Taï monkeys vary widely in their densities (Shultz & Noë 2002, Zuberbuhler & Jenny 2002). If organisms at naturally low densities are more susceptible to extirpation than those at higher densities (e.g. Brashares 2002, Davies *et al.* 2000) due to the relative effects of removing individuals from different sized pools, then animals at higher densities should be better able to withstand losses than species at lower densities. Primates found at low population densities tend to be large-bodied and, as noted above, hunters tend to prefer large-bodied prey over smaller forms.

Density disparities combined with differences in group size mean that some species are intrinsically rarer than others. This is an important consideration for hunters trying to minimize time spent looking for monkeys; monkeys that are encountered sooner and more frequently are at greater risk than those found in smaller, more dispersed groups (Mittermeier 1987, Peres 1991, Vickers 1991). Among the *arboreal* monkeys at Taï, there is a strong correlation between population density and group size (corr coeff $= 0.99$): monkeys found at the highest densities are those with the largest groups. Assuming hunters will try to maximize off-take in the shortest period of time (particularly when hunting in prohibited areas), the most vulnerable animals are those that hunters encounter first. On average, the primates encountered first will be those with the largest group sizes and greatest densities. There is one exception: *Cercocebus atys* is the only terrestrial monkey species at Taï and it has the largest average group size (~ 100 individuals). Because its average home range (~ 500 ha) is five times that of the next nearest species, the density of sooty mangabeys at Taï is comparatively low (25 ind./km^2) and the odds of encountering a mangabey group based on density alone are relatively small.

The densities of the Taï cercopithecids are reported in Table 11.1. Based on these data, I scored the density of each Taï cercopithecid as high (1), moderate (2), and low (3) relative to one another.

Relative vulnerability of the Taï monkey fauna
Red colobus
Red colobus are the monkey species most vulnerable to hunting and, ultimately, extirpation at Taï (see Table 11.2). They are large monkeys that live in large, noisy groups. Red colobus typically flee to the tops of trees and sit immobile when threatened by poachers (Gartlan 1975, Struhsaker 1999). This may be an effective strategy against natural predators such as eagles, leopards and, in some cases, chimpanzees but it is disastrous against

Table 11.1. *Group size, density, and home range size of cercopithecids in Taï National Park (from Schultz & Noë 2002, Zuberbuhler & Jenny 2002)*

Species	X Group size	Density (ind./km^2)	Home range (ha)
Procolobus badius	52.9	123.8	57
Cercopithecus diana	20.2	48.2	62
Colobus polykomos	15.4	35.2	55
Cercopithecus campbelli	10.8	24.4	50
Cercocebus atys	69.7	11.9	500
Cercopithecus petaurista	17.5	29.3	100
Procolobus verus	6.7	17.3	62

Table 11.2. *Relative vulnerability to extirpation of Taï Forest monkeys*

	Habitat sensitivity	Body size	Substrate preference	Response to humans	Natural density	Total
Procolobus badius	3	3	1	2	1	10
Cercopithecus diana	3	2	1	1	2	9
Cercocebus atys	1	3	2	1	2	9
Colobus polykomos	1	3	1	1	2	8
Cercopithecus petaurista	1	1	1	1	3	7
Procolobus verus	1	1	1	1	3	7
Cercopithecus campbelli	1	1	1	1	2	6

Habitat Sensitivity: High (3), Moderate (2) Low (1)
Body Size: Large (3), Medium (2), Small (1)
Substrate Preference: Terrestrial (2), Arboreal (1)
Response to Humans: Good (1), Poor (2)
Density: High (1), Moderate (2), Low (3)

hunters armed with shotguns. Struhsaker (1999) believes that red colobus are the easiest primates to hunt in Africa. "Hunters claim that they could kill most of an entire group in one day if they wanted and had enough ammunition. After shooting one, the hunter simply hides behind a tree, waits a few minutes, then steps out and shoots another one and so forth ... red colobus seem poorly adapted to human hunting pressure and unable to modify their behavior accordingly." (Struhsaker 1999: 291–2)

The densities of red colobus vary significantly in areas of heavy poaching compared with areas where humans tend to avoid hunting (Refisch & Kone 2005a, 2005b). In the latter regions (e.g. western portion of the park including our study site), red colobus densities are by far the highest of any primate and over twice that of the next most common monkey species (see Table 11.1). Despite their inability to effectively counter pressure from human hunters, the sheer number of naturally occurring red colobus monkeys has probably provided a safety net against local elimination, at least until recent years. As humans apply increasing pressure on the park, a greater number of hunters will contact remaining red colobus populations sooner and more easily. It is questionable whether Taï red colobus can withstand human hunting pressure even with their high but rapidly diminishing densities. Data from markets around the park reveal that red colobus monkeys comprise the greatest percentage of monkey bushmeat and that offtake rates are not sustainable (Refisch & Kone 2005a, 2005b).

Red colobus monkeys are ecologically sensitive primates. Harcourt *et al.* (2002) show that specialization − not rarity − is the factor most responsible for increasing a species' risk of extinction. Although the red colobus radiation includes more than a dozen forms across Africa (Grubb *et al.* 2003), each appears to be quite specialized. Most populations are at least endangered (Oates 1996) and 38.9 per cent are threatened with extinction in the near future (Struhsaker 2005). Red colobus are dependent on high canopy forest[3] and are known to suffer dramatic reductions in population density and increases in infant mortality when habitats are sufficiently altered (Davies 1987, Decker 1994, Oates *et al.* 2000). Red colobus feed on a diverse array of plant items and although the diets of several subspecies are among the most well-studied of any primate (e.g. Struhsaker 1975, Korstjens 2001, Chapman *et al.* 2002, Chapman & Chapman 2002), authorities have been unable to replicate them in captivity (Collins & Roberts 1978, Oates & Davies 1994). This obstacle precludes any controlled breeding program and there are no red colobus in captivity.

In sum, their large body size, large group sizes, noisy behavior, and reluctance to flee from hunters make red colobus a favorite prey for poachers. Their naturally high densities have provided some buffer against extirpation, however it is likely that this species could be the first to be eliminated in heavily hunted areas since such densities make them easy to locate and their large body sizes make them a profitable return on

---

[3] The Zanzibar red colobus (*Procolobus kirkii*) is known to occur in a variety of forest types including secondary forest, mangrove swamp, and ground-water forest (Siex & Struhsaker 1999).

a shotgun shell investment. These factors combined with their ecological fragility, put Taï red colobus in extremely vulnerable positions. Red colobus are classified as *Endangered* by the International Union for the Conservation of Nature (IUCN 2007).

Diana monkey

The Diana monkey *Cercopithecus diana* is probably the second most endangered species in the Taï Forest. Like the red colobus, Diana monkeys are restricted to high canopy forest and Taï is almost certainly the last stronghold for *C. diana diana* in Ivory Coast. To the east, *C. diana roloway* is classified as *Critically Endangered* and populations of Roloways have recently been eliminated from a number of Ghanaian forest reserves (Struhsaker & Oates 1995, Oates 2006). At Taï, densities of Diana monkeys are second only to those of Red colobus monkeys (Table 11.1). Unlike its congeners (*C. campbelli* and *C. petaurista*), Diana monkeys do not fare well in disturbed or degraded forest and are restricted to the park's interior away from human influences. Although red colobus monkeys seek out Diana monkeys because of their abilities to detect monkey hunting chimpanzees (Noë & Bshary 1997), the latter are also the noisiest and most raucous monkeys in Taï: they are easily located by their vocalizations, coloring, and active movement throughout all canopy layers.

Populations of *C. diana roloway* that we located in the heavily hunted forests of eastern Ivory Coast utter barely audible contact calls and we heard adult males give few, if any, loud calls (McGraw 1998b). It appears that populations of Roloway monkeys in eastern Ivory Coast that have eluded hunters to date have adjusted their behavior by remaining silent or nearly so. These observations are consistent with results from studies at Taï suggesting that Diana monkeys in different regions of the park have adapted their behavior in response to human hunting pressure (Bshary 2001, Chapter 10). Diana monkeys are not normally found near forest edges near villages and hunters must venture well within the park to kill them. These striking monkeys are adept at detecting humans and for this reason, Diana monkeys are prized as bushmeat which could add to their vulnerability. In sum, Diana monkeys are found at high densities in undisturbed forest where poachers can locate them easily. The combination of these factors makes Diana monkeys highly vulnerable to extirpation. The Diana monkey is classified as *Endangered* by the International Union for the Conservation of Nature (IUCN 2007).

Sooty mangabey

Sooty mangabeys *Cercocebus atys atys* are at risk because they are the largest and only terrestrial monkeys in the Taï Forest. Although their densities are much lower than either red colobus or Diana monkeys, sooty mangabey groups are large, noisy, and easy to locate. They are hunted with dogs throughout West Africa (Struhsaker 1971). Mangabeys frequently feed on hard nuts and fruits and the sounds of large groups cracking open these food items can be heard from considerable distances. Although Sooty mangabeys are excellent at detecting some ground predators (e.g. leopards) from a distance (McGraw & Bshary 2002), it is possible that they are less successful against cautious hunters with shotguns. A common response of sooty mangabeys when alarmed is to flee into the trees: a behavior that makes a hunter's task easier.

Kingdon (1997) reports that two West African mangabeys – both *C. a. atys* and *C. a. lunulatus* (Grubb *et al.* 2003) – are not threatened within their ranges, but this is not accurate. The white-crowned mangabey *C. a. lunulatus* is becoming increasingly scarce in eastern Ivory Coast, may be locally extinct in a number of Ghanaian forest reserves (Struhsaker & Oates 1995, Oates 1996, McGraw 1998b) and was recently placed on the list of the 25 most endangered primates (McGraw *et al.* 2006). Little information exists on the abundance of *C. a. atys* elsewhere in Ivory Coast but it is likely that the majority of individuals left in the country are found at Taï. *C. a. atys* is classified as *Lower Risk: Near Threatened* by the International Union for the Conservation of Nature (IUCN 2007), but it is likely that its status will be changed to vulnerable in the near future.

Black and white colobus

The black and white colobus *Colobus polykomos* is at risk because of its large body size. Black and white colobus are probably hunted less often than red colobus due to the ease of locating and killing the latter: red colobus live in large noisy groups whereas black and white colobus live in smaller, more cryptic societies (Korstjens 2001) at lower densities (Table 11.1). As red colobus become scarcer, the frequency that black and white colobus are taken by poachers will undoubtedly increase. During recent surveys in eastern Ivory Coast and western Ghana, we encountered no living populations of the local form of red colobus (*Procolobus badius waldroni*) and groups of the local black and White colobus (*Colobus vellerosus*) were quite rare (McGraw 1998b,

Oates *et al.* 2000). Nevertheless, we encountered one poaching party in southeastern Ivory Coast preparing 11 recently killed primates for transport back to their village; six of these were black and white colobus. Primate skeletal remains collected from several poaching camps in eastern Ivory Coast and western Ghana included two guenon species, mangabeys (*C. a. lunulatus*) and many *Colobus vellerosus* bones. Thus, there is evidence that as the number of sympatric species declines, black and white colobus will become increasingly threatened. On an encouraging note, black and white colobus (unlike red colobus) are able to survive in degraded primary or secondary forest and do well in captive environments. These factors may provide some safety net as preferred habitats dwindle and hunting pressure increases. At present, *Colobus polykomos* is classified as *Lower Risk: Near Threatened* by the International Union for the Conservation of Nature (IUCN 2007).

Olive colobus

Although many adaptations (e.g. habitat use, coloration, grouping behavior) of the olive colobus *Procolobus verus* make it difficult to locate and kill, the major factors putting the smallest African colobine at risk are its low densities and small group sizes (Table 11.1). This cryptic species is permanently associated with Diana monkeys and enjoys the protection the latter provide in the form of predator-detection capabilities (Oates & Whitesides 1990, Korstjens 2001). Nevertheless, if hunters kill only a few individuals, the cost to the group's social organization and local population genetics is proportionately greater than that of other species. The olive colobus is able to exploit areas of secondary forests and was encountered in several heavily poached forest reserves in eastern Ivory Coast where larger colobine species were absent (McGraw 1998b). Thus, despite its small body size, small group sizes, and ability to exploit disturbed forest, olive colobus are at risk because of their naturally low population densities. Olive colobus are classified as *Lower Risk: Near Threatened* by the International Union for the Conservation of Nature (IUCN 2007).

Campbell's monkey

The Campbell's monkey *Cercopithecus campbelli* is one of the most common monkeys in West Africa because of its ability to survive in disturbed habitats (Jones 1950, Fimbel 1994, McGraw *et al.* 1999). During surveys in eastern Ivory Coast and western Ghana we encountered Campbell's monkeys in nearly every forest reserve we studied (McGraw 1998b). Because Campbell's monkeys live in small, cryptic groups, they are probably more difficult to hunt than Diana monkeys or red colobus.

However, the fact that this species is able to exploit areas of disturbed forest puts them in closer proximity to humans where they are killed near villages or agricultural fields. Villagers in The Gambia claimed that Campbell's monkeys were the worst crop raiders and that farmers never hesitated to kill them because they were pests (Starin 1989). Other individuals interviewed by Starin claimed that Campbell's monkey was the best tasting. For these reasons, Campbell's monkeys were nearly extirpated in The Gambia over 20 years ago. Similar trends in and around Taï National Park could eventually threaten the survival of Campbell's monkey although current populations appear to be under only moderate threat. Campbell's monkeys are classified as *Not Endangered* by the International Union for the Conservation of Nature (IUCN 2007).

Lesser spot-nosed monkey
In many ways, the lesser spot-nosed monkey *Cercopithecus petaurista* is ecologically similar to Campbell's monkey and the factors responsible for the success and vulnerability of the two species are the same. Lesser spot-nosed monkeys are able to survive and may in fact thrive in secondary forest (Mackenzie 1952, Fimbel 1994, McGraw 1998b, McGraw *et al.* 1999). This small guenon is among the most abundant monkeys in Ivory Coast despite their naturally low densities and small group sizes. We encountered *C. petaurista* in all the forest reserves we surveyed in eastern Ivory Coast (McGraw 1998b). The ability of this monkey to exploit degraded forest near villages makes it an easy target for hunters and it is one of the most common monkeys found in bushmeat markets. In addition to the interior of Taï National Park, lesser spot-nosed monkeys are found throughout the degraded forest on the park's perimeter and in secondary forest of regenerating agricultural fields. Hunters in the Taï region need not travel far to kill *C. petaurista*. As monkeys living near villages are eliminated and larger species within the park become scarcer, populations in and around our study area will become increasingly threatened. Lesser spot-nosed monkeys are classified as *Not Endangered* by the International Union for the Conservation of Nature (IUCN 2007).

Discussion
A thorough treatment of the vulnerability of the Taï cercopithecids demands a rigorous, comparative approach using additional criteria (Cowlishaw 1999, Harcourt 1998, Harcourt *et al.* 2002). The criteria used in this exercise are based on previous studies (Mittermeier 1987) as well as information collected from hunters in the Taï region and

elsewhere in West Africa[4]. We know that red colobus monkeys are at highest risk because of the ease that hunters encounter and kill them; red colobus are large, noisy primates found at high densities that are not well adapted to avoiding poachers. However, variables putting certain Taï primates at risk do not necessarily indicate increased vulnerability of other taxa in other regions. For example, Purvis *et al.* (2000) argued that primates at low densities were more likely to die out due to the effects of local catastrophes, slow rates of adaptation or inbreeding. Red colobus appear to violate the abundance rule, however Purvis *et al.* (2000) also show that large-bodied primates tend to be found at low densities. As large-bodied monkeys found at high densities, Red colobus are unusual and highlight the perhaps idiosyncratic nature of the Taï primate fauna. It is likely that since the smaller monkeys are found at lower densities, their larger cercopithecid relatives could be eliminated earlier because of the increased likelihood of encountering them sooner. This possibility is supported by Bshary (2001) who found that the three monkeys most commonly killed by hunters from six villages bordering the Taï Forest were red colobus, Diana monkey and black and white colobus, respectively. These taxa represent three of the four largest monkeys at Taï and are also those found at the highest densities.

Other authors foresee a bleak future for the monkeys of western Ivory Coast unless immediate conservation measures are enacted. Struhsaker (1999) predicted that the African primates most likely to go extinct in the near future (1) are reliant on old growth forest, (2) weigh at least 4–5 kg, (3) are partly terrestrial, (4) are loud and conspicuous, and (5) live in areas with increasing human population densities and hunting pressures. These criteria apply to several of our study species and three of the six African taxa identified by Struhsaker as being at greatest risk of extinction are found at Taï: *Procolobus badius*, *Cercopithecus diana*, and

---

[4] An additional consideration is a species' acquired rarity. As a monkey becomes less abundant, it becomes more difficult to locate and kill. At the same time, its value as a prestige item may increase. Hunters who must travel further to kill monkeys unable to live near forest edges might seek greater compensation for their efforts. The ability to afford less abundant, and therefore higher-priced bushmeat, is viewed by some local residents as a sign of prestige. The result is that species that have already become scarce may be pursued with greater fervor because of their acquired rarity. During surveys in eastern Ivory Coast, we learned from local peoples that the least abundant primates (*Cercopithecus diana roloway*, *Cercocebus atys lunulatus*, and *Procolobus badius waldroni*) were also the most prized by villagers. We suspect, though cannot yet confirm, that a similar mentality might exist in the villages surrounding Taï National Park. Although acquired rarity is not used as a criterion in this paper, it is possible that this phenomenon may add to the vulnerability of species such as *Procolobus badius* and *Cercopithecus diana* that are restricted to the interior of the park.

Cercocebus atys. Based on our experience in the forest and the likely increases in human pressure around it, we have every reason to believe Struhsaker's predictions are correct and that Ivory Coast could soon lose a significant portion of its primate diversity (Cowlishaw 1999).

The principal threat to the remaining monkeys is hunting, spurred by an increasing human population. Although commercial hunting is concentrated in the eastern region of the park, monkeys are poached everywhere within the forest. It is possible that one of the few areas left inside the nearly 4,500 km^2 park where primates are found at natural densities is within or immediately adjacent to our small study area. Even here, poachers threaten the animals. Our study animals have been fired at by persons with shotguns while members of our research team were observing them. The sound of gunshots around our study grid and research station is an almost daily occurrence. The ongoing civil war in Ivory Coast has occasionally disrupted our research in the forest and there have been times when we have been forced to evacuate our field site. After one such period in 2003, we learned that 27 individuals from a red colobus group under continuous study since 1992 were killed by a local poacher, most likely during a single excursion into our study grid. If nothing else, our continued presence in the forest provides protection for a small number of the park's inhabitants.

Monkey densities vary significantly throughout the park as a function of poaching pressure. Even if all hunting stopped today, the uneven distribution of monkeys in the forest could have already led to at least two detrimental changes. For example, decreases in suitable and safe habitat are barriers to normal dispersal and reproductive behaviors in different groups. Changes in migration patterns (e.g. opportunities for group transfer, increased distances between neighboring groups, etc.) could negatively impact the nature of each species' gene pool. At present, we lack information on the genetic diversity of the cercopithecid populations in the park, but considering the number of primates already culled from the park's periphery, it is possible that the viability of some species may soon be jeopardized. Unfortunately, the severity of these losses may not be known for some time. Cowlishaw (1998) argued that a time lag of up to 100 years ("extinction debt") exists between the end of deforestation and the disappearance of primate species owing, in part, to reduction in a species' gene pool. Cowlishaw further predicted that Ivory Coast would eventually suffer the highest loss of primate species (based on moderate forest loss) in West Africa, eventually losing between 40–50 per cent of its primate species. The disturbing footnote to these conclusions is that they are based solely on habitat loss and do not factor in a burgeoning human population and increased hunting pressure. Considering these influences, the rate of predicted loss is most likely higher.

The reduction or elimination of primates in the park could also affect the forest itself (e.g. Dominy & Duncan 2005). Chapman and Chapman (1997) estimate that upwards of 60 per cent of all fruiting trees could be lost as a consequence of removing seed-dispersing frugivores. Loss of primates has already severely affected the biology of other forested areas in West Africa (Maisels et al. 2001). Refisch's (2001) study of the role of primates as seed dispersers in poached and non-poached areas showed that portions of the park with high poaching rates could suffer dramatic regeneration problems from loss of dispersal agents. Just as the Taï primates can ill afford to have their habitat reduced further, the remaining forest is equally dependent on healthy primate populations.

What is not debatable
There are at least two issues being debated by biologists working in tropical forests. The first concerns the sustainability of traditional (subsistence) hunting (Cowlishaw et al. 2005, Damania et al. 2005). Excellent studies on subsistence hunting have been carried out in East (Carpento & Fusari 2000), Central (Wilkie et al. 1998, Muchaal & Ngandjui 1999, Wilkie & Carpenter 1999), and West Africa (Caspary et al. 2001, Newing 2001, Refisch & Kone 2005a, 2005b) and it has been shown that in some areas, traditional hunting methods can be compatible with wildlife populations if humans do not put excessive pressure on natural resources. Such cases tend to be exceptions, however, and the prevailing notion is that hunting among traditional cultures is more unsustainable than not (see studies in Robinson & Bennett 2000). Accepting this, various authors have offered suggestions to control hunting practices when off-take rates jeopardize wildlife populations. These include logging restrictions, hunting seasons, snare regulations, species-kill quotas, taxes on bushmeat, a multi-actor approach (Bowen-Jones & Pendrey 1999, Davies 2002, Cowlishaw et al. 2004).

Both subsistence and commercial hunting is already widespread at Taï and it is clear that the human population adjacent to the park will continue to put even greater pressure on the forest and its wildlife. Most hunting of primates is carried out with non-traditional weapons (i.e. shotguns) and is done on a large (non-subsistence) scale. Even in areas where monkey hunting is less commercial, off-take rates are non-sustainable (Refisch & Kone 2005a, 2005b). Many measures needed to control hunting and the trade of bushmeat trade are − theoretically − in place and questions about the sustainability of hunting should be non-issues since hunting is prohibited throughout Côte d'Ivoire, let alone in the crown jewel of its national park system. Regulations meant to curtail the illegal killing of Taï's monkeys are falling well short of their mark.

A second issue concerns linking wildlife protection with community-based conservation and development schemes (Oates 1995, 1999, Terborgh 1999, Schwartzman *et al.* 2000, Adams & Hulme 2001, Kiss 2004, Odling-Smee 2005). Many authors believe that any attempt to preserve wildlife in areas of high human population density must include provisions for harvesting alternate sources of protein (e.g. Bowen-Jones & Pendrey 1999, Brashares *et al.* 2004). Others maintain that community-based conservation is heavily skewed towards human interests, that these costly programs hasten the disappearance of biodiversity or that, for example, successful development programs ultimately put increased pressure on local wildlife by attracting more migrants to areas already buckling under high human pressure (Attwell & Cotterill 2000, Oates *et al.* 2004). If the human population around the park continues to grow at its present rate, conservation strategies must eventually include measures that provide for alternate sources of protein: there will be too many people to resist the temptation of bushmeat within the park. However unless immediate action is taken to protect the remaining wildlife, debates about integrating conservation goals with human welfare will not be necessary since little wildlife will be left.

As a high profile national park with a long history of international support, Taï should enjoy the best possible protection. The reality is quite different and primates in the park continue to be killed in large numbers. A strong taste for wild meat fuels the bushmeat trade on both subsistence and commercial levels and aggressive education programs are needed to change the general attitude that wildlife is a limitless food source. Ultimately, any hope of conserving wildlife will require this kind of grass-roots change but in the meantime − and until poachers decide not to head into the forest − the most important measure that could be taken to decrease the vulnerability of the Taï primates is to increase policing within the park so that poachers are effectively discouraged from venturing inside (Oates *et al.* 2004, Struhsaker *et al.* 2005). Current levels of law enforcement by the Ivorian forest police are far below those necessary to prevent hunters from conducting major poaching exercises throughout the park. Response times by authorities to reports of gunfire are slow due to inadequate transportation, faulty communication networks or lack of motivation. On several occasions, days elapsed by the time forest police responded to our reports of poaching activity in our primary study grid immediately adjacent to our research station. Even if development projects were to supply sufficient alternatives to bushmeat, why should hunters stop poaching when there is little chance of being caught?

Of course, no proximate solution to Taï's problems will be successful unless pressure from an increasing human population is eased. Butynski (2002) notes that although Africa's population growth rate (between 2 and 3 per cent) is by far the highest in the world, no African nation has a population control policy. Further, although most authorities agree that an expanding human population is at the root of many conservation conflicts, few know how to address it in a realistic way. In reviewing a collection of studies on the sustainability of hunting in the tropics, Struhsaker (2001b) noted, "Although the majority of authors recognized human population growth as an issue, none considered it important enough to warrant discussion about how it should be reduced." Given the complexity and delicate nature of the problem, this is probably no accident. Nevertheless, some authors have been explicit about the relationship between human population growth and the biodiversity crisis (e.g. Smail 1997, 2003, Pimm 2001, McKee 2003). These authors have offered sobering statistics and potential remedies. An unflinching and honest assessment must inevitably conclude that any hope of preserving non-human primates will require measures to slow human population growth.

Concluding remarks

Conservation journals are filled with reports on the amount of dwindling habitat, number of species at risk, extent of the bushmeat trade and potential remedies for conserving biodiversity (e.g. Brooks *et al.* 2002). Hotspots are identified and debated, threats to wildlife discussed, priorities made, surveys conducted and allocated funds acknowledged in summary reports (Mittermeier *et al.* 1998, Orme *et al.* 2005). What is clear is that (1) Taï National Park is a highly significant forest in terms of providing a refuge for West African wildlife, (2) the park has received considerable domestic and international support for its protection, (3) escalating human activities are putting increased pressure on the remaining wildlife and, for these reasons, (4) the park's wildlife is in serious trouble.

A recent report stated, "(The Taï) sanctuary faces a growing danger not just from poachers finding easy targets in the abundant wildlife here, but even more seriously from expanding human settlements on the boundaries of the park. For people who desperately need more land to plant crops, the Taï forest looks like a vast empty territory just waiting for the farmer's plow. In a poor developing country like Ivory Coast, it is hard for many to understand why a place like this should be protected. People see no economic benefit. And in the years to come, with an even bigger human population demanding more agricultural land, it might be politically impossible for any government to continue protecting this forest (Streiker 1997)." Considering that the specter of continued civil war and the fragmentation of their country occupy the attention of most Ivorians, the political obstacles to conservation are even greater.

In the spirit of the passages introducing this chapter, an honest assessment of Taï National Park is that its primates are vulnerable, some extremely so; arguments to the contrary are naive and misleading. The more time spent debating what kinds of development projects could best ease the human demand for protein without addressing the immediate problem of monkeys being killed more rapidly than they can reproduce, the more difficult it will be to salvage any viable populations. The adage that it is hard to practice conservation on an empty stomach is true, however conservation rhetoric will not stop hunters from continuing to kill monkeys when the chances of getting caught are slim. As biologists concerned with the future of all wildlife, we believe that preserving the Taï Forest and its inhabitants is critical because so much surrounding biodiversity has not been. We recognize that if we are unable to protect this national park, the future of wildlife in less "protected" regions of Africa is bleak.

We therefore view the conservation of the Taï primates as a two-tiered process: the proximate issue is to enhance protection of the park's residents by increasing the efficacy of police in the forest because laws protecting Taï's wildlife are ignored and are difficult to enforce. Ultimately, changing attitudes towards wildlife and bringing Africa's population growth under control will undoubtedly prove far more challenging. In the meantime, the single most important step that can be taken to ensure the survival of the Taï monkeys is to immediately increase the number and effectiveness of the forest police charged with protecting them (Butynski 2002, Rowcliffe *et al.* 2004, Struhsaker *et al.* 2005). Conservation money in carefully managed funds should go directly to expanding the number of forest police in the park, improving equipment, providing better transportation, and strengthening the motivation of park police. Conservation will be greatly enhanced when poachers are promptly and actively pursued, brought to trial, convicted, and sentenced. The effectiveness of law enforcement personnel can be increased with real incentive programs to reward employees. These are obtainable objectives: park authorities have successfully controlled illegal timber exploitation and, to a lesser extent, illegal plantations within the park's boundaries (Refisch & Kone, Chapter 10, Butynski 2002). On paper, many elements needed to ensure the safety of the park's residents already exist: a complete ban on hunting, absence of logging, few visitors, etc. However, unless improved forest police can bring poaching under control in the short term, long-term changes in attitudes towards wildlife will not be necessary.

Summary

1. Taï National Park is the last stronghold for a number of West African primates. Despite its status as a national park, poaching is widespread throughout the forest leaving some of the park's residents — including primates — vulnerable to extirpation.

2. Five criteria are used to assess the vulnerability to extirpation of seven cercopithecid monkeys within the park. Based on the combined effect of these variables, the relative vulnerability (from most to least vulnerable) and predicted order of disappearance is *Procolobus badius, Cercopithecus diana, Cercocebus atys, Colobus polykomos, Cercopithecus petaurista, Procolobus verus,* and *Cercopithecus campbelli.*

3. The interaction of factors contributing to increased vulnerability is important. For example, examining densities alone might lead to the erroneous conclusion that the red colobus is the least vulnerable monkey while the olive colobus is most at risk. This analysis demonstrates that additional characteristics such as group size, body size, and anti-predator behaviors must also be considered.

4. In the short term, dramatic improvements in the effectiveness of anti-poaching patrols within the park are necessary to prevent further loss of the Taï monkey fauna. Ultimately, the future of the Taï primate fauna rests on reducing human pressure on wild resources and changing attitudes about local wildlife from one consisting exclusively of "monkeys as food" to one of monkeys as income for food.

5. Discussions of Ivory Coast's primate fauna have taken a back seat to questions about the country's future. As this book goes to press (January 2007), peace has not returned to the Ivory Coast and the future of Taï's primate community continues to hang in the balance.

References

Adams, W. M. and Hulme, D. (2001). If community conservation is the answer in Africa, what is the question? *Oryx,* **35**, 193–200.

Anonymous. (2002). World deforestation rate slows down – outside Africa. Afrol.com/categories/environment/env055_fao_deforestation.htm

Attwell, C. A. M. and Cotterill, F. P. D. (2000). Postmodernism and African conservation science. *Biodiversity and Conservation*, **9**, 559–77.

Beier, P., van Drielend, M. and Kankam, B. O. (2002). Avifaunal collapse in West African forest fragments. *Conservation Biology*, **16**, 1097–111.

Boesch, C. and Boesch-Achermann, H. (2000). *The Chimpanzees of the Taï Forest*. Oxford: Oxford University Press.

Bowen-Jones, E. and Pendry, S. (1999). The threat to primates and other mammals from the bushmeat trade in Africa, and how this threat could be diminished. *Oryx*, **33**, 233–46.

Brashares, J. S. (2002). Ecological, behavioural and life-history correlates of mammal extinctions in West Africa. *Conservation Biology*, **17**, 733–43.

Brashares, J. S., Arcese, P. and Sam, M. K. (2001). Human demography and reserve size predict wildlife extinction in West Africa. *Proceedings of the Royal Society of London*, B, **268**, 2473–8.

Brashares, J. S., Arcese, P., Sam, M. K., Coppolillo, P. B., Sinclair, A. R. E. and Balmford, A. (2004). Bushmeat hunting, wildlife declines, and fish supply in West Africa. *Science*, **306**, 1180–3.

Brooks, T. M., Mittermeier, R. A. and Mittermeier, C. G. (2002). Habitat loss and extinction in the hotspots of biodiversity. *Conservation Biology*, **16**, 909–23.

Bshary, R. (2001). Diana monkeys, *Cercopithecus diana*, adjust their anti-predator response behavior to human hunting strategies. *Behavioural Ecology and Sociobiology*, **50**, 251–6.

Butynski, T. M. (2002). Conservation of the guenons: An overview of status, threats and recommendations. In *The Guenons: Diversity and Adaptation in African Monkeys*, ed. M. E. Glenn and M. Cords. New York: Kluwer Academic/Plenum Publishers, pp. 411–24.

Carpaneto, G. M. and Fusari, A. (2000). Subsistence hunting and bushmeat exploitation in central-western Tanzania. *Biodiversity and Conservation*, **9**, 1571–2000.

Caspary, H. U., Konee, I., Prouot, C. and de Pauw, M. (2001). *La chasse et la filiere viande de broussee dans l'espace Taï, Côte d'Ivoire*. Tropenbos – Côte d'Ivoire serie 2, Tropenbos – Côte d'Ivoire, Abidjan.

Chapman, C. A. and Chapman, L. J. (1997). Forest regeneration in logged and unlogged forests of Kibale National Park, Uganda. *Biotropica*, **29**, 396–412.

Chapman, C. A. and Chapman, L. J. (2002). Foraging challenges of red colobus monkeys: influence of nutrients and secondary compounds. *Comparative Biochemistry and Physiology Part A*, **133**, 861–75.

Chapman, C. A., Chapman, L. J. and Gillespie, T. R. (2002). Scale issues in the study of primate foraging: red colobus of Kibale National Park. *American Journal of Physical Anthropology*, **117**, 349–63.

Collins, L. and Roberts, M. (1978). Arboreal folivores in captivity – maintenance of a delicate minority. In *The Ecology of Arboreal Folivores*, ed. G. G. Montgomery. Washington, DC: Smithsonian Institute Press, pp. 5–12.

Cowlishaw, G. (1999). Predicting the pattern of decline of African primate diversity: an extinction debt from historical deforestation. *Conservation Biology*, **14**, 1183–93.

Cowlishaw, G. and Dunbar, R. (2000). *Primate Conservation Biology.* Chicago: The University of Chicago Press.

Cowlishaw, G., Mendelson, S. and Rowcliffe, J. M. (2004). Structure and operation of a bushmeat commodity chain in Southwestern Ghana. *Conservation Biology*, **19**, 139–49.

Cowlishaw, G., Mendelson, S. and Rowcliffe, J. M. (2005). Evidence for post-depletion sustainability in a mature bushmeat market. *Journal of Applied Ecology*, **42**, 460–8.

Crockford, C., Herbinger, I., Vigilant, L. and Boesch, C. (2004). Wild chimpanzees produce group-specific calls: A case for vocal learning? *Ethology*, **110**, 221–43.

Damania, R., Milner-Gulland, E. J. and Crookes, D. J. (2005). A bioeconomic analysis of bushmeat hunting. *Proceedings of the Royal Society*, B, **272**, 259–66.

Davies, G. (1987). Conservation of primates in the Gola Forest Reserves, Sierra Leone. *Primate Conservation*, **8**, 151–3.

Davies, G. (2002). Bushmeat and international development. *Conservation Biology*, **16**, 587–9.

Davies, K. F., Margules, C. R. and Lawrence, J. F. (2000). Which traits of species predict population declines in experimental forest fragments? *Ecology*, **81**, 1450–61.

Decker, B. S. (1994). Effects of habitat disturbance on the behavioral ecology and demographics of the Tana River red colobus (*Colobus badius rufomitratus*). *International Journal of Primatology*, **15**, 703–37.

Dominy, N. J. and Duncan, B. W. (2005). Seed-spitting primates and the conservation and dispersion of large-seeded trees. *International Journal of Primatology*, **26**, 631–49.

Fairhead, J. and Leach, M. (1998). *Reframing Deforestation. Global Analyses and Local Realities: Studies in West Africa.* Routledge: New York.

Fimbel, C. (1994). The relative use of abandoned clearings and old forest habitats by primates and a forest antelope at Tiwai, Sierra Leone, West Africa. *Biological Conservation*, **70**, 277–86.

Fischer, F., Gross, M. and Kunz, B. (1999–2000). Primates of the Comoe National Park, Ivory Coast. *African Primates*, **4**, 10–15.

Fischer, F. and Linsenmair, K. E. (2001). Decreases in ungulate population densities. Examples from the Comoe National Park, Ivory Coast. *Biological Conservation*, **101**, 131–5.

Freese, C. H., Heltne, P. G., Castro, R. N. and Whitesides, G. H. (1982). Patterns and determinants of monkey densities in Peru and Bolivia, with notes on distributions. *International Journal of Primatology*, **3**, 53–90.

Gartlan, J. S. (1975). The African coastal rain forest and its primates – threatened resources. In *Primate Utilization and Conservation*, ed. G. Bermant and D. Linburg. New York: Wiley and Sons, pp. 67–82.

Grubb, P., Butynski, T. M., Oates, J. F. *et al.* (2003). Assessment of the diversity of African primates. *International Journal of Primatology*, **24**(6), 1301–57.

Hacker, J. E., Cowlishaw, G. and Williams, P. H. (1998). Patterns of African primate diversity and their evaluation for the selection of conservation areas. *Biological Conservation*, **84**, 251–62.

Harcourt, A. H. (1998). Ecological indicators of risk for primates, as judged by species' susceptibility to logging. In *Behavioral Ecology and Conservation Biology*, ed. T. Caro. New York: Oxford University Press, pp. 56–79.

Harcourt, A. H., Coppeto, S. A. and Parks, S. A. (2002). Rarity, specialization and extinction in primates. *Journal of Biogeography*, **29**, 445–56.

IUCN. (2007). IUCN Red List of Threatened Species. (www.iucnredlist.org <http://www.iucnredlist.org>).

Jones, T. S. (1950). Notes on the monkeys of Sierra Leone. *Sierra Leone Agricultural Notes* 22.

Johns, A. D. and Skorupa, J. P. (1987). Responses of rainforest primates to habitat disturbance: a review. *International Journal of Primatology*, **8**, 157–91.

Kingdon, J. (1997). *The Kingdon Field Guide to African Mammals*. San Diego: Academic Press.

Kiss, A. (2004). Is community based ecotourism a good use of biodiversity conservation funds? *Trends in Ecology and Evolution*, **19**, 232–7.

Korstjens, A. H. (2001). The mob, the secret sorority and the phantoms: an analysis of the socio-ecological strategies of the three colobines of Taï. Ph.D. Dissertation, University of Utrecht, Netherlands.

Mackenzie, A. F. (1952). *Proceedings of the Zoological Society of London*, **122**, 541.

Maisels, F., Keming, E., Kemei, M. and Toh, C. (2001). The extirpation of large mammals and implications for montane forest conservation: the case of the Kilum-Ijim Forest, North-west Province, Cameroon. *Oryx*, **35**, 322–31.

McGraw, W. S. (2005). Update on the search for Miss Waldron's red colobus (*Procolobus badius waldroni*). *International Journal of Primatology*, **26**(3), 605–19.

McGraw, W. S. (1998a). Comparative locomotion and habitat use of six monkeys in the Taï Forest, Ivory Coast. *American Journal of Physical Anthropology*, **105**, 493–510.

McGraw, W. S. (1998b). Three monkeys nearing extinction in the forest reserves of eastern Côte d'Ivoire. *Oryx*, **32**, 233–6.

McGraw, W. S. and Bshary, R. (2002). Association of terrestrial mangabeys (*Cercocebus atys*) with arboreal monkeys: experimental evidence for the effects of reduced ground predator pressure on habitat use. *International Journal of Primatology*, **23**, 311–25.

McGraw, W. S., Magnuson, L., Kormos, R. and Konstant, W. R. (2006). White-naped mangabey *Cercocebus atys lunulatus*. In Primates in Peril: The World's Most Endangered Primates 2004–2006, eds. R. A. Mittermeier *et al.* *Primate Conservation*, **20**, 1–28.

McGraw, W. S., Monah, I. T. and Abedi-Lartey, M. (1999). Survey of endangered primates in the forest reserves of eastern Côte d'Ivoire. *African Primates*, **3**, 22–5.

McGraw, W. S. and Oates, J. F. (2002). Evidence for a surviving population of Miss Waldron's red colobus. *Oryx*, **36**(3), 223.

McKee, J. K. (2003). *Sparing Nature*. New Brunswick: Rutgers University Press.

Mittermeier, R. A. (1987). Effects of hunting on rain forest primates. In *Primate Conservation in the Tropical Rainforest*, ed. C. Marsh and R. Mittermeier. New York: Alan R. Liss, pp. 109–46.

Mittermeier, R. A., Myers, N. and Thomsen, J. B. (1998). Biodiversity hotspots and major tropical wilderness areas: approaches to setting conservation priorities. *Conservation Biology*, **12**, 516–20.

Muchaal, P. K. and Ngandjui, G. (1999). Impact of village hunting on wildlife populations in the western Dja Reserve, Cameroon. *Conservation Biology*, **13**, 385–96.

Myers, N., Mittermeier, R. A., Mittermeier, C. G., da Fonseca, G. A. B. and Kent, J. (2000). Biodiversity hotspots for conservation priorities. *Nature*, **403**, 853–8.

Newing, H. (2001). Bushmeat hunting and management: implications of duiker ecology and interspecific competition. *Biodiversity and Conservation*, **10**, 99–118.

Noë, R. and Bshary, R. (1997). The formation of red colobus-Diana monkey associations under predation pressure from chimpanzees. *Proceedings of the Royal Society of London*, **64**, 253–9.

Oates, J. F. (1995). The dangers of conservation by rural development – a case-study from the forests of Nigeria. *Oryx*, **29**, 115–22.

Oates, J. F. (1996). *African Primates. Status Survey and Conservation Action Plan. Revised Edition*. Gland, Switzerland: IUCN.

Oates, J. F. (1999). *Myth and Reality in the Rain Forest: How Conservation Strategies are Failing in West Africa*. Los Angeles: University of California Press.

Oates, J. F. (2006). *Primate Conservation in the Forests of Western Ghana: Field Survey Results, 2005–2006*. Report to the Wildlife Division, Forestry Commission, Ghana.

Oates, J. F., Abedi-Lartey, M., McGraw, W. S., Struhsaker, T. T. and Whitesides, G. H. (2000). Extinction of a West African colobus monkey. *Conservation Biology*, **14**, 1526–32.

Oates, J. F., Bergl, R. A. and Linder, L. M. (2004). Africa's Gulf of Guinea forests: biodiversity patterns and conservation priorities. *Advances in Applied Biodiversity Science, No. # 6*. Washington, DC: Conservation International.

Oates, J. F. and Davies, A. G. (1994). Conclusions: the past, present and future of the colobines. In *Colobine Monkeys: Their Ecology, Behavior and Evolution*, ed. A. G. Davies and J. F. Oates. Cambridge: Cambridge University Press, pp. 347–58.

Oates, J. F. and Whitesides, G. (1990). Association between olive colobus (*Procolobus verus*), Diana guenons (*Cercopithecus diana*) and other forest monkeys in Sierra Leone. *American Journal of Primatology*, **21**, 129–46.

Oates, J. F., Whitesides, G. H., Davies, A. G. *et al.* (1990). Determinants of variation in tropical forest primate biomass: New evidence from West Africa. *Ecology*, **71**, 328–43.

Odling-Smee, L. (2005). Dollars and sense. *Nature*, **437**, 614–16.

Orme, C. D. L., Davies, R. G., Burgess, M. *et al.* (2005). Global hotspots of species richness are not congruent with endemism or threat. *Nature*, **436**, 1016–19.

Peres, C. (1991). Humboldt's woolly monkeys decimated by hunting in Amazonia. *Oryx*, **25**, 89–95.

Pimm, S. (2001). *The World According to Pimm: A Scientist Audits the Earth*. New York: McGraw-Hill.

Purvis, A., Gittleman, J. L., Cowlishaw, G. and Mace, G. M. (2000). Predicting extinction risk in declining species. *Proceedings of the Royal Society of London*, B, **267**, 1947–52.

Refisch, J. (2001). What would tropical rainforests become without primate seed dispersers? *Folia Primatologica*, **72**(3), 179–80.

Refisch, J. and Kone, I. (2005a). Impact of commercial hunting on monkey populations in the Taï region, Côte d'Ivoire. *Biotropica*, **37**(1), 136–44.

Refisch, J. and Kone, I. (2005b). Market hunting in the Taï region, Côte d'Ivoire and implications for monkey populations. *International Journal of Primatology*, **26**(3), 621–9.

Robinson, J. G. and Bennett, E. L. ed. (2000). *Hunting for Sustainability in Tropical Forests*. New York: Columbia University Press.

Robinson, J. G. and Bennett, E. L. (2004). Having your wildlife and eating it too: an analysis of hunting sustainability across tropical ecosystems. *Animal Conservation*, **7**, 397–408.

Rosser, A. M. and Mainka, S. A. (2002). Overexploitation and species extinctions. *Conservation Biology*, **16**(33), 584–6.

Rowcliffe, J. M., de Merode, E. and Cowlishaw, G. (2004). Do wildlife laws work? Species protection and the applications of a prey choice model to poaching decisions. *Proceedings of the Royal Society of London*, B, **271**, 2631–6.

Said, M. Y., Chunge, R. N., Craig, G. C., Thouless, C. R., Barnes, R. F. W. and Dublin, H. T. (1995). *African Elephant Database 1995*. Gland, Switzerland: IUCN.

Schaller, G. B. (2000). Forward. In *Hunting for Sustainability in Tropical Forests*, ed. J. G. Robinson and E. L. Bennett. New York: Columbia University Press, xv–xviii.

Schwartzman, S., Moreira, A. and Nepstad, D. (2000). Rethinking tropical forest conservation: perils and parks. *Conservation Biology*, **14**, 1351–7.

Shultz, S. (2001). Interactions between crowned eagles and monkeys in Taï National Park, Ivory Coast. *Folia Primatologica*, **72**, 248–50.

Shultz, S. and Noë, R. (2002). The consequences of crowned eagle central-place foraging on predation risk in monkeys. *Proceedings of the Royal Society of London*, B, **269**, 1797–802.

Siex, K. S. and Struhsaker, T. T. (1999). Ecology of the Zanzibar red colobus monkey: demographic variability and habitat stability. *International Journal of Primatology*, **20**, 163–92.

Smail, J. K. (1997). Beyond population stabilization: the case for dramatically reducing global human numbers. *Politics and the Life Sciences*, **16**, 183.

Smail, J. K. (2003). Remembering Malthus III: implementing global population reduction. *American Journal of Physical Anthropology*, **122**(3), 295–300.

Starin, E. D. (1989). Threats to the monkeys of The Gambia. *Oryx*, **23**, 208–14.

Streiker, G. (1997). Human settlement threatens spectacle of life in Ivory Coast rainforest. *CNN* May **6**, 1997.

Struhsaker, T. T. (1971). Notes on *Cercocebus a. atys*. In Senegal, West Africa. *Mammalia*, **35**, 343–4.

Struhsaker, T. T. (1975). *The Red Colobus Monkey*. Chicago: University of Chicago Press.

Struhsaker, T. T. (1999). Primate communities in Africa: The consequence of long-term evolution or the artifact of recent hunting. In *Primate Communities*, ed. J. G. Fleagle, C. Janson and K. Reed. Cambridge: Cambridge University Press, pp. 289−94.

Struhsaker, T. T. (2001a). *Africa's Rain Forest Protected Areas: Problems and Possible Solutions.* Report to the Center for Applied Biodiversity Science, Conservation International.

Struhsaker, T. T. (2001b). Unsustainable hunting in tropical forests. *Trends in Ecology and Evolution*, **16**, 163−4.

Struhsaker, T. T. (2005). Conservation of red colobus and their habitats. *International Journal of Primatology*, **26**, 525−38.

Struhsaker, T. T. and Oates, J. F. (1995). The biodiversity crisis in Southwestern Ghana. *African Primates*, **1**, 5−6.

Struhsaker, T. T., Struhsaker, P. J. and Siex, K. S. (2005). Conserving Africa's rain forests: problems in protected areas and possible solutions. *Biological Conservation*, **123**, 45−54.

Terborgh, J. (1999). *Requiem for Nature*. Island Press: Washington DC.

Tockman, J. (2002). Côte d'Ivoire: IMF, cocoa, coffee, logging and mining. *World Rainforest Movement's Bulletin 54* (January).

Vickers, W. T. (1991). Hunting yields and game composition over ten years in an Amazon Indian territory. In *Neotropical Wildlife and Conservation*, ed. J. G. Robinson and K. H. Redford. Chicago: University of Chicago Press, pp. 53−81.

Wilkie, D. S. and Carpenter, J. F. (1999). Bushmeat hunting in the Congo Basin: an assessment of impacts and options for mitigation. *Biodiversity and Conservation*, **8**, 927−55.

Wilkie, D. S., Curran, B., Tshombe, R. and Morelli, G. A. (1998). Managing bushmeat hunting in Okapi Wildlife Reserve, Democratic Republic of Congo. *Oryx*, **32**, 131−44.

Zuberbuhler, K. and Jenny, D. (2002). Leopard predation and primate evolution. *Journal of Human Evolution*, **43**, 873−86.

Zuberbuhler, K., Jenny, D. and Bshary, R. (1999). The predator deterrence function of primate alarm calls. *Ethology*, **105**, 477−90.

Zuberbuhler, K., Noë, R. and Seyfarth, R. M. (1997). Diana monkey long-distance calls: messages for conspecifics and predators. *Animal Behaviour*, **53**, 589−604.

Appendix

Here we present data on polyspecific associations for several social groups within our primary study grid. The focal subjects were two groups of *Procolobus badius* (bad1 and bad2) and two groups of *Cercopithecus diana* (dia1 and dia2). Association data on the *Procolobus badius* groups were collected between June 1996 and May 1998 while association data on the *Cercopithecus diana* groups were collected between April 1997 and May 1998. On most observation days we followed a focal group from 7:00 h to 17:30 h. Data on association behavior were collected at every one-hour interval and one hourly group scan represents a single data point. During each hourly scan, we recorded the number of species "in association" with the focal species. Groups of different species were considered associated when the distance between imaginary polygons connecting the most outermost members of each group was less than 25 m. Association rates were calculated by dividing the time each group was seen "in association" with another species by the total observation time. In late 1997 both *Procolobus badius* groups divided into subgroups, each of which had a preferred *Cercopithecus diana* association partner.

Table 1. *Time* Procolobus badius *group 1 (bad1) spent in association with other Taï cercopithecids from June 1996−May 1998*

	% Association	N
Cercopithecus diana	33.58	586
Cercopithecus campbelli	0.74	13
Cercopithecus petaurista	13.06	228
Colobus polykomos	2.8	49
Procolobus verus	12.03	210
Cercocebus atys	2.6	47
Total sample		1745

Table 2. *Time* Procolobus badius *group 2 (bad2) spent in association with other Taï cercopithecids from June 1996—May 1998*

	% Association	N
Cercopithecus diana	46.4	739
Cercopithecus campbelli	41.2	656
Cercopithecus petaurista	23	367
Colobus polykomos	1.12	18
Procolobus verus	11.2	180
Cercocebus atys	3.6	57
Total sample		1593

Table 3. *Time* Cercopithecus diana *group 1 (dia1) spent in association with other Taï cercopithecids from April 1997—May 1998*

	% Association	N
Procolobus badius	37.3	237
Cercopithecus campbelli	4.6	29
Cercopithecus petaurista	37.6	239
Colobus polykomos	24.4	155
Procolobus verus	83.5	530
Cercocebus atys	15.1	96
Total sample		635

Table 4. *Time* Cercopithecus diana *group 2 (dia2) spent in association with other Taï cercopithecids from April 1997—May 1998*

	% Association	N
Procolobus badius	71.1	683
Cercopithecus campbelli	85.6	823
Cercopithecus petaurista	62.2	598
Colobus polykomos	15.1	145
Procolobus verus	16.6	160
Cercocebus atys	12.3	118
Total sample		961

Table 5. *Percent of observations two neighboring red colobus Procolobus badius groups (bad1 and bad2) were found in association with/without Diana monkeys (Cercopithecus diana) and with/without other species. All tests used the Bonferroni corrections for adjusted significance levels (p divided by 5, $N_{bad1} = 1745$, $N_{bad2} = 1602$)*

		Procolobus badius group 1 (bad1)		**Procolobus badius** group 2 (bad2)		Fisher's Exact Test
		Cercopithecus diana present (N=586)	*Cercopithecus diana* absent (N=1159)	*Cercopithecus diana* present (N=739)	*Cercopithecus diana* absent (N=855)	
Cercopithecus campbelli	present	1.4% (8)	0.4% (5)	85.8% (634)	2.5% (21)	The red colobus groups differ in the way they associate with Diana monkeys
	absent	98.6% (578)	99.6% (1154)	14.1% (104)	97.5% (834)	
Cercopithecus petaurista	present	28.8% (169)	5.1% (59)	47.2% (349)	2.1% (18)	$P < 0.001$
	absent	71.2% (417)	94.9% (1100)	52.6% (389)	97.9% (837)	
Colobus polykomos	present	4.6% (27)	1.9% (22)	1.8% (13)	0.6% (5)	$P < 0.001$
	absent	95.4% (559)	98.1% (1137)	98.1% (725)	99.4% (850)	
Procolobus verus	present	35.4% (208)	0.2% (2)	23.5% (174)	0.7% (6)	$P < 0.001$
	absent	64.5% (378)	99.8% (1157)	76.3% (564)	99.3% (849)	
Cercocebus atys	present	6.7% (39)	0.69% (8)	7.2% (53)	0.5% (4)	$P < 0.001$
	absent	93.3% (547)	99.3% (1151)	92.7% (685)	99.5% (851)	

Index

Printed in the United States
by Baker & Taylor Publisher Services